INNOVATIONS IN MULTIVARIATE STATISTICAL ANALYSIS

## Advanced Studies in Theoretical and Applied Econometrics

**Volume 36**

---



The titles published in this series are listed at the end of this volume.

# Innovations in Multivariate Statistical Analysis

## A Festschrift for Heinz Neudecker

*Edited by*

R.D.H. Heijmans, D.S.G. Pollock and A. Satorra

**Springer-Science+Business Media, B.V.**

**Library of Congress Cataloging-in-Publication Data**

Innovations in multivariate statistical analysis: a festschrift for Heinz Neudecker / edited by R.D.H. Heijmans, D.S.G. Pollock and A. Satorra

p. cm. – (Advanced studies in theoretical and applied econometrics; v. 36)

Includes index.

ISBN 978-1-4613-7080-2 ISBN 978-1-4615-4603-0 (eBook)

DOI 10.1007/978-1-4615-4603-0

1. Multivariate analysis. I. Neudecker, Heinz. II. Heijmans, Risto. III. Pollock, D.S.G. IV. Satorra, A. V. Series.

QA278.I48 2000

519.5'35–dc21 99-088162



**Originally published by Kluwer Academic Publishers in 2000**

**Softcover reprint of the hardcover 1st edition 2000**



Typeset by SigmaPi (D.S.G. Pollock), London

E-mail <stephen_pollock@sigmapi.u-net.com>

*Printed on acid-free paper*

# Contents

# List of Contributors


1. G. ALPARGU
   Department of Mathematics and Statistics, McGill University, Québec, Canada

2. H. AMMAN
   Department of Economics, University of Amsterdam, The Netherlands

3. P. BEKKER
   Department of Economics, University of Groningen, The Netherlands

4. H.P. BOSWIJK
   Department of Econometrics, University of Amsterdam, The Netherlands

5. C.M. CUADRAS
   Departament d'Estadística, Universitat de Barcelona, Spain

6. F.J.H. DON
   CPB Netherlands Bureau for Economic Policy Analysis, The Hague, and University of Amsterdam

7. R.W. FAREBROTHER
   Department of Economic Studies, Victoria University of Manchester, U.K.

8. J. FORTIANA
   Departament d'Estadística, Universitat de Barcelona, Spain

9. M.J. GREENACRE
   Departament d'Economia i Empresa, Universitat Pompeu Fabra, Spain

10. P.C.N GROENWALD
    Department of Mathematical Statistics, University of the Orange Free State, South Africa

11. R.D.H. HEIJMANS
    Department of Econometrics, University of Amsterdam, The Netherlands

12. D.A. KENDRICK
    University of Texas, U.S.A.

13. T. KOLLO
    University of Tartu, Estonia

14. R.H. KONING
    Department of Economics, University of Groningen, The Netherlands

15. S.J. LEYBOURNE
    Department of Economics, University of Nottingham, U.K.

16. S. LIU
Institute of Statistics and Econometrics, University of Basel, Switzerland

17. J.R. MAGNUS
CentER for Economic Research, Tilburg University, The Netherlands

18. B.P.M. McCABE
Faculty of Commerce, University of British Columbia, Canada

19. D.G. NEL
Statistical Research, Clover, South Africa

20. W. POLASEK
Institute of Statistics and Econometrics, University of Basel, Switzerland

21. D.S.G. POLLOCK
Department of Economics, Queen Mary College, University of London, U.K.

22. G. RIDDER
Department of Economics, John Hopkins University, Baltimore, U.S.A.

23. T.J. ROTHENBERG
University of California at Berkeley, U.S.A.

24. A. SATORRA
Departament d'Economia i Empresa, Universitat Pompeu Fabra, Spain

25. G.P.H. STYAN
Department of Mathematics and Statistics, McGill University, Québec, Canada

26. G. TRENKLER
Department of Statistics, University of Dortmund, Germany

27. M. van de VELDEN
Department of Econometrics, University of Amsterdam, The Netherlands

28. T. WANSBEEK
Department of Economics, University of Groningen, The Netherlands

29. F. WINDMEIJER
Institute for Fiscal Studies, London, U.K.

# Introduction to the Book

R.D.H. HEIJMANS, D.S.G. POLLOCK and A. SATORRA

The three decades which have followed the publication of Heinz Neudecker's seminal paper "Some theorems on matrix differentiation with special reference to Kronecker products" in the *Journal of the American Statistical Association,* (1969) have witnessed the growing influence of matrix analysis in many scientific disciplines. Amongst these are the disciplines to which Neudecker has contributed directly—namely econometrics, economics, psychometrics and multivariate analysis.

This book aims to illustrate how powerful the tools of matrix analysis have become as weapons in the statistician's armoury. The majority of its chapters are concerned primarily with theoretical innovations, but all of them have applications in view, and some of them contain extensive illustrations of the applied techniques.

Several of the contributors to this volume are concerned with the theory of matrix inequalities to which Neudecker has also contributed over the years. These can be very powerful tools in distribution theory, in the theory of statistical estimation and in many other applications besides. Chapter 1 provides an extensive bibliography of matrix inequalities. The chapters 1, 2 and 3, which deal with this topic, present several new inequalities together with some new applications of known inequalities.

Two of the chapters deal with the covariance matrices which are associated with vectors of random variables. Chapter 4, which has been inspired by a fascinating problem in electronics, is concerned with the covariance matrix of the multinomial distribution which may become rank deficient. Chapter 5 is concerned with the information-matrix equality (IME) which entails the covariance matrix of a vector of estimated quantities. This chapter considers various strategies for testing whether the parameters of certain models are fixed constants or whether they should be regarded as random variables. Several alternative representations of the IME are derived.

One multivariate technique which has been growing in popularity in recent years is correspondence analysis. This is a technique which provides a visual rendition of categorical data. Chapter 6 clarifies the connections and differences between correspondence analysis and the closely related technique of dual scaling when these techniques are applied to rank order data. Chapter 7 extends correspondence analysis to a continuous bivariate distribution. The results are applied to a generalisation of the Farlie–Gumbel–Morgensten family of bivariate distributions.

The economic hypothesis of utility maximisation imposes certain restrictions on the equations of consumer demand. The analogues of these conditions for discrete-choice models are developed in Chapter 8 and they are used to assess the degree of rationality in choices of the mode of payment.

Sometimes the variables of an econometric model cannot all be observed directly, or else the observations of the variables are afflicted by errors. In such cases, the Bayesian version of the method of Gibbs sampling, which is the topic of Chapter 9, can be used. This chapter deals specifically with latent variables in autoregressive and vector autoregressive models.

In autoregressive models, the assumptions about the initial conditions can have major influences on the statistical estimates when some of the roots of the characteristic polynomial are close to the boundary of the unit circle. Chapter 10 explores the effects upon the sampling distributions of least-squares estimates and test statistics of having roots which are close to the boundary.

The generalised method of moments (GMM) estimator for dynamic panel data models, which is the subject of Chapter 11, combines moment conditions for the differenced version of a dynamic behavioural equation with moment conditions for the equation in levels. It is common practice to use the inverse of the moment matrix of the instruments as the initial weight matrix in the criterion function of the estimation. An assessment is made of the potential efficiency loss resulting from the use of this matrix.

Two of the chapters are concerned with the stability of linear dynamic systems. Chapter 12 considers a system of first-order linear differential equations with a transition matrix which may be singular. For this system, a rank condition is established; and it is shown that a rank and an order condition are jointly sufficient for the existence of a non-explosive solution. Chapter 13 is concerned with a class of reflection and permutation matrices which are related to the Samuelson–Wise conditions for the stability of linear difference equations. It also discusses a class of permutation matrices which can be used to generate pseudo-random numbers.

Chapter 14 concerns the issue of conditioning in multivariate time-series models. The concept of weak exogeneity is contrasted with various forms of ancillarity. It is shown that only ancillarity allows for conditional inference.

Asymptotic theory is needed in statistics whenever finite sample-theory does not deliver tractable expressions for the distributions of the estimators. Four chapters, numbers 15–18, are devoted to this topic.

In principal component analysis, it is sometimes desired to know whether the columns of a given matrix lie in the subspace generated by the eigenvectors of the sample covariance matrix. To verify this, a test is developed in Chapter 15 which is based on the asymptotic properties of eigenprojections of covariance matrices.

One of the most important matrices in multivariate analysis is the Wishart matrix. A Fisher–Cornish type expansion of Wishart matrices is extensively studied in Chapter 16.

Moment structure analysis is widely used in behavioural, social and economic studies to analyse structural relations among variables, some of which can be latent or unobservable. The concern of Chapter 17 is to make corrections to the score, the difference and the Wald test statistics which can be used in testing the validity of certain parametric restrictions on structural relations which impact upon the moment structures

Chapter 18, which is the last in the group of five, concerns the asymptotic properties of sums of powers of residuals in the classic linear regression model.

Nonlinear dynamic optimisation models are widely used in theoretical and empirical economic modelling, especially in the field of optimal growth and inter temporal modelling. In solving the equation of these models, matrix methods play a crucial *rôle*. Chapter 19 of the book proposes new methods for solving the Ricatti equation which are of major interest in the theory of linear filtering.

The concluding chapter of the book takes a fresh look at the multilinear algebra which lies at the heart of the techniques of matrix differential calculus to which Heinz Neudecker has made such substantial contributions. The chapter proposes a notation which reveals the structure of the derivatives more clearly than does the conventional matrix notation. By adopting this notation, one can make light of the task of deriving and memorising some of the essential results.

## Aims and Scope of the Book

This book aims to provide research workers and graduate students with a cross section of innovative work in the fields of matrix methods and multivariate statistical analysis. The book should be of interest to students and practitioners in a wide range of subjects which rely upon modern methods of statistical analysis.

The contributors to the book are themselves practitioners of a wide range of subjects including econometrics, psychometrics, educational statistics, computation methods and electrical engineering, but they find a common ground in the methods which are represented in the book. It is envisaged that the book will serve as an important work of reference and as a source of inspiration for some years to come.

## The Editors

The editors divided their tasks so that each was performing a set of largely distinct functions. The managing editor was Risto Heijmans of the University of Amsterdam. Matters of book production and typography were handled by Stephen Pollock of the University of London who used the LaTeX2$\epsilon$ package in compiling the book. The role of copy editor was assumed by Albert Satorra of the Pompeu Fabra University.

CHAPTER 1

# Some Comments and a Bibliography on the Frucht–Kantorovich and Wielandt Inequalities

Gülhan ALPARGU and George P.H. STYAN

## 1.1 Introduction and Mise-en-scène: The Frucht–Kantorovich Inequality

The well-known inequality

$$\frac{x'Ax \cdot x'A^{-1}x}{(x'x)^2} \leq \frac{(\lambda_1 + \lambda_n)^2}{4\lambda_1\lambda_n} \tag{1}$$

is usually attributed (cf. e.g., [35], [36]) to Nobel Laureate and Academician Leonid Vitalévich Kantorovich (1912–1986) for the inequality he established in 1948 in a long survey article (in Russian) on "Functional Analysis and Applied Mathematics" ([111], pp. 142–144; [112], pp. 106–107). In (1) $x$ is a real $n \times 1$ vector and $A$ is a real $n \times n$ symmetric positive definite matrix, with $\lambda_1$ and $\lambda_n$, respectively, its (fixed) largest and smallest, necessarily positive, eigenvalues. (All matrices and vectors in this paper are real.)

Another way of expressing (1) is in the "normalised reduced" form:

$$\sum_{i=1}^{n} \lambda_i z_i^2 \cdot \sum_{i=1}^{n} \frac{1}{\lambda_i} z_i^2 \leq \frac{(\lambda_1 + \lambda_n)^2}{4\lambda_1\lambda_n}, \qquad \text{with} \quad \sum_{i=1}^{n} z_i^2 = 1, \tag{2}$$

which follows by putting $A = P\Lambda P'$ and then $P'x/\sqrt{x'x} = z = \{z_i\}$ in (1), where $P$ is orthogonal and $\Lambda = \operatorname{diag}\{\lambda_i\}$.

As observed by Watson, Alpargu and Styan [266], the inequality (1) was established, however, already five years earlier in 1943 by the graph theorist Roberto Frucht [Wertheimer] (1906–1997) in Spanish [71] (published in Argentina)[1] and so we now name (1) the Frucht–Kantorovich inequality. An

[1] We are very grateful to Josip E. Pečarić for drawing our attention (in September 1996) to this paper by Frucht (see also Beckenbach [27] and Mitrinović, Pečarić and Fink [174], pp. 125, 132).

English translation of [71] appears in [266].

The paper by Frucht [71] built on the solution of the following problem: Given positive numbers $x_1, x_2, \ldots, x_n$ such that $\sum_{i=1}^{n} x_i = k$, establish the inequalities:

$$\sum_{i=1}^{n} x_i^2 \geq \frac{k^2}{n}, \qquad \sum_{i=1}^{n} \frac{1}{x_i} \geq \frac{n^2}{k}. \tag{3}$$

This problem (3) was posed [anonymously, but presumably by Beppo Levi (1875–1961)] in 1942 and solved (also in 1942) by Abraham H. Bender [30] and by Ernesto M. Saleme [229]; English translations of these solutions are also included in [266]. Both inequalities in (3) are easily proved using the well-known Cauchy–Schwarz inequality; see also (13) below.

Equality holds in the Frucht–Kantorovich inequality (1) when $x = \frac{1}{2}(q_1 \pm q_n)$, where $q_1$ and $q_n$ are eigenvectors of $A$ corresponding to the eigenvalues $\lambda_1$ and $\lambda_n$. Equality holds in (2) when $z_1 = z_n = 1/\sqrt{2}$ and $z_2 = \cdots = z_{n-1} = 0$. When the eigenvalues $\lambda_1$ and $\lambda_n$ each have multiplicity one then these conditions are also necessary.

Our purpose in this paper is to show, with a historical perspective, how the Frucht–Kantorovich inequality (1) is related to several other inequalities. We also present an extensive bibliography, essentially doubling our earlier bibliography [4]; some biographical articles are also cited. References to *Current Mathematical Publications* are denoted by CMP, *Jahrbuch für die Fortschritte der Mathematik* by JFM, *Mathematical Reviews* by MR, and to *Zentralblatt für Mathematik* by Zbl.

Our favourite proof of the Frucht–Kantorovich inequality may be the following simple probabilistic proof given by T. W. Anderson ([8], Lemma 10.2.5, p. 569); see also Bühler [40], Marshall and Olkin [161], and Watson [265]. For other proofs see, e.g., Alpargu [2], Pták [219], and Styan [244].

We may write the left-hand side of (2) as

$$\sum_{i=1}^{n} \lambda_i z_i^2 \cdot \sum_{i=1}^{n} \frac{1}{\lambda_i} z_i^2 = E(T) \cdot E\left(\frac{1}{T}\right), \tag{4}$$

the product of the expected value of a random variable $T$ and the expected value of its reciprocal $1/T$, where $T$ assumes the values $\lambda_i \in [m, M]$ with probabilities $p_i = z_i^2$ $(i = 1, \ldots, n)$; $0 < m = \lambda_n < \lambda_1 = M < \infty$.

For $m \leq T \leq M$, we see that

$$0 \leq (M - T)(T - m) = (M + m - T)T - mM \tag{5}$$

and so with $m > 0$ we have

$$\frac{1}{T} \leq \frac{m + M - T}{mM}, \tag{6}$$

which implies

$$\begin{aligned} E(T)\cdot E\left(\frac{1}{T}\right) &\le E(T)\cdot\frac{m+M-E(T)}{mM} \\ &= \frac{(m+M)^2}{4mM}-\frac{1}{mM}[E(T)-\frac{1}{2}(m+M)]^2 \\ &\le \frac{(m+M)^2}{4mM}, \end{aligned} \tag{7}$$

and our proof is complete.

Equality holds in (2) if and only if equality holds in (5) and (7) and this happens if and only if $E(T)=\frac{1}{2}(m+M)$ and $T=m$ or $T=M$. Hence $P(T=m)=P(T=M)=\frac{1}{2}$ and so equality holds in (2) when $z_1=z_n=1/\sqrt{2}$ and $z_2=\cdots=z_{n-1}=0$. When the eigenvalues $\lambda_1$ and $\lambda_n$ each have multiplicity one then this condition is also necessary.

## 1.2 The Wielandt Inequality

In 1960, Friedrich L. Bauer and Alston S. Householder [24] showed that for any two non-null vectors $x$ and $y$ and positive definite matrix $A$, the inequalities

$$\frac{(x'y)^2}{x'x\cdot y'y}\le\cos^2\phi \quad\text{and}\quad 0\le\phi\le\frac{\pi}{2} \tag{8}$$

together imply that

$$\frac{(x'Ay)^2}{x'Ax\cdot y'Ay}\le\cos^2\theta,$$

where $\cot^2(\theta/2)=\kappa\cot^2(\phi/2)$ and the condition number $\kappa=\lambda_1/\lambda_n$, with, as earlier, $\lambda_1\ge\cdots\ge\lambda_n$ the, necessarily positive, eigenvalues of $A$.

When $\phi=\pi/2$ the vectors $x$ and $y$ must be orthogonal, cf. (8), and then

$$\frac{(x'Ay)^2}{x'Ax\cdot y'Ay}\le\left(\frac{\kappa-1}{\kappa+1}\right)^2=\left(\frac{\lambda_1-\lambda_n}{\lambda_1+\lambda_n}\right)^2, \qquad\text{with}\quad x'y=0, \tag{9}$$

since now $\kappa=\cot^2(\theta/2)$.

We will refer to (9) as the Wielandt inequality[2], since Bauer and Householder [24] credit (9) to Helmut Wielandt (1953) [268] "and also private communication" (we have found it difficult to deduce (9) from the results in [268]). Earl R. Barnes and Alan J. Hoffman (1994) [19] and Henry Wolkowicz (1981) [271] refer to the right-hand side of (9) as the Kantorovich ratio.[3]

[2]This is not the only inequality so called: we found 11 references in *Mathematical Reviews* to the Wielandt inequality but only two are about (9).

[3]We have not found any other references to the Kantorovich ratio.

Morris L. Eaton [68] rediscovered (9) in 1976, cf. Olkin [203], while much more recently (1995 to date) the Wielandt inequality (9) has been studied by Chiang [52], Fujii, Katayama and Nakamoto [75], Gustafson [91], [92], and by Yeh [284].

Householder seems to be the first person to observe [in his well-known book *The Theory of Matrices in Numerical Analysis* ([105], p. 83)] that when

$$y = (x'x)A^{-1}x - (x'A^{-1}x)x$$

then the Wielandt inequality (9) becomes the Frucht–Kantorovich inequality (1).

To prove the Wielandt inequality (9) we follow Householder ([105], pp. 81–85, §3.4) and start by introducing the $2 \times 2$ matrix

$$G = (u : v)'A(u : v) = \begin{pmatrix} u'Au & u'Av \\ v'Au & v'Av \end{pmatrix},$$

where $u = x/\sqrt{x'x}$ and $v = y/\sqrt{y'y}$, and

$$(u : v)'(u : v) = \begin{pmatrix} u' \\ v' \end{pmatrix}(u : v) = \begin{pmatrix} 1 & 0 \\ 0 & 1 \end{pmatrix} = I_2,$$

the $2 \times 2$ identity matrix.

Let $\gamma_1 \geq \gamma_2$ denote the eigenvalues of $G$ and so the trace $\mathrm{tr}G = \gamma_1 + \gamma_2$ and the determinant $\det G = \gamma_1\gamma_2$. Then by the Poincaré Separation Theorem, cf. e.g., Scott and Styan (1985) [235], where some historical information may also be found,

$$\lambda_1 \geq \gamma_1 \geq \gamma_2 \geq \lambda_n, \tag{10}$$

and so

$$\frac{4\gamma_1\gamma_2}{(\gamma_1 + \gamma_2)^2} \geq \frac{4\lambda_1\lambda_n}{(\lambda_1 + \lambda_n)^2} = \frac{4\kappa}{(\kappa + 1)^2}, \tag{11}$$

with $\kappa = \lambda_1/\lambda_n$. Applying (11) yields

$$\begin{aligned} 1 - \frac{(x'Ay)^2}{x'Ax \cdot y'Ay} &= 1 - \frac{(u'Av)^2}{u'Au \cdot v'Av} \\ &= \frac{4\det G}{(\mathrm{tr}G)^2 - (u'Au - v'Av)^2} \\ &= \frac{4\gamma_1\gamma_2}{(\gamma_1 + \gamma_2)^2 - (u'Au - v'Av)^2} \\ &\geq \frac{4\gamma_1\gamma_2}{(\gamma_1 + \gamma_2)^2}, \end{aligned}$$

with equality if and only if $u'Au = v'Av$. Hence

$$\frac{(x'Ay)^2}{x'Ax \cdot y'Ay} \leq 1 - \frac{4\kappa}{(\kappa+1)^2} = \left(\frac{\kappa-1}{\kappa+1}\right)^2 = \cos^2\theta,$$

and the Wielandt inequality (9) is established.

Equality holds in the Wielandt inequality (9) whenever $x = p_1 + p_n$ and $y = p_1 - p_n$, where $p_1$ and $p_n$ are normalised eigenvectors corresponding to the eigenvalues $\lambda_1$ and $\lambda_n$ of the matrix $A$.

## 1.3 The Schweitzer Inequality

The oldest inequality related to the Frucht–Kantorovich inequality appears to be:

$$\frac{1}{n}(x_1 + \cdots + x_n) \cdot \frac{1}{n}\left(\frac{1}{x_1} + \cdots + \frac{1}{x_n}\right) \leq \frac{(m+M)^2}{4mM}, \tag{12}$$

where $0 < m \leq x_i \leq M \quad (i = 1, \ldots, n)$. The inequality (12) was established in 1914 by Pál Schweitzer [234] (in Hungarian).[4]

The complementary inequality

$$1 \leq \frac{1}{n}(x_1 + \cdots + x_n) \cdot \frac{1}{n}\left(\frac{1}{x_1} + \cdots + \frac{1}{x_n}\right), \tag{13}$$

which follows at once from the Cauchy–Schwarz inequality, is the well-known arithmetic-harmonic mean inequality, cf. e.g., Mitrinović ([172], pp. 27–28, 206–207); see also (3) above.

The arithmetic-geometric-harmonic mean inequality follows at once from the equivalence of the geometric-harmonic mean inequality[5] with the well-known arithmetic-geometric mean inequality. The proof of this by Cauchy (1821) (cf. [47], pp. 375–377) is reprinted (in the original French) in Pólya and Szegö ([216], pp. 50–51 and [217], p. 64).

Beckenbach and Bellman [29] refer to the arithmetic-geometric mean inequality as a result of "singular elegance" (p. 3) and present twelve proofs (pp. 4–19), while Mitrinović ([172], pp. 27–28) observes that "it is likely that the Pythagoreans [*fl.* *c.* 6th cent. BC] knew of the inequality $\sqrt{ab} \leq (a+b)/2$, but there is no doubt that it was proved by Euclid (*fl.* *c.* 300 BC) (cf. [69], Book V, Proposition 25 and Commentary [vol. II, pp. 185–186]). The first, and one of the most beautiful proofs of the arithmetic-geometric mean inequality, was certainly the one given by Cauchy."

---

[4] Translated into English in Alpargu [2] and Watson, Alpargu and Styan [266]. We believe that Pál Schweitzer died in 1941 (cf. Mitrinović ([172], p. 394) but we do not know when he was born.

[5] Cf. Bouniakowsky (1859), p. 3.

Equality holds in the Schweitzer inequality (12) only when $n$ is even and then if

$$x_1 = \cdots = x_{\frac{n}{2}} = M \quad \text{and} \quad x_{\frac{n}{2}+1} = \cdots = x_n = m; \tag{14}$$

when $n$ is even and $x_n \leq \cdots \leq x_1$ then (14) is also necessary for equality to hold in the Schweitzer inequality (12).

In 1972 Alexandru Lupaş [149] obtained the following version of the Schweitzer inequality, which is stronger than the (original) Schweitzer inequality (12) when $n$ is odd (and identical to (12) when $n$ is even):

$$\sum_{i=1}^{n} x_i \cdot \sum_{i=1}^{n} \frac{1}{x_i} \leq \frac{1}{mM}\left(\left[\frac{n}{2}\right]M + \left[\frac{n+1}{2}\right]m\right)\left(\left[\frac{n+1}{2}\right]M + \left[\frac{n}{2}\right]m\right), \tag{15}$$

where $[\cdot]$ denotes the integral part. Equality is attained when the smallest $[n/2]$ of the numbers $x_1, \ldots, x_n$ are equal to $m$ and the $[n/2]$ largest are equal to $M$, and when $n$ is odd the "middle" $x_i$ is equal to either $m$ or $M$.

The Schweitzer inequality (12) is the special case of the Frucht–Kantorovich inequality (2) with $\lambda_i = x_i$, $z_i = 1/\sqrt{n}$, $\lambda_1 = M$ and $\lambda_n = m$, and so the Frucht–Kantorovich inequality is a "weighted" version of the Schweitzer inequality.

Surprisingly, the Schweitzer inequality also implies the Frucht–Kantorovich inequality, as was shown in 1961 by Peter Henrici [101].[6] To see this, we put $\lambda_i = x_i$ in (2), and let

$$\text{LSI} = \frac{1}{n}\sum x_i \cdot \frac{1}{n}\sum \frac{1}{x_i} \quad \text{and} \quad \text{LKI} = \sum x_i z_i^2 \cdot \sum \frac{1}{x_i} z_i^2,$$

so that LSI denotes the left-hand side of the Schweitzer inequality (12) and LKI the left-hand side of the Frucht–Kantorovich inequality (2). It suffices to prove that LSI = LKI for all $z_i^2$ rational with $\sum_{i=1}^{n} z_i^2 = 1$.

We choose $n$ to be "very large" so that each $x_i$ occurs "many times," and write

$$x_{(1)} < x_{(2)} < \cdots < x_{(d)}$$

for the $d$ distinct $x$'s with multiplicities $m_1, m_2, \ldots, m_d$ and $\sum_{h=1}^{d} m_h = n$. Then

$$\text{LSI} = \frac{\sum_{h=1}^{d} m_h x_{(h)}}{\sum_{h=1}^{d} m_j} \cdot \frac{\sum_{h=1}^{d} m_h / x_{(h)}}{\sum_{h=1}^{d} m_h} = \sum_{h=1}^{d} z_h^2 x_{(h)} \cdot \sum_{h=1}^{d} z_h^2 \frac{1}{x_{(h)}} = \text{LKI},$$

with $z_h^2 = m_h / \sum_{h=1}^{d} m_h$, and so the Frucht–Kantorovich inequality (2) is "essentially equivalent" to the Schweitzer inequality (12).

---

[6] A similar technique was proposed already in 1934 by Hardy, Littlewood, and Pólya ([99], p. 13). See also Makai [152] for a different approach.

## 1.4 The Pólya–Szegö Inequality

The inequality:

$$\frac{\sum_{i=1}^n a_i^2 \cdot \sum_{i=1}^n b_i^2}{(\sum_{i=1}^n a_i b_i)^2} \leq \frac{(ab+AB)^2}{4abAB}, \tag{16}$$

where

$$0 < a \leq a_i \leq A, \quad 0 < b \leq b_i \leq B \quad (i = 1, \ldots, n),$$

was proved in 1925 by George Pólya (1887–1985) and Gábor Szegö (1895–1985) in their well-known book *Aufgaben und Lehrsätze aus der Analysis, Band I: Reihen, Integralrechnung, Funktiontheorie* [216], pp. 57, 213–214 (see also [217], pp. 71–72, 253–255).

Pólya and Szegö also showed that equality holds in (16) if and only if

$$p = n\frac{A}{a} \Big/ \left(\frac{A}{a} + \frac{B}{b}\right) \quad \text{and} \quad q = n\frac{B}{b} \Big/ \left(\frac{A}{a} + \frac{B}{b}\right)$$

are integers and if $p$ of the numbers $a_1, \ldots, a_n$ are equal to $a$ and $q$ of these numbers are equal to $A$, and if the corresponding $b_i$ are equal to $B$ and $b$, respectively.

If we put $a_i^2 = 1/{b_i}^2 = x_i, b = 1/A, B = 1/a, M = A^2$ and $m = a^2$ in the Pólya–Szegö inequality (16) then it becomes the Schweitzer inequality (12). In [266] we show that the Schweitzer inequality also implies the Pólya–Szegö inequality.

The continuous version of the special case of the Pólya–Szegö inequality (16) with $a = b$ and $A = B$:

$$\frac{\int_c^d f^2(x)\,dx \int_c^d g^2(x)\,dx}{[\int_c^d f(x)g(x)\,dx]^2} \leq \frac{(a^2+A^2)^2}{4a^2A^2},$$

was already posed in 1914 as a "Problem" [122][7] by József Kürschák (1864–1933).

## 1.5 The Cassels, Krasnosel′skiĭ–Kreĭn and Greub–Rheinboldt Inequalities

There are three other inequalities closely related to the Frucht–Kantorovich inequality. The Cassels inequality:

$$\frac{\sum_{i=1}^n a_i^2 w_i \cdot \sum_{i=1}^n b_i^2 w_i}{(\sum_{i=1}^n a_i b_i w_i)^2} \leq \frac{(m+M)^2}{4mM}, \tag{17}$$

[7] In the same journal and volume as Schweitzer [234] but just over a hundred pages later! As far as we know there was no published solution *per se* to this "Problem" [122].

where $a_i > 0,\ b_i > 0$ and $w_i \geq 0$ $(i = 1, \ldots, n)$, with

$$m = \min_i \frac{a_i}{b_i} \quad \text{and} \quad M = \max_i \frac{a_i}{b_i}, \tag{18}$$

appears as an appendix [45] by John William Scott Cassels (1922–) in Watson's 1951 PhD thesis [260], cf. also [46] in [261].

Another closely related inequality was established in 1952 by Mark Aleksandrovich Krasnosel′skiĭ and Selim Griogor'evich Kreĭn in a study [119] of "iteration processes with minimal residuals" (in Russian):

$$\frac{\sum_{i=1}^n \lambda_i^2 u_i^2 \cdot \sum_{i=1}^n u_i^2}{(\sum_{i=1}^n \lambda_i u_i^2)^2} \leq \frac{(\lambda_1 + \lambda_n)^2}{4\lambda_1\lambda_n}, \tag{19}$$

where, as in (1), $0 < \lambda_n \leq \lambda_i \leq \lambda_1$ fixed, $i = 1, \ldots, n$. A continuous version of (19) was given already by Frucht (1943), cf. p. 38 in [266]. A matrix version of (19), along the lines of the Frucht–Kantorovich inequality (1), is:

$$\frac{x'A^2x \cdot x'x}{(x'Ax)^2} \leq \frac{(\lambda_1 + \lambda_n)^2}{4\lambda_1\lambda_n}, \tag{20}$$

where, as before, $x$ is a real $n \times 1$ vector and $A$ is a real $n \times n$ symmetric positive definite matrix, with $\lambda_1$ and $\lambda_n$, respectively, its (fixed) largest and smallest, necessarily positive, eigenvalues.

Indeed, as pointed out in 1960 by Andreas H. Schopf [231], cf. also Householder ([105], p. 83),

$$\frac{x'A^{h+1}x \cdot x'A^{h-1}x}{(x'A^hx)^2} \leq \frac{(\lambda_1 + \lambda_n)^2}{4\lambda_1\lambda_n}, \tag{21}$$

where $v$ is an integer. [To establish (21) we put $x = t$ and then $t = A^{h/2}x = (A^{1/2})^h x$ in the Frucht–Kantorovich inequality (1).] Moreover Schopf [231] shows that (21) remains valid for any $h$ when $A$ is complex Hermitian positive definite and $x$ is complex (with $x'$ its conjugate transpose). Clearly $h = 0$ in (21) yields the Frucht–Kantorovich inequality (1) while $h = 1$ yields the Krasnosel′skiĭ–Kreĭn inequality (20).

In 1959 Greub and Rheinboldt [89] obtained this "weighted" version of the Pólya–Szegö inequality (16):

$$\frac{\sum_{i=1}^n a_i^2 w_i \cdot \sum_{i=1}^n b_i^2 w_i}{\left(\sum_{i=1}^n a_i b_i w_i\right)^2} \leq \frac{(ab + AB)^2}{4abAB}, \tag{22}$$

where

$$0 < a \leq a_i \leq A, \quad 0 < b \leq b_i \leq B \quad (i = 1, \ldots, n),$$

Watson, Alpargu and Styan show in [266] that the six inequalities Frucht–Kantorovich (1), Schweitzer (12), Pólya–Szegö (16), Cassels (17), Krasnosel′skiĭ–Kreĭn (19), and Greub–Rheinboldt (22) are all "essentially equivalent".

## 1.6 The Bloomfield–Watson–Knott Inequality

The inequality:

$$\frac{|X'X|^2}{|X'AX| \cdot |X'A^{-1}X|} \geq \prod_{i=1}^{\min(k,\, n-k)} \frac{4\lambda_i \lambda_{n-i+1}}{(\lambda_i + \lambda_{n-i+1})^2}, \tag{23}$$

where the $n \times k$ matrix $X$ has full column rank $k$, while, as earlier, $\lambda_1 \geq \cdots \geq \lambda_n$ are the, necessarily positive, eigenvalues of the $n \times n$ positive definite matrix $A$, was originally conjectured in 1955 by James Durbin (cf. Watson [261], p. 331) but first established for $k > 1$ only twenty years later in 1975 by Peter Bloomfield and Geoffrey S. Watson [33] and Martin Knott [118]; see also Alpargu, Drury and Styan [3], Khatri and Rao [115] and Yang [275]. The interest of these authors centered on the efficiency of the ordinary least squares estimator in the usual Gauss–Markov linear model with known positive definite dispersion matrix $A$ and model (design) matrix $X$ of full column rank.

When $k = 1$ the Bloomfield–Watson–Knott inequality (23) reduces, with the matrix $X$ replaced by the vector $x$, to:

$$\frac{(x'x)^2}{x'Ax \cdot x'A^{-1}x} \geq \frac{4\lambda_1\lambda_n}{(\lambda_1 + \lambda_n)^2},$$

which is equivalent to the Frucht–Kantorovich inequality (1).

Alpargu, Drury and Styan [3] present a proof of the Bloomfield–Watson–Knott inequality (23), based closely on the proof given by Bloomfield and Watson [33] but which avoids using Lagrange multipliers, and show that the "brief proof" given in 1990 by Yang [275] is incomplete.

In 1981 Khatri and Rao [115] (see also [116]) generalised the Bloomfield–Watson–Knott inequality (23) in several directions. Let $T$ and $U$ be $n \times k$ matrices such that $T'T = U'U = I_k$ and let $F$ be an $n \times n$ nonsingular matrix with (fixed) singular values $\sigma_1 \geq \cdots \geq \sigma_n > 0$. Then

$$|T'FU| \cdot |U'F^{-1}T| \leq \prod_{i=1}^{\min(k,n-k)} \frac{(\sigma_i + \sigma_{n-i+1})^2}{4\sigma_i\sigma_{n-i+1}} \tag{24}$$

and

$$\begin{aligned} \operatorname{tr}(T'FU \cdot U'F^{-1}T) &\leq \sum_{i=1}^{k} \frac{(\sigma_i + \sigma_{n-i+1})^2}{4\sigma_i\sigma_{n-i+1}} && \text{if} \quad 2k \leq n \\ &\leq \sum_{i=1}^{n-k} \frac{(\sigma_i + \sigma_{n-i+1})^2}{4\sigma_i\sigma_{n-i+1}} + 2k - n && \text{if} \quad 2k > n. \end{aligned} \tag{25}$$

When $k = 1$ then (24) and (25) reduce to the inequality obtained in 1960 by W. Gilbert Strang [243],

$$t'Fu \cdot u'F^{-1}t \leq \frac{(\sigma_1 + \sigma_n)^2}{4\sigma_1\sigma_n}, \qquad \text{with} \quad t't = u'u = 1, \tag{26}$$

where the matrices $T$ and $U$ are now the $n \times 1$ vectors $t$ and $u$. When $F = A$, symmetric positive definite, and $t = u = x/\sqrt{x'x}$ then (26) becomes the Frucht–Kantorovich inequality (1).

Let $B$ and $C$ be symmetric $n \times n$ nonsingular matrices such that $BC = CB$ is positive definite, let $T$ be an $n \times k$ matrix of rank $k$ and let $\lambda_1 \geq \cdots \geq \lambda_n$ be the eigenvalues of $BC^{-1}$. Then, cf. [115] and [116],

$$\frac{|T'B^2T| \cdot |T'C^2T|}{|T'BCT|^2} \leq \prod_{i=1}^{\min(k,n-k)} \frac{(\lambda_i + \lambda_{n-i+1})^2}{4\lambda_i\lambda_{n-i+1}} \tag{27}$$

and

$$\begin{aligned} &\operatorname{tr}[T'B^2T(T'BCT)^{-1}T'C^2T(T'BCT)^{-1}] \\ &\leq \sum_{i=1}^{k} \frac{(\lambda_i + \lambda_{n-i+1})^2}{4\lambda_i\lambda_{n-i+1}} \qquad \text{if} \quad 2k \leq n \\ &\leq \sum_{i=1}^{n-k} \frac{(\lambda_i + \lambda_{n-i+1})^2}{4\lambda_i\lambda_{n-i+1}} + 2k - n \qquad \text{if} \quad 2k > n. \end{aligned}$$

## 1.7 Some Other Related Inequalities

Let us define the variance (with divisor $n$) of the real numbers $\lambda_1 \geq \cdots \geq \lambda_n$ with $n \geq 2$:

$$\sigma^2 = \frac{1}{n}\left\{\sum \lambda_i^2 - \frac{1}{n}\left(\sum \lambda_i\right\}^2\right) = \frac{1}{n}\sum \lambda_i^2 - \left(\frac{1}{n}\sum \lambda_i\right)^2. \tag{28}$$

Then

$$\sigma^2 \leq \frac{1}{4}(\lambda_1 - \lambda_n)^2 \leq \frac{n}{2}\sigma^2. \tag{29}$$

As pointed out in 1959 by Brauer and Mewborn [38], the second inequality in (29) was established (almost certainly for the first time) in 1918 by Julius von Szökefalvi Nagy in [195] and the first inequality (possibly for the first time) in 1935 by Tiberiu Popoviciu (1906–1975) in [218].

When $n = 2$ the inequality string (29) collapses to equality throughout. For $n \geq 3$ the second inequality in (29) is sharp: equality holds if and only if $\lambda_2 = \cdots = \lambda_{n-1} = (\lambda_1 + \lambda_n)/2$. The first inequality in (29) is sharp only when $n$ is even: equality then holds if and only if $\lambda_1 = \cdots = \lambda_k$ and $\lambda_{k+1} = \cdots \lambda_n$ with

$n = 2k$. When $n$ is odd then Popoviciu [218] showed that the first inequality in (29) may be strengthened to:

$$\frac{n^2}{n^2-1}\sigma^2 \le \frac{1}{4}(\lambda_1 - \lambda_n)^2, \tag{30}$$

which is sharp: equality holds if and only if $\lambda_1 = \cdots = \lambda_{k+1}$ and $\lambda_{k+2} = \cdots \lambda_n$ with $n = 2k + 1$.

In statistics the difference $\lambda_1 - \lambda_n$ is known as the "range" of the $\lambda_i$ and the ratio $(\lambda_1 - \lambda_n)/\sigma$, or $\sqrt{(n-1)/n}\,(\lambda_1 - \lambda_n)/\sigma$, as the "Studentised range", cf. e.g., David [58]. The second inequality in (29) was established in 1948, probably for the first time in a statistical context, by Keshavan Raghavan Nair (1910–) [196], while the first inequality in (29) and the inequality (30) were established in 1959 by Hugh Daniel Brunk [39] (these inequalities were also "given" in 1955 by George W. Thomson [248]); see also Nair (1959) [197], as well as the recent surveys by Arnold and Balakrishnan [15], §3.2, Jensen and Styan [107], and Olkin [204].

When the $\lambda_i$ are eigenvalues of a matrix $A$ then Mirsky [170] introduced the term "spread of a matrix" for $\lambda_1 - \lambda_n$ in 1956 and noted that the variance of the $\lambda_i$:

$$\sigma^2 = \frac{1}{n}\sum \lambda_i^2 - \left(\frac{1}{n}\sum \lambda_i\right)^2 = \frac{1}{n}\mathrm{tr}A^2 - \left(\frac{1}{n}\mathrm{tr}A\right)^2, \tag{31}$$

when, for example, the matrix $A$ is symmetric, cf. also Mirsky [171], Brauer and Mewborn [38], Johnson, Kumar and Wolkowicz [110], Marcus and Minc [160], pp. 167–168, and Wolkowicz and Styan [272].

Let $z$ be an $n \times 1$ vector with $z'z = 1$, and let $A$ be an $n \times n$ symmetric matrix, not necessarily nonnegative (or positive) definite. Then the inequality:

$$z'A^2z - (z'Az)^2 \le \frac{1}{4}(\lambda_1 - \lambda_n)^2, \tag{32}$$

was established in 1983 by Styan [244]; here, as before, $\lambda_1$ and $\lambda_n$ are, respectively, the largest and the smallest eigenvalues of the symmetric matrix $A$. Already in 1975 Bloomfield and Watson [33], however, proved that

$$\mathrm{tr}[Z'A^2Z - (Z'AZ)^2] \le \frac{1}{4}\sum_{i=1}^{p}(\lambda_i - \lambda_{n-i+1})^2, \tag{33}$$

where $Z$ is an $n \times p$ matrix with $Z'Z = I_p$ and $n \ge 2p$, and where, as before, $\lambda_1 \ge \cdots \ge \lambda_n$ are the ordered eigenvalues of the symmetric matrix $A$. When $p = 1$ then (33) reduces to (32). In 1996 Jia [109] reestablished (33) and referred to (32) as the Styan inequality (even in the title of [109]). We note that the matrix $A$ in (32) and (33) need not be nonnegative (or positive) definite since both sides of these inequalities remain unchanged if we replace $A$ by $A + kI$, where $k$ is a scalar (positive or negative).

The first inequality in (29) follows at once from (32) by putting $z = Pe/\sqrt{n}$, where $e$ is the $n \times 1$ vector with each element equal to 1, and the $n \times n$ orthogonal matrix $P$ diagonalises $A$ so that the diagonal matrix $\Lambda = \operatorname{diag}\{\lambda_i\} = P'AP$. We may use the "Henrici trick", which, as we saw above, shows that the Schweitzer inequality implies the Frucht–Kantorovich inequality, to show that the inequality (32) also implies the lower bound in (29) and so they are "essentially equivalent".

Now let $A$ be an $n \times n$ symmetric positive definite matrix with necessarily positive eigenvalues $\lambda_1 \geq \cdots \geq \lambda_n > 0$, and again let $Z$ be an $n \times p$ matrix such that $Z'Z = I_p$. Then Liu [133] posed the problem (solved by Bebiano, da Providencia and Li [25] and by Liu [134]): Show that provided $n \geq 2p$:

$$\frac{\operatorname{tr} Z'A^2Z}{\operatorname{tr}(Z'AZ)^2} \leq \frac{\sum_{j=1}^{p}(\lambda_j + \lambda_{n-j+1})^2}{4\sum_{j=1}^{p}\lambda_j\lambda_{n-j+1}}. \tag{34}$$

When $p = 1$ then the inequality (34) reduces to the Krasnosel'skiĭ–Kreĭn inequality (20).

Another extension of (32) was obtained by Liu and Neudecker [140], who showed that:

$$A_1^2 \odot A_2^2 - (A_1 \odot A_2)^2 \leq_{\mathrm{L}} \frac{1}{4}(\lambda_1 - \lambda_n)^2 I_p,$$

where $\odot$ denotes the elementwise Hadamard (or Schur) product and $\leq_{\mathrm{L}}$ the Löwner or nonnegative definite partial ordering. The matrices $A_1$ and $A_2$ are $n \times n$ and positive definite with $\lambda_1$ and $\lambda_n$, respectively, the largest and smallest eigenvalues of the Kronecker product $A_1 \otimes A_2$.

A similar extension was given by Pečarić, Puntanen and Styan [214], who proved that

$$T'A^2T - (T'AT)^2 \leq_{\mathrm{L}} \frac{1}{4}(\lambda_1 - \lambda_r)^2 T'AA^{\dagger}T,$$

where $A$ is an $n \times n$ nonnegative definite matrix of rank $r$ with $r$ positive eigenvalues $\lambda_1 \geq \cdots \geq \lambda_r > 0$, $A^{\dagger}$ is the Moore–Penrose inverse of $A$, and $T$ is an $n \times k$ matrix such that $AA^{\dagger}T$ is a partial isometry, i.e., $T'AA^{\dagger}T$ is (symmetric) idempotent.

For further closely related results see e.g. Baksalary and Puntanen [17], Liu and Neudecker [140], and Mond and Pečarić [180], [181], [182].

## Acknowledgements

Our recent research associated with the Frucht–Kantorovich inequality began at the Fourth International Workshop on Matrix Methods for Statistics (Montréal, July 1995) with the presentation and discussion of the papers [142] by Shuangzhe Liu and Heinz Neudecker and [266] by the late Geoffrey S. Watson. We are especially grateful to the late Roberto Frucht Wertheimer for his comments on some of the results in this paper and for providing us with copies of Bender [30] and Saleme [229]. Many thanks go to Heinz Neudecker, as well as to Rajendra Bhatia, S. W. Drury, Gene H. Golub, Shane T. Jensen, Shuangzhe Liu, Jorma Kaarlo Merikoski, Bertram Mond, Josip E. Pečarić, Simo Puntanen, and Geoffrey S. Watson, for helpful discussions.

Much of the biographical information was obtained by visiting the excellent O'Connor–Robertson Internet website [202], while web access to the databases MathSciNet (for *Mathematical Reviews*) and MATH Database (for *Zentralblatt für Mathematik*) has been of great help in compiling our bibliography.

Preliminary versions of this paper were presented at the Conference in Honor of Shayle R. Searle (Cornell University, August 1996), cf. [3], and at the Sixth International Workshop on Matrix Methods for Statistics (İstanbul, August 1997). Much of this paper is part of the MSc thesis by the first author [2]. This research was supported in part by a scholarship from Harran University, ŞanlıUrfa, Turkey (to the first author), and by a research grant from the Natural Sciences and Engineering Research Council of Canada (to the second author).

## References

## [and Bibliography]

[1] M. Alić, B. Mond, J. Pečarić, and V. Volenec. The arithmetic–geometric–harmonic-mean and related matrix inequalities. *Linear Algebra and Its Applications* 264:55-62 (1997) [Zbl 980.01234].

[2] Gülhan Alpargu. *The Kantorovich Inequality, With Some Extensions and With Some Statistical Applications.* MSc thesis, Dept. of Mathematics and Statistics, McGill University, Montréal (1996). (Includes English translations of Chen [51], Lin [126] and Schweitzer [234].)

[3] Gülhan Alpargu, S. W. Drury and George P. H. Styan. Some remarks on the Bloomfield–Watson–Knott Inequality and on some other inequalities related to the Kantorovich Inequality. In *Proceedings of the Conference in Honor of Shayle R. Searle, August 9–10, 1996*, Biometrics Unit, Cornell University, Ithaca, New York, pp. 125–143 (1998).

[4] Gülhan Alpargu and George P. H. Styan. Some remarks and a bibliography on the Kantorovich inequality. In *Multidimensional Statistical Analysis and Theory of Random Matrices: Proceedings of the Sixth Eugene Lukacs Symposium, Bowling Green, OH, USA, March 29–30, 1996* (Arjun K. Gupta and Vyacheslav L. Girko, eds.), VSP International Science Publishers, Utrecht & Zeist (The Netherlands), pp. 1–13 (1996) [MR 98h:15033, Zbl 879.60015].

[5] Gülhan Alpargu and George P. H. Styan. Research papers in Chinese on the Kantorovich and related inequalities. *Image: Bulletin of the International Linear Algebra Society* 17:12–13 (1996).

[6] Gülhan Alpargu and George P. H. Styan. A third bibliography on the Frucht-Kantorovich inequality. In *Three Bibliographies and a Guide* (George P. H. Styan, ed.), Prepared for the Seventh International Workshop on Matrices and Statistics, (Fort Lauderdale, Florida, December 1998), Dept. of Mathematics and Statistics, McGill University, pp. 17–26 (1998).

[7] T[heodore] W[ilbur] Anderson. On the theory of testing serial correlation. *Skandinavisk Aktuarietidskrift* 31:88–116 (1948) [MR 10:312f, Zbl 033.08001]. (Reprinted, with corrections and commentary, in *The Collected Papers of T. W. Anderson: 1943–1985* (George P. H. Styan, ed.), Wiley, New York, vol. 1, pp. 61–89 (1990) [MR 91j:01064].)

[8] T[heodore] W[ilbur] Anderson. *The Statistical Analysis of Time Series.* Wiley Classics Library [Reprint Edition]. Wiley, New York (1994) [Zbl 835.62074]. (Cf. §10.2, pp. 560–571. Original version: 1971 [MR 44:1169, Zbl 225.62108].)

[9] Tsuyoshi Ando. Concavity of certain maps on positive definite matrices and applications to Hadamard products. *Linear Algebra and Its Applications* 26:203–241 (1979) [MR 80f:15023, Zbl 495.15018].

[10] Tsuyoshi Ando. On the arithmetic–geometric–harmonic-mean inequalities for positive definite matrices. *Linear Algebra and Its Applications* 52/53:31–37 (1983) [MR 84j:15016, Zbl 516.15011].

[11] Tsuyoshi Ando. Quasi-orders on the positive cone of a $C^*$-algebra. *Linear and Multilinear Algebra* 41:81–94 (1996) [Zbl 870.46035].

[12] Tsuyoshi Ando. *Literatures on Matrix Inequalities and Hadamard Products.* Technical Report, Faculty of Economics, Hokusei Gakuen University, Sapporo, Japan, ii + 21 pp., March 1998.

[13] Tsuyoshi Ando. *Operator-Theoretic Methods for Matrix Inequalities.* Technical Report, Faculty of Economics, Hokusei Gakuen University, Sapporo, Japan, iv + 77 pp., March 1998.

[14] D. Andrica and C. Badea. Grüss' inequality for positive linear functionals. *Periodica Mathematica Hungarica (Budapest)* 19:155–167 (1988) [MR 89d:26018, Zbl 619.26011].

[15] Barry C. Arnold and Narayanaswamy Balakrishnan. *Relations, Bounds and Approximations for Order Statistics.* Lecture Notes in Statistics, vol. 53. Springer-Verlag, Berlin (1989) [MR 90i:62061, Zbl 703.62064].

[16] Zhaojun Bai and Gene H. Golub. Bounds for the trace of the inverse and the determinant of symmetric positive definite matrices. *Annals of Numerical Mathematics (Amsterdam)* 4:29–38 (1997) [MR 97k:65074, Zbl 883.15013].

[17] Jerzy K. Baksalary and Simo Puntanen. Generalised matrix versions of the Cauchy-Schwarz and Kantorovich inequalities. *Aequationes Mathematicae* 41:103–110 (1991) [MR 91k:15038, Zbl 723.15017].

[18] Jerzy K. Baksalary, Simo Puntanen and George P. H. Styan. On T. W. Anderson's contributions to solving the problem of when the ordinary least-squares estimator is best linear unbiased and to characterizing rank additivity of matrices. In *The Collected Papers of T. W. Anderson: 1943–1985* (George P. H. Styan, ed.), Wiley, New York, vol. 2, pp. 1579–1591 (1990).

[19] Earl R. Barnes and Alan J. Hoffman. Bounds for the spectrum of normal matrices. *Linear Algebra and Its Applications* 201:79–90 (1994) [MR 95c:15037, Zbl 803.15016].

[20] Flavio C. Bartmann and Peter Bloomfield. Inefficiency and correlation. *Biometrika* 68:67–71 (1981) [MR 83c:62093, Zbl 472.62073].

[21] D. M. Bătineţu-Giurgiu. În legătură cu inegalitatea L. V. Kantorovici [in Romanian: "Connections with the inequality of L. V. Kantorovich"]. *Gazeta Matematică (Bucharest)* 99(2):51–61 (1994) [Zbl 805.26014].

[22] D. M. Bătineţu-Giurgiu, Maria Bătineţu-Giurgiu and Valentin Gârban. *Analiza Matematică–Exerciţii si Probleme* [in Romanian: *Mathematical Analysis–Exercises and Problems*]. Editura Militară, Bucharest (1992). (Cf. Problem 83. Cited as [7] in Bătineţu-Giurgiu [21].)

[23] Friedrich L. Bauer. A further generalization of the Kantorovič inequality. *Numerische Mathematik* 3:117–119 (1961) [Zbl 099.24801].

[24] Friedrich L. Bauer and Alston S. Householder. Some inequalities involving the euclidean condition of a matrix. *Numerische Mathematik* 2:308–311 (1960) [Zbl 104:34502].

[25] Natália Bebiano, João da Providencia and Chi-Kwong Li. Solution 18-4.1 to Problem 18-4: Bounds for a ratio of matrix traces (posed by Liu [133]). *Image: Bulletin of the International Linear Algebra Society* 20:23–25 (1998).

[26] Eugen Beck. Komplementäre Ungleichungen bei vergleichbaren Mittelwerten [in German]. *Monatshefte für Mathematik* 73:289–308 (1969) [MR 41:3691, Zbl 181.33101].

[27] Edwin F. Beckenbach. Review of Frucht [71]. *Mathematical Reviews* 4:286 (1943).

[28] Edwin F. Beckenbach. On the inequality of Kantorovich. *American Mathematical Monthly* 71:606–619 (1964) [MR 29:5971, Zbl 126.28003].

[29] Edwin F. Beckenbach and Richard Bellman. *Inequalities*. Ergebnisse der Mathematik und ihrer Grenzgebiete, vol. 30, Fourth [Revised] Printing. Springer-Verlag, Berlin (1983) [MR 33:236, Zbl 513.26003]. (Cf. pp. 44–45. Original version: 1961 [MR 28:1266, Zbl 097.26502].)

[30] Abraham H. Bender. Problema N° 21.–Siendo $x_1, x_2, x_3, \ldots x_n$ números positivos con la condición $\sum x_i = k$, demostrar las desigualdades $\sum_{i=1}^{n} x_i^2 \geq k^2/n$, $\sum_{i=1}^{n} 1/x_i \geq n^2/k$ : 1° Solución [in Spanish]. *Mathematicæ Notæ–Boletin del Instituto de Matemática (Rosario)* 2:195–197 (1942). (Translated into English, together with Frucht [71], Levi [123] and Saleme [229], as Appendix A of Watson, Alpargu and Styan [266].)

[31] Rajendra Bhatia and Chandler Davis. More matrix forms of the arithmetic-geometric mean inequality. *SIAM Journal on Matrix Analysis and Applications* 14:132–136 (1993) [MR 94b:15017, Zbl 767.15012].

[32] Mieczysław Biernacki, H. Pidek and Czesław Ryll-Nardzewski. Sur une inégalité entre des intégrales définies [in French]. *Annales Universitatis Mariae Curie-Skłodowska, Sectio A: Mathematica (Lublin)* 4:1–4 (1950) [MR 13:118a, Zbl 040.31904].

[33] Peter Bloomfield and Geoffrey S. Watson. The inefficiency of least squares. *Biometrika* 62:121–128 (1975) [MR 51:9377, Zbl 308.62056].

[34] James V. Bondar. Comments on and complements to: *Inequalities: Theory of Majorization and its Applications* [Academic Press, New York, 1979] by A. W. Marshall and I. Olkin [162]. *Linear Algebra and Its Applications* 199:115–130 (1994) [MR 95c:00001, Zbl 793.26014].

[35] Ephraim J. Borowski and Jonathan M. Borwein. *Dictionary of Mathematics.* With the assistance of J. F. Bowers, A. Robertson and M. McQuillan. HarperCollins, London (1989) [MR 94b:00008].

[36] Jonathan [M.] Borwein, Carolyn Watters, and Ephraim [J.] Borowski. *The MathResource Interactive Math Dictionary on CD-ROM.* MathResources Inc., Halifax, Nova Scotia, and Springer-Verlag, Berlin (1997) [Zbl 884.00004].

[37] V[iktor Yakovlevich] Bouniakowsky [Buniakovski, Bunyakovsky]. Sur quelques inégalités concernant les intégrales ordinaires et les intégrales aux différences finies [in French]. *Mémoires de l'Académie Impériale des Sciences de St.-Pétersbourg, Septième Série*, vol. 1, no. 9, pp. 1–18 (1859). (Cf. pp. 3–4.)

[38] Alfred Brauer and A. C. Mewborn. The greatest distance between two characteristic roots of a matrix. *Duke Mathematical Journal* 26:653–661 (1959) [MR 22:10997, Zbl 095.01202].

[39] H[ugh] D[aniel] Brunk. Note on two papers of K. R. Nair. *Journal of the Indian Society of Agricultural Statistics*, 11:186–189 (1959).

[40] Wolfgang J. Bühler. Two proofs of the Kantorovich inequality and some generalizations [in English]. *Revista Colombiana de Matemáticas* 21:147–154 (1987) [MR 89k:15027, Zbl 656.60030].

[41] Peter S. Bullen. A chapter on inequalities. *Mathematical Medley (Singapore)* 21(2):48–69 (1993) [MR 95e:26021, Zbl 805.26016].

[42] Peter S. Bullen, Dragoslav S. Mitrinović and Petar M. Vasić. *Means and Their Inequalities* [in English]. Revised and Updated Edition. Mathematics and Its Applications: East European Series, vol. 31. D. Reidel, Dordrecht (1988) [MR 89d:26003, Zbl 687.26005]. (Cf. pp. 201–209. Original version: *Sredine i sa Njima Povezane Nejednakosti* [in Serbo-Croatian] 1977 [MR 80b:26001, Zbl 422.26009].)

[43] G. T. Cargo. An elementary, unified treatment of complementary inequalities. In *Inequalities-III: Proceedings of the Third Symposium on Inequalities held at the University of California, Los Angeles, September 1–9, 1969* (Oved Shisha, ed.), Academic Press, New York, pp. 39–63 (1972) [Zbl 306:26010].

[44] G. T. Cargo and Oved Shisha. Bounds on ratios of means. *Journal of Research of the National Bureau of Standards: Section B, Mathematics and Mathematical Physics* 66B:169–170 (1962) [Zbl 113:27102].

[45] J[ohn] W[illiam] S[cott] Cassels. Appendix 1: A proof of the inequality (3.3.2). In Watson [260], pp. 138–139 (1951). (Appendix published as Cassels [46].)

[46] J[ohn] W[illiam] S[cott] Cassels. Appendix: A proof of the inequality (3.2). In Watson [261], pp. 340–341 (1955). (Original version: Cassels [45].)

[47] [Baron] Augustin-Louis Cauchy. Note II: Sur les formules qui résultent de l'emploi du signe > ou <, et sur les moyennes entre plusieurs quantités [in French]. In *Cours d'Analyse de l'École Royale Polytechnique, Première Partie: Analyse Algébrique.* L'Imprimerie Royale Chez Debure Frères, Libraires du Roi et de la Bibliothèque du Roi, Paris, pp. 360–377 (1821). (Cf. pp. 373–374. Reprinted in *Œuvres Complètes d'Augustin Cauchy, Publiées sous la direction scientifique de l'Académie des Sciences et sous les auspices de M. le Ministre de l'Instruction Publique, Seconde Série: Mémoires Divers et Ouvrages*, Gauthier-Villars, Paris, vol. 3 (Mémoires publiés en corps d'ouvrage), pp. 360–377 (1897).)

[48] Sin-Chung Chang. On the product of convex combinations and the Kantorovich inequality. *Linear Algebra and Its Applications* 93:9–38 (1987) [MR 88d:15023, Zbl 626.15013].

[49] Yu-Chi Chang and Chun-Tu Lin. A generalization of Kantorovich's inequality with application to the measurement of inefficiency of least squares. Report B-84-6, Institute of Statistics, Academia Sinica, Taipei, Taiwan (1984).

[50] Jianbao Chen and T. Chen. The error ratio efficiency of the mean square in the general Gauss–Markov model. In *Proceedings of the Fourth China–Japan Symposium on Statistics* (Kunming, China), vol. 4, pp. 17–19 (year not known).

[51] Yong-Lin Chen. A simpler proof of Kantorovich's inequality and of a generalization of it [in Chinese]. *Shuxue de Shijian yu Renshi / Mathematics in Practice and Theory (Beijing)* 1987(4):78–79 (1987) [MR 89c:26033]. (Translated into English as Appendix C of [2].)

[52] Chiou-Wen Chiang. On a generalization of the Wielandt inequality [in English]. *Soochow Journal of Mathematics (Taipei)* 21:117–120 (1995) [MR 95m:47029, Zbl 824.15018].

[53] Achim Clausing. Kantorovich-type inequalities. *American Mathematical Monthly* 89:314 & 327–330 (1982) [MR 83g:26026, Zbl 491.26009].

[54] Salvatore Coen. Beppo Levi: la vita [in Italian]. In *Seminari di Geometria 1991–1993* (Salvatore Coen, ed.), Dipartimento di Matematica, Università degli Studi di Bologna, pp. 193–232 (1994) [MR 95c:01034, Zbl 795.01022].

[55] Noel Cressie. M-estimation in the presence of unequal scale. *Statistica Neerlandica* 34:19–32 (1980) [MR 81i:62067, Zbl 428.62002].

[56] Daniel Culea. Inegalitāţile lui P. Schweitzer şi L. B. Kantorovici, aplicaţii [in Romanian: "Inequalities of P. Schweitzer and L. V. Kantorovich: applications"]. *Gazeta Matematicā (Bucharest)* "*Seria Metodicá* no. 2, pp. 62–70 (1991)". (Cited as [9] in Bătineţu-Giurgiu [21].)

[57] Malaya Das Gupta and Kali Charan Das. A closer look at the Kantorovich inequality. *Journal of Mathematical and Physical Sciences (Madras)* 27:335–346 (1993) [MR 95m:47030, Zbl 817.47021].

[58] Herbert A. David. Studentised range. In *Encyclopedia of Statistical Sciences, Volume 9: Strata Chart to Zyskind–Martin Models; Cumulative Index, Volumes 1–9* (Samuel Kotz, Norman L. Johnson and Campbell B. Read, eds.), Wiley, New York, pp. 39–43.

[59] Chandler Davis. Extending the Kantorovič inequality to normal matrices. *Linear Algebra and Its Applications* 31:173–177 (1980) [MR 81g:15020, Zbl 434.15004].

[60] Joaquín Basilio Diaz, Alan J. Goldman and Frederic T. Metcalf. Equivalence of certain inequalities complementing those of Cauchy–Schwarz and Hölder. *Journal of Research of the National Bureau of Standards: Section B, Mathematics and Mathematical Physics* 68B:147–149 (1964) [MR 31:295, Zbl 125.03001].

[61] Joaquín Basilio Diaz and Frederic T. Metcalf. Stronger forms of a class of inequalities of G. Pólya–G. Szegö, and L. V. Kantorovich. *Bulletin of the American Mathematical Society* 69:415–418 (1963) [MR 26:3846, Zbl 129.26904].

[62] Joaquín Basilio Diaz and Frederic T. Metcalf. Complementary inequalities I: Inequalities complementary to Cauchy's inequality for sums of real numbers. *Journal of Mathematical Analysis and Applications* 9:59–74 (1964) [MR 30:4879, Zbl 135.34702]. (Published in part as [66].)

[63] Joaquín Basilio Diaz and Frederic T. Metcalf. Complementary inequalities, II: Inequalities complementary to the Buniakowsky-Schwarz inequality for integrals. *Journal of Mathematical Analysis and Applications* 9:278–293 (1964) [MR 30:4880, Zbl 135.34702].

[64] Joaquín Basilio Diaz and Frederic T. Metcalf. Complementary inequalities, IV: Inequalities complementary to Cauchy's inequality for sums of complex numbers. *Rendiconti del Circolo Matematico di Palermo, Serie II* 13:291–328 (1964) [MR 33:1419, Zbl 151.17704]. (Part III published in 1965.)

[65] Joaquín Basilio Diaz and Frederic T. Metcalf. Complementary inequalities, III: Inequalities complementary to Schwarz's inequality in Hilbert space. *Mathematische Annalen* 162:120–139 (1965) [MR 32:6235, Zbl 135.34702].

[66] Joaquín Basilio Diaz and Frederic T. Metcalf. Inequalities complementary to Cauchy's inequality for sums of real numbers. In *Inequalities: Proceedings of a Symposium held at Wright-Patterson Air Force Base, Ohio, August 19–27, 1965* (Oved Shisha, ed.), Academic Press, New York, pp. 73–77 (1967) [MR 36:5280]. ("Based on the talk 'Four inequalities in search of two authors' and on Diaz and Metcalf [62]".)

[67] Zdeněk Dostál. Rennie's generalization of Kantorovich's inequality in a Hilbert space [in English]. *Sborník Prací Přírodovědečke Fakulty University Palackého v Olomouci, Matematika (Prague)* 19:101–102 (1980) [MR 82j:47025, Zbl 481.47015].

[68] Morris L. Eaton. A maximisation problem and its applications to canonical correlation. *Journal of Multivariate Analysis* 6:422–425 (1976) [MR 54:5272, Zbl 332.15008].

[69] Euclid [of Alexandria]. *The Thirteen Books of Euclid's Elements.* Translated from the text of Heiberg with Introduction and Commentary by Sir Thomas L[ittle] Heath. Second Edition, Revised with Additions. Three volumes, Dover, New York (1956). (Cf. Book V, Proposition 25 and Commentary: vol. II, pp. 185–186.)

[70] Ky Fan. Some matrix inequalities. *Abhandlungen aus dem Mathematischen Seminar der Hamburgischen Universität* 29:185–196 (1966) [MR 35:2915, Zbl 145.25303].

[71] Roberto Frucht [Wertheimer]. Sobre algunas desigualdades: Observación relativa a la solución del Problema N° 21, indicada por el Ing. Ernesto M. Saleme [in Spanish]. *Mathematicæ Notæ–Boletin del Instituto de*

*Matemática (Rosario)* 3:41–46 (1943) [MR 4:286, Zbl 060.14901]. (This paper, which includes an untitled appendix by Levi [123], builds on the solution and generalization (of Problem No. 21) by Saleme [229]. Translated into English, together with Bender [30], Levi [123] and Saleme [229], as Appendix A of Watson, Alpargu and Styan [266].)

[72] Masatoshi Fujii, Takayuki Furuta, Ritsuo Nakamoto, and Sin-Ei Takahasi. Operator inequalities and covariance in noncommutative probability. *Mathematica Japonica* 46:317–320 (1997) [MR 98j:46072, Zbl 902.47014].

[73] Masatoshi Fujii, S. Izumino, Ritsuo Nakamoto, and Yuki Seo. Operator inequalities related to Cauchy–Schwarz and Hölder–McCarthy inequalities. *Nihonkai Mathematical Journal* 8:117–122 (1997) [MR 98j:47037].

[74] Masatoshi Fujii, Eizaburo Kamei and Akemi Matsumoto. Parameterised Kantorovich inequality for positive operators. *Nihonkai Mathematical Journal* 6:129–134 (1995) [MR 96i:47032].

[75] Masatoshi Fujii, Yoshikazu Katayama and Ritsuo Nakamoto. Generalizations of the Wielandt theorem. Unpublished manuscript, 6 pp., Submitted for publication (1997).

[76] Takayuki Furuta. Extensions of Hölder–McCarthy and Kantorovich inequalities and their applications. *Proceedings of the Japan Academy Series A, Mathematical Sciences* 73:38–41 (1997) [MR 98e:26020, Zbl 885.47005].

[77] Takayuki Furuta. Operator inequalities associated with Hölder–McCarthy and Kantorovich inequalities. *Journal of Inequalities and Applications* 2:137–148 (1998) [Zbl 980.31729].

[78] Takayuki Furuta. Extensions of Mond–Pečarić generalization of Kantorovich inequality. Preprint. (Cited in Mićić *et al.* [167].)

[79] Takayuki Furuta. Two extensionms of Ky Fan generalization and Mond–Pečarić matrix version generalization of Kantorovich inequality Preprint. (Cited by [167].)

[80] Dao-De Gao. Estimation of the deviation between the least squares and the best linear unbiased estimators of the mean vector in a linear model [in Chinese]. *Yingyong Gailü Tongji / Chinese Journal of Applied Probability and Statistics (Shanghai)* 8(4):391–397 (1992) [MR 93m:62150].

[81] Dao-De Gao. A new class of Kantorovich-type inequalities and their applications in statistics [in Chinese]. *Xitong Kexue yu Shuxue / Journal of Systems Science and Mathematical Sciences (Beijing)* 13(4):331–337 (1993) [MR 94k:62102, Zbl 790.62064].

[82] Dao-De Gao. Some extensions of the Kantorovich inequality and their applications in statistics [in Chinese with English summary]. *Gaoxiao Yingyong Shuxue Xuebao / A Journal of Chinese Universities, Series A (Hangzhou)* 10(2):183–188 (1995) [MR 96k:15009, Zbl 841.62049].

[83] Dao-De Gao and Jing-Long Wang. Efficiency of generalised least square estimates [in Chinese]. *Xitong Kexue yu Shuxue / Journal of Systems Science and Mathematical Sciences (Beijing)* 10(2):125–130 (1990) [MR 96c:62059].

[84] Şerban A. Gheorghiu. Note sur une inégalité de Cauchy [in French]. *Bulletin Mathématique de la Société Roumaine des Sciences (Bucharest)* 35:117–119 (1933) [Zbl 008:34504].

[85] Alan J. Goldman. A generalization of Rennie's inequality. *Journal of Research of the National Bureau of Standards: Section B, Mathematics and Mathematical Physics* 68B:59–63 (1964) [MR 29:5970, Zbl 136.34202].

[86] Allen A. Goldstein. A modified Kantorovich inequality for the convergence of Newton's method. In *Mathematical Developments arising from Linear Programming: Proceedings of the Joint Summer Research Conference, Bowdoin College, Brunswick, Maine, June 25–July 1, 1988* (Jeffrey C. Lagarias and Michael J. Todd, eds.), Contemporary Mathematics, vol. 114, pp. 285–294 (1990) [MR 92d:90078, Zbl 725.90088].

[87] Gene H. Golub. Comparison of the variance of minimum variance and weighted least squares regression coefficients. *The Annals of Mathematical Statistics* 34:984–991 (1963) [MR 27:5336, Zbl 203.21404].

[88] Carlos Gonzalez de la Fuente. Roberto W. Frucht: The mathematician, the teacher, the man. *Scientia, Series A: Mathematical Sciences, Universidad Técnica Federico Santa María (Valparaíso)* 1:iii–v (1988) [Zbl 679.01006].

[89] Werner Greub and Werner Rheinboldt. On a generalization of an inequality of L. V. Kantorovich. *Proceedings of the American Mathematical Society* 10:407–415 (1959) [MR 21:37744, Zbl 093.12405].

[90] Gerhard Grüss. Über das Maximum des absoluten Betrages von $\frac{1}{b-a}\int_a^b f(x)g(x)dx - \frac{1}{(b-a)^2}\int_a^b f(x)dx \int_a^b g(x)dx$ [in German]. *Mathematische Zeitschrift* 39:215–226 (1934) [Zbl 010:01602].

[91] Karl E. Gustafson. Commentary on "Topics in the analytic theory of matrices, Section 23: Singular angles of a square matrix". In *Helmut Wielandt: Mathematische Werke / Mathematical Works–Volume 2: Linear Algebra and Analysis* (Bertram Huppert and Hans Schneider, eds.), Walter de Gruyter, Berlin, pp. 356–368.

[92] Karl E. Gustafson. *Lectures on Computational Fluid Dynamics, Mathematical Physics, and Linear Algebra*. Edited and with a foreword by T. Abe and K. Kuwahara. With comments by Abe, Kuwahara, Izumi Ojima and Mitsuharu Ôtani. World Scientific, Singapore (1997) [MR 99b:00008].

[93] Karl E. Gustafson. The geometrical meaning of the Kantorovich–Wielandt inequalities. *Linear Algebra and Its Applications*, in press (1999).

[94] Shelby J. Haberman. How much do Gauss–Markov and least-square estimates differ? A coordinate-free approach. *The Annals of Statistics* 3:982–990 (1975) [MR 51:14437, Zbl 311.62031].

[95] Edward James Hannan. *Time Series Analysis.* Methuen, London (1960) [MR 22:5105, Zbl 095.13204]. (Cf. pp. 111–113.)

[96] Edward James Hannan. *Multiple Time Series.* Wiley, New York (1970) [MR 43:5673, Zbl 211.49804]. (Cf. pp. 420–423.)

[97] Zhi-Chuan Hao. Generalization and applications of Kantorovich inequality [in English]. Economic Information Department of Gui Zhou Finance and Economic Institute of China, Preprint 13 pp. (ca. 1992). (Results published, in part, in Mitrinović, Pečarić and Fink [174], pp. 122–123.)

[98] Frank Harary. Homage to Roberto Frucht. *Journal of Graph Theory* 6:97–99 (1982) [MR 83i:01065, Zbl 492.05033].

[99] Godfrey Harold Hardy, John Edensor Littlewood and George Pólya. *Inequalities.* Second Edition. Cambridge University Press (1952) [MR 89d:26016, Zbl 634.26008]. (Cf. # 71, p. 62. Original version: 1934; first paperback version of the Second Edition (reprinted without any changes): 1988, reprinted: 1994.)

[100] John Z. Hearon. A generalised matrix version of Rennie's inequality. *Journal of Research of the National Bureau of Standards: Section B, Mathematics and Mathematical Physics* 71B:61–64 (1967) [MR 37:232, Zbl 178.03102].

[101] Peter Henrici. Two remarks on the Kantorovich inequality. *American Mathematical Monthly* 68:904–906 (1961) [MR 25:91, Zbl 102.04001].

[102] . Grant H. Hillier and Maxwell L. King. Linear regression with correlated errors: bounds on coefficient estimates and $t$-values. In *Specification Analysis in the Linear Model (In Honour of Donald Cochrane* (Maxwell L. King and David E. A. Giles, eds.), Routledge & Kegan Paul, London, pp. 74–80 (1987).

[103] Roger A. Horn and Charles R. Johnson. *Matrix Analysis.* Corrected Reprint Edition. Cambridge University Press (1990) [MR 91i:15001, Zbl 801.15001]. (Cf. pp. 441–445, 451–452. Original version: 1985 [MR 87e:15001, Zbl 729.15001].)

[104] Alston S. Householder. The Kantorovich and some related inequalities. *SIAM Review* 7:463–473 (1965) [MR 32:6656, Zbl 161.03001].

[105] Alston S. Householder. *The Theory of Matrices in Numerical Analysis.* Second [Reprint] Edition. Dover, New York (1975) [MR 51:14539, Zbl 329.65003]. (Cf. §3.4, pp. 81–85. Original version: Blaisdell, New York, 1964 [MR 30:5475, Zbl 161.12101].)

[106] Shane Tyler Jensen. *The Laguerre-Samuelson Inequality with Extensions and Applications in Statistics and Matrix Theory.* MSc thesis, Dept. of Mathematics and Statistics, McGill University, Montréal (1999).

[107] Shane T. Jensen and George P. H. Styan. Some comments and a bibliography on the Laguerre–Samuelson inequalities with extensions and applications to statistics and matrix theory. *Analytic and Geometric Inequalities and Their Applications* (Hari M. Srivastava and Themistocles M. Rassias, eds.), Kluwer Academic Publishers, in press (1999).

[108] Shane T. Jensen and George P. H. Styan. Some comments and a bibliography on the von Szökefalvi Nagy–Popoviciu and Nair–Thomson inequalities with extensions and applications in statistics and matrix theory. In preparation for presentation at The Eighth International Workshop on Matrices and Statistics (Tampere, Finland: August 1999).

[109] Zhongzhen Jia. An extension of Styan inequality [in Chinese with English summary]. *Gongcheng Shuxue Xuebao / Journal of Engineering Mathematics (Xi'an)* 13(1):122–126 (1996) [MR 97i:15021].

[110] Charles R. Johnson, Ravinder Kumar and Henry Wolkowicz. Lower bounds for the spread of a matrix. *Linear Algebra and Its Applications* 71:161–173 (1985).

[111] L[eonid] V[ital′evich] Kantorovich. Funkcional'nyi analiz i prikladnaya matematika [in Russian: Functional analysis and applied mathematics]. *Uspekhi Matematičeskiĭ Nauk, Novaya Seriya* 3(6/28):89–185 (1948). (Cf. pp. 142–144. Translated into English as [112].)

[112] Leonid Vital′evich Kantorovich. *Functional Analysis and Applied Mathematics* [in English]. Translated from the Russian [111] by Curtis D. Benster and edited by George E. Forsythe. NBS Report no. 1509 (1101-10-5100), U. S. Dept. of Commerce, National Bureau of Standards, Los Angeles (1952) [MR 14:766d]. (Cf. pp. 106–109.)

[113] Louis Kates, Serge Kruk and Henry Wolkowicz. Solution No. 1 to Problem 16-4: Algebraic reverse of a convex matrix inequality (posed by Liu [132]). *Image: Bulletin of the International Linear Algebra Society* 17:31 (1996).

[114] C[hinubhai] G[helabhai] Khatri. An inequality concerning the difference of variances of two different estimates. *Mathematics Today (Ahmedabad)* 2:29–32 (1984) [MR 87g:62104, Zbl 573.62049].

[115] C[hinubhai] G[helabhai] Khatri and C. Radhakrishna Rao. Some extensions of the Kantorovich inequality and statistical applications. *Journal of Multivariate Analysis* 11:498–505 (1981) [MR 84e:15006, Zbl 482.15010].

[116] C[hinubhai] G[helabhai] Khatri and C. Radhakrishna Rao. Some generalizations of Kantorovich inequality. *Sankhyā Series A* 44:91–102 (1982) [MR 85m:15005, Zbl 586.62095].

[117] Konrad Knopp. Über die maximalen Abstände und Verhältnisse verschiedener Mittelwerte [in German]. *Mathematische Zeitschrift* 39:768–776 (1935) [Zbl 011:01203].

[118] Martin Knott. On the minimum efficiency of least squares. *Biometrika* 62:129–132 (1975) [MR 51:9378, Zbl 308.62057].

[119] Mark Aleksandrovich Krasnosel'skiĭ and Selim Griogor'evich Kreĭn. An iteration process with minimal residuals [in Russian]. *Matematičeskiĭ Sbornik, Novaya Seriya* 31(73):315–334 (1952) [MR 14:692d]. (Cf. pp. 323–325.)

[120] Hsu-Tung Ku, Mei-Chin Ku and Xin-Min Zhang. Inequalities for symmetric means, symmetric harmonic means, and their applications. *Bulletin of the Australian Mathematical Society* 56:409–420 (1997) [Zbl 894.26010].

[121] Hiroshi Kurata and Takeaki Kariya. Least upper bound for the covariance matrix of a generalised least squares estimator in regression with applications to a seemingly unrelated regression model and a heteroscedastic model. *The Annals of Statistics* 24:1547–1559 (1996) [Zbl 868.62060].

[122] [József] Kürschák. Kitűzött feladat [in Hungarian: "Posed problem"]. *Matematikai és Physikai Lapok* 23:378 (1914).

[123] B[eppo] L[evi]. Untitled appendix [in Spanish], in Frucht [71], pp. 45–46 (1943). (Translated into English, together with Bender [30], Frucht [71] and Saleme [229], in Watson, Alpargu and Styan [266].)

[124] Chi-Kwong Li and Roy Mathias. Matrix inequalities involving a positive linear map. *Linear and Multilinear Algebra* 41:221–231 (1996) [MR 97k:15042, Zbl 885.15013].

[125] Chun-Tu Lin. Extrema of quadratic forms and statistical applications. *Communications in Statistics–Theory and Methods* 13:1517–1520 (1984) [MR 86a:62083, Zbl 554.62055].

[126] Chun-Tu Lin. An extension of the Kantorovich inequality and its application to estimating parameters in linear models [in Chinese]. *Xitong Kexue yu Shuxue–Journal of Systems Science and Mathematical Sciences* 6(3):217–220 (1986) [MR 88a:62176, Zbl 628.15014]. (Translated into English by Ming-Yan Venus Chiu as Appendix B of [2].)

[127] Chun-Tu Lin. Extremes of determinants and optimality of canonical variables. *Communications in Statistics–Simulation and Computation* 19:1415–1430 (1990) [MR 91m:62091, Zbl 850.62483].

[128] Chun-Tu Lin. Extremes of ratios of determinants and canonical correlation variables [in English]. *Acta Mathematicae Applicatae Sinica, English Series)* 7(3):272–278 (1991) [MR 93e:62154, Zbl 735.62058].

[129] Erkki P. Liski, Simo Puntanen and Song-Gui Wang. Bounds for the trace of the difference of the covariance matrices of the OLSE and BLUE. *Linear Algebra and Its Applications* 176:121–130 (1992) [MR 94d:62167, Zbl 753.62033].

[130] Shuangzhe Liu. *Contributions to Matrix Calculus and Applications in Econometrics.* Tinbergen Institute Research Series, no. 106. Thesis Publishers, Amsterdam (1995). (Cf. pp. 25–58.)

[131] Shuangzhe Liu. Cauchy and Kantorovich inequalities involving sums of matrices. Unpublished paper, Institute of Actuarial Science and Econometrics, University of Amsterdam, 5 pp. (ca. 1995).

[132] Shuangzhe Liu. Problem 16-4: Reverse of a convex matrix inequality. *Image: Bulletin of the International Linear Algebra Society* 16:32 (1996). (Solved by Kates, Kruk and Wolkowicz [113] and by Olkin [205].)

[133] Shuangzhe Liu. Problem 18-4: Bounds for a ratio of matrix traces. *Image: Bulletin of the International Linear Algebra Society* 18:32 & 19:27 (1997). (Solved by Bebiano, da Providencia and Li [25] and by Liu [134].)

[134] Shuangzhe Liu. Solution 18-4.2 to Problem 18-4: Bounds for a ratio of matrix traces (posed by Liu [133]). *Image: Bulletin of the International Linear Algebra Society* 20:25–26 (1998).

[135] Shuangzhe Liu. On matrix trace Kantorovich-type inequalities. Chapter 2 in *Innovations in Multivariate Statistical Analysis: A Festschrift for Heinz Neudecker* (R. J. Heijmans, D. S. G. Pollock & A. Satorra. eds.), Kluwer Academic Publishers, pp. 39–50 (1999).

[136] Shuangzhe Liu. Efficiency comparisons between the OLSE and BLUE in a singular linear model. *Journal of Statistical Planning and Inference*, in press (1999).

[137] Shuangzhe Liu. Efficiency comparisons between the OLSE and the BLUE based on matrix determinant Kantorovich-type inequalities, Institute of Statistics and Econometrics, Universität Basel, Preprint 8 pp. (1999).

[138] Shuangzhe Liu and Heinz Neudecker. Kantorovich inequalities and efficiency of estimators useful in panel data analysis. Technical Report no. AE 17/94, Institute of Actuarial Science and Econometrics, University of Amsterdam, 14 pp. (1994).

[139] Shuangzhe Liu and Heinz Neudecker. Matrix-trace Cauchy–Schwarz inequalities and applications in canonical correlation analysis. *Statistical Papers* 36:287–298 (1995) [MR 97k:15043, Zbl 838.62047].

[140] Shuangzhe Liu and Heinz Neudecker. Several matrix Kantorovich-type inequalities *Journal of Mathematical Analysis and Applications* 197:23–26 (1996) [MR 97d:26023, Zbl 853.15014].

[141] Shuangzhe Liu and Heinz Neudecker. Kantorovich and Cauchy–Schwarz inequalities involving positive semidefinite matrices, and efficiency comparisons for a singular linear model. *Linear Algebra and Its Applications* 259:209–221 (1997) [MR 98c:15056, Zbl 881.15020].

[142] Shuangzhe Liu and Heinz Neudecker. Kantorovich inequalities and efficiency comparisons for several classes of estimators in linear models. *Statistica Neerlandica* 51:345–355 (1997) [MR 99b:62106, Zbl 887.62066].

[143] Shuangzhe Liu and Heinz Neudecker. A survey of Cauchy–Schwarz and Kantorovich-type matrix inequalities, *Statistical Papers*, 40:55–73 (1999).

[144] Shuangzhe Liu, Wolfgang Polasek and Heinz Neudecker. Equality conditions for matrix Kantorovich-type inequalities. *Journal of Mathematical Analysis and Applications* 212:517–528 (1997) [MR 98e:15012, Zbl 882.15018].

[145] G. Loizou. Bounds for the inequality of Wielandt. *Numerische Mathematik* 10:142–146 (1967) [MR 36:1098].

[146] C. J. Lombard. Union–intersection tests for sphericity. *South African Statistical Journal* 17:165–175 (1983) [MR 86b:62089, Zbl 548.62036].

[147] Yu-Tong Lou. The generalization of Schweitzer's inequality and Grüss's inequality and their relation [in Chinese with English summary]. *Qufu Shifan Daxue Xuebao Ziran Kexue Ban / Journal of Qufu Normal University, Natural Sciences Edition (Qufu)* 17(4):24–28 (1991) [Zbl 751.26008].

[148] Chang-yu Lu. Generalised matrix versions of the Wielandt inequality with some statistical applications [in English]. Dept. of Mathematics, Northeast Normal University, Changchun, Jilin, Preprint, 8 pp. (1999).

[149] Alexandru Lupaş. A remark on the Schweitzer and Kantorovich inequalities. *Publikacije Elektrotehničkog Fakulteta Univerziteta u Beogradu, Serija Matematika i Fizika / Publications de la Faculté d'Électrotechnique de l'Université à Belgrade, Série Mathématiques et Physique* 383:13–15 (1972) [MR 49:5278, Zbl 248.26016].

[150] T. A. Magness and J. B. McGuire. Comparison of least squares and minimum variance estimates of regression parameters. *The Annals of Mathematical Statistics* 33:462–470 (1962) [Zbl 212.22601].

[151] Jan R. Magnus and Heinz Neudecker. *Matrix Differential Calculus with Applications in Statistics and Econometrics.* Revised Edition. Wiley, Chichester (1999) [Original version, 1988: MR 89g:15001, Zbl 651.15001].

[152] Endre Makai. On generalizations of an inequality due to Pólya and Szegő [in English]. *Acta Mathematica Hungarica* 12:189–191 (1961) [MR 27:251, Zbl 098.26405].

[153] Valeriĭ Leonidovich Makarov and Sergeĭ L'vovich Sobolev. Academician L. V. Kantorovich (19 January 1912 to 7 April 1986) [in English: translated from the Russian by Lev J. Leifman]. In *Functional Analysis, Optimization, and Mathematical Economics: A Collection of Papers Dedicated to the Memory of Leonid Vital'evich Kantorovich* (Lev J. Leifman, ed.), Oxford University Press, pp. 1–7 (1990) [MR 91i:01098].

[154] Timo Mäkeläinen. A specification analysis of the general linear model. *Commentationes Physico-Mathematicae Societatis Scientiarum Fennicae* 38:55–100 (1970) [Zbl 207.49601].

[155] S. M. Malamud. On perturbation of positive definite operators with fixed bounds of the spectrum. *Dopovīdī Natsīonaĺnoï Akademī ï Nauk Ukraïni, Matematika Prirodoznavstvo Tekhnīchnī Nauki* 1997(4):38–42 (1997) [MR 98k:47041].

[156] Marvin Marcus. A unified exposition of some classical matrix theorems. *Linear and Multilinear Algebra* 25:137–147 (1989) [MR 91a:15004, Zbl 687.15016].

[157] Marvin Marcus and Afton H. Cayford. Further results on the Kantorovich inequality [abstract]. *Notices of the American Mathematical Society* 9:300 (1962). (Paper published as Marcus and Cayford [158].)

[158] Marvin Marcus and Afton H. Cayford. Equality in certain inequalities. *Pacific Journal of Mathematics* 13:1319–1329 (1963) [MR 28:168]. (Abstract: Marcus and Cayford [157].)

[159] Marvin Marcus and Nisar A. Khan. Some generalizations of Kantorovich's inequality. *Portugaliae Mathematica* 20:33–38 (1961) [MR 24:A127].

[160] Marvin Marcus and Henryk Minc. *A Survey of Matrix Theory and Matrix Inequalities.* [Second Corrected] Reprint Edition. Dover, New York (1992). (Cf. §3.5.2, p. 110; §4.3, p.117. Original version: Allyn and Bacon, Boston, 1964 [MR 29:112, Zbl 126.02404]; reprinted (with corrections): Prindle, Weber and Schmidt, Boston, 1969.)

[161] Albert W. Marshall and Ingram Olkin. Reversal of the Lyapunov, Hölder, and Minkowski inequalities and other extensions of the Kantorovich inequality. *Journal of Mathematical Analysis and Applications* 8:503–514 (1964) [MR 30:4881].

[162] Albert W. Marshall and Ingram Olkin. *Inequalities: Theory of Majorization and its Applications.* Mathematics in Science and Engineering, vol. 143, Academic Press, New York (1979) [MR 81b:00002, Zbl 437.26007]. (Cf. # D.3.a, p. 71. "Comments on and complements to" by Bondar [34].)

[163] Albert W. Marshall and Ingram Olkin. Matrix versions of the Cauchy and Kantorovich inequalities. *Aequationes Mathematicae* 40:89–93 (1990) [MR 91f:15045, Zbl 706.15019].

[164] Motosaburo Masuyama. An upper bound for CV. *TRU Mathematics* 18(1):25–28 [MR 83k:26016].

[165] Motosaburo Masuyama. A refinement of the Pólya–Szegö inequality and a refined upper bound of CV. *TRU Mathematics* 21:201-205 (1985) [MR 87k:26023, Zbl 645.26014].

[166] Günter Meinardus. Über eine Verallgemeinerung einer Ungleichung von L. V. Kantorowitsch [in German]. *Numerische Mathematik* 5:14–23 (1963) [MR 28:3525, Zbl 114.32001].

[167] Jadranka Mićić, Yuki Seo, Sin-Ei Takahasi, and Masaru Tominaga. Inequalities of Furuta and Mond–Pečarić. *Mathematical Inequalities & Applications* 2:83–111 (1999).

[168] Lucia Migliaccio and Luciana Nania. The Kantorovich inequality under integral constraints. *Journal of Mathematical Analysis and Applications* 181:524–530 (1994) [MR 94m:26021, Zbl 805.26013].

[169] Kenneth S. Miller. *Some Eclectic Matrix Theory.* Robert E. Kreiger, Malabar, Florida (1987) [MR 88h:15002, Zbl 654.15002]. (Cf. pp. 29–31.)

[170] L[eonid] Mirsky. The spread of a matrix. *Mathematika (London)* 3:127–130 (1956) [MR 18:460c, Zbl 073.00903].

[171] L[eonid] Mirsky. Inequalities for normal and Hermitian matrices. *Duke Mathematics Journal*, 24:591–599 (1957) [MR 19:832c, Zbl 081.25101].

[172] Dragoslav S. Mitrinović. *Analytic Inequalities.* In cooperation with Petar M. Vasić. Die Grundlehren der mathematischen Wissenschaften in Einzeldarstellungen mit besonderer Berücksichtigung der Anwendungsgebiete, vol. 165, Springer-Verlag, New York (1970) [MR 43:448, Zbl 199.38101]. (Cf. pp. 59–66. Translated into Serbo-Croatian and updated: *Analitičke Nejednakosti.* University of Belgrade Monographs, vol. 2. Gra̅devinska Knjiga, Belgrade.)

[173] Dragoslav S. Mitrinović and Josip E. Pečarić. Comments on an inequality of M. Masuyama. *SUT Journal of Mathematics* 27(1):89–91 (1991) [MR 94a:26048, Zbl 751.26009].

[174] Dragoslav S. Mitrinović, Josip E. Pečarić and Arlington M. Fink (1993). *Classical and New Inequalities in Analysis.* Mathematics and Its Applications: East European Series, vol. 61, Kluwer, Dordrecht (1993) [MR 94c:00004, Zbl 771.26009]. (Cf. pp. 121–125, 222, 684–685.)

[175] Dragoslav S. Mitrinović and Petar M. Vasić. Addenda to the monograph *Analytic Inequalities* by Mitrinović [172]—I [in English]. *Publikacije Elektrotehničkog Fakulteta Univerziteta u Beogradu, Serija Matematika i Fizika / Publications de la Faculté d'Électrotechnique de l'Université à Belgrade, Série Mathématiques et Physique* 577–598:3–10 (1977) [MR 56:5818, Zbl 369.26010].

[176] Bertram Mond. A matrix version of Rennie's generalization of Kantorovich's inequality. *Proceedings of the American Mathematical Society* 16:1131 (1965) [MR 32:2423, Zbl 145.25302].

[177] Bertram Mond. A matrix inequality including that of Kantorovich. *Journal of Mathematical Analysis and Applications* 13:49–52 (1966) [MR 32:2423, Zbl 131.01503].

[178] Bertram Mond. An inequality for operators in a Hilbert space. *Pacific Journal of Mathematics* 18:161–163 (1966) [MR 33:7857, Zbl 146.12504].

[179] Bertram Mond. Generalised inverse extensions of matrix inequalities. *Linear Algebra and Its Applications* 2:393–399 (1969) [MR 40:1410, Zbl 184.05704].

[180] Bertram Mond and Josip E. Pečarić. Converses of Jensen's inequality for linear maps of operators. *Analele Universităţii din Timişoara, Seria Matematică-Informatică* 31:223–228 (1993) [MR 97i:47028, Zbl 866.47010].

[181] Bertram Mond and Josip E. Pečarić. Inequalities with weights for powers of generalised inverses. *Bulletin of the Australian Mathematical Society* 48:7–12 (1993) [MR 94d:15002, Zbl 782.15009].

[182] Bertram Mond and Josip E. Pečarić. Matrix versions of some means inequalities. *Australian Mathematical Society Gazette* 20:117–120 (1993) [MR 94h:26025].

[183] Bertram Mond and Josip E. Pečarić. Inequalities with weights for powers of generalised inverses–II. *Linear Algebra and Its Applications* 210:265-272 (1994) [MR 95h:15031, Zbl 812.15011]. (Reprinted in *Analele Universităţii din Timişoara, Seria Matematică-Informatică* 32:51-58 (1994) [MR 97i:15023, Zbl 865.15015]).

[184] Bertram Mond and Josip E. Pečarić. A matrix version of the Ky Fan generalization of the Kantorovich inequality. *Linear and Multilinear Algebra* 36:217–221 (1994) [MR 95j:15017, Zbl 791.15015].

[185] Bertram Mond and Josip E. Pečarić. Converses of Jensen's inequality for several operators. *Revue d'Analyse Numérique et de Théorie de l'Approximation (Bucharest)* 23:179–183 (1994) [MR 97m:47019, Zbl 847.47016].

[186] Bertram Mond and Josip E. Pečarić. Inequalities involving powers of generalised inverses. *Linear Algebra and Its Applications* 199:293–303 (1994) [MR 95a:15004].

[187] Bertram Mond and Josip E. Pečarić. Generalisation of a matrix inequality of Ky Fan. *Journal of Mathematical Analysis and Applications* 190:244–247 (1995) [MR 95m:15029, Zbl 829.15015].

[188] Bertram Mond and Josip E. Pečarić. Reverse forms of a convex matrix inequality. *Linear Algebra and Its Applications* 220:359–364 (1995) [MR 96d:15027, Zbl 824.15019].

[189] Bertram Mond and Josip E. Pečarić. A matrix version of the Ky Fan generalisation of the Kantorovich inequality, II. *Linear and Multilinear Algebra* 38:309–313 (1995) [MR 96h:15021, Zbl 835.15011].

[190] Bertram Mond and Josip E. Pečarić. Some matrix inequalities of Ky Fan type. *Tamkang Journal of Mathematics* 26:321–326 (1995) [MR 97e:15015, Zbl 866.15012].

[191] Bertram Mond and Josip E. Pečarić. Matrix inequalities for convex functions. *Journal of Mathematical Analysis and Applications* 209:147–153 (1997) [Zbl 879.15014].

[192] Bertram Mond and Oved Shisha. A difference inequality for operators in Hilbert space. *Blanch Anniversary Volume*, Aerospace Research Laboratories (Office of Aerospace Research, United States Air Force), Wright-Patterson Air Force Base, Ohio, pp. 269–275 (February 1967) [MR 35:5958, Zbl 165.15101].

[193] Bertram Mond and Oved Shisha. Ratios of means and applications. In *Inequalities: Proceedings of a Symposium held at Wright-Patterson Air Force Base, Ohio, August 19–27, 1965* (Oved Shisha, ed.), Academic Press, New York, pp. 191–197 (1967) [MR 39:2930a].

[194] Bertram Mond and Oved Shisha. Difference and ratio inequalities in Hilbert space. In *Inequalities-II: Proceedings of the Second Symposium on Inequalities held at the United States Air Force Academy, Colorado, August 14–22, 1967* (Oved Shisha, ed.), Academic Press, New York, pp. 241–249 (1970) [Zbl 217.45401].

[195] J[ulius] v[on] Sz[ökefalvi] Nagy. Über algebraische Gleichungen mit lauter reellen Wurzeln [in German]. *Jahresbericht der Deutschen Mathematiker-Vereinigung* 27:37–43 (1918) [JFM 46:125].

[196] K[eshavan] R[aghavan] Nair. Certain symmetrical properties of unbiased estimates of variance and covariance. *Journal of the Indian Society of Agricultural Statistics*, 1:162–172 (1948) [MR 11:448a].

[197] K[eshavan] R[aghavan] Nair. A tail-piece to Brunk's paper [39]. *Journal of the Indian Society of Agricultural Statistics*, 11:189–190 (1959).

[198] Ritsuo Nakamoto and Masahiro Nakamura. Operator mean and Kantorovich inequality. *Mathematica Japonica* 44:495–498 (1996) [MR 97j:47029, Zbl 866.47011].

[199] Masahiro Nakamura. A remark on a paper of Greub and Rheinboldt. *Proceedings of the Japan Academy, Series A: Mathematical Sciences (Tokyo)* 36:198–199 (1960) [MR 22:12188, Zbl 099.31804].

[200] Heinz Neudecker and Shuangzhe Liu. Alternative matrix versions of Kantorovich inequalities. Technical Report no. AE 5/94, Institute of Actuarial Science and Econometrics, University of Amsterdam, 7 pp., 15 January 1994.

[201] Morris Newman. Kantorovich's inequality. *Journal of Research of the National Bureau of Standards: Section B, Mathematics and Mathematical Physics* 64B:33–34 (1959) [Zbl 101.25502].

[202] John J. O'Connor and Edmund F. Robertson. MacTutor History of Mathematics Archive Web site: `http://www-history.mcs.st-andrews.ac.uk/history/BiogIndex.html`.

[203] Ingram Olkin. Review of Eaton [68]. *Mathematical Reviews*, review no. 5272, 54:752 (1977).

[204] Ingram Olkin. A matrix formulation on how deviant an observation can be. *The American Statistician* 46:205–209 (1992) [CMP 1:183-071]

[205] Ingram Olkin. Solution No. 2 to Problem 16-4: Algebraic reverse of a convex matrix inequality (posed by Liu [132]). *Image: Bulletin of the International Linear Algebra Society* 17:31 (1996).

[206] Nobuo Ozeki. On the estimation of the inequalities by the maximum or minimum values [in Japanese]. *Journal of the College of Arts and Sciences, Chiba University* 5(2):199–203 (1968) [MR 40:7408].

[207] Zsolt Páles. On complementary inequalities. *Publicationes Mathematicae (Debrecen)* 30:75–88 (1983) [MR 83h:26022, Zbl 544.26009].

[208] Josip E. Pečarić. On an inequality of G. Grüss [in English]. *Matematicki Vesnik (Belgrade)* 35:59–64 (1983).

[209] Josip E. Pečarić. Power matrix means and related inequalities. *Mathematical Communications (Osijek)* 1:91–110 (1996) [MR 98a:15039, Zbl 871.15017].

[210] Josip E. Pečarić and Behdžet A. Mesihović. On some complementary inequalities [in English]. *Contributions: Section of Mathematical and Technical Sciences, Macedonian Academy of Sciences and Arts (Skopje)* 14:49–54 (1993) [CMP 1:374-935].

[211] Josip E. Pečarić and Bertram Mond. A matrix inequality including that of Kantorovich–Hermite, II. *Journal of Mathematical Analysis and Applications* 168:381–384 (1992) [MR 93m:15031, Zbl 762.15010].

[212] Josip E. Pečarić and Bertram Mond. An inequality for operators in a Hilbert space, III. *Journal of Mathematical Analysis and Applications* 183:385–390 (1994) [MR 95g:47029, Zbl 804.47024].

[213] Josip E. Pečarić and Bertram Mond. The arithmetic mean, the geometric mean and related matrix inequalities. In *General Inequalities 7, Proceedings of the 7th International Conference: Oberwolfach, Germany, November 13–18, 1995* (C. Bandle, W. N. Everitt, L. Losonczi, and W. Walter, eds.), International Series in Numerical Mathematics, vol. 123, Birkhäuser, Basel, pp. 77–91 (1997) [Zbl 886.15019].

[214] Josip E. Pečarić, Simo Puntanen and George P. H. Styan. Some further matrix extensions of the Cauchy–Schwarz and Kantorovich inequalities, with some statistical applications. *Linear Algebra and Its Applications* 237/238:455–476 (1996) [MR 97c:15035, Zbl 860.15021].

[215] W. V. Petryshyn. Direct and iterative methods for the solution of linear operator equations in Hilbert space. *Transactions of the American Mathematical Society* 105:136–175 (1962) [MR 26:3180].

[216] George Pólya and Gábor Szegö. *Aufgaben und Lehrsätze aus der Analysis, Band I: Reihen, Integralrechnung, Funktionentheorie* [in German]. Fourth Edition. Heidelberger Taschenbücher, vol. 73. Springer-Verlag, New York (1970) [Zbl 201.38102]. (Cf. # 92–93, pp. 57, 213–214. Original version: Julius Springer, Berlin, 1925. Reprint: Dover, New York, 1945 [MR7:418e]. Translated into English as Pólya and Szegö [217] 1972.)

[217] George Pólya and Gábor Szegö. *Problems and Theorems in Analysis, Volume I: Series, Integral Calculus, Theory of Functions* [in English]. Translated from the German [216] by Dorothee Aeppli. Corrected printing of the revised translation of the fourth German edition. Springer-Verlag, New York, 1972 [Zbl 236.00003]. (Cf. # 92–93, pp. 71–72, 253– 255.)

[218] Tiberiu Popoviciu. Sur les équations algébriques ayant toutes leurs racines réeles [in French]. *Mathematica (Cluj)* 9:129–145 (1935) [Zbl 014.10003].

[219] Vlastimil Pták. The Kantorovich inequality. *American Mathematical Monthly* 102:820–821 (1995) [MR 96f:26025, Zbl 856.26011].

[220] Simo Puntanen. On the relative goodness of ordinary least squares estimation in the general linear model. *Acta Universitatis Tamperensis, Series A*, vol. 216, vi + 132 pp. (1987) [MR 88d:62115, Zbl 652.62050]. (Cf. Paper [1], §2.)

[221] Simo Puntanen and George P. H. Styan. The equality of the ordinary least squares estimator and the best linear unbiased estimator [with discussion]. *The American Statistician* 43:153–164 (1989) [MR 92e:62125].

[222] M. Raghavachari. A linear programming proof of Kantorovich's inequality. *The American Statistician* 40:136–137 (1986).

[223] C. Radhakrishna Rao. The inefficiency of least squares: extensions of the Kantorovich inequality. *Linear Algebra and Its Applications* 70:249–255 (1985) [MR 87d:62131, Zbl 594.62073].

[224] C. Radhakrishna Rao and M. Bhaskara Rao. *Matrix Algebra and Its Applications to Statistics and Econometrics.* World Scientific, Singapore, 1998.

[225] Basil C. Rennie. An inequality which includes that of Kantorovich. *American Mathematical Monthly* 70:982 (1963) [Zbl 127.28003].

[226] Basil C. Rennie. On a class of inequalities. *Journal of the Australian Mathematical Society* 3:442–448 (1963) [MR 29:3590, Zbl 137.03101].

[227] Peter D. Robinson and Andrew J. Wathen. Variational bounds on the entries of the inverse of a matrix. *IMA Journal of Numerical Analysis* 12:463–486 (1992) [MR 93g:65044, Zbl 759.15016].

[228] Masahiko Sagae and Kunio Tanabe. Upper and lower bounds for the arithmetic-geometric-harmonic means of positive definite matrices. *Linear and Multilinear Algebra* 37:279–282 (1994) [MR 95m:15030, Zbl 816.15017].

[229] Ernesto M. Saleme. Problema N° 21.—Siendo $x_1, x_2, x_3, \ldots, x_n$ números positivos con la condición $\sum x_i = k$, demostrar las desigualdades $\sum_{i=1}^{n} x_i^2 \geq k^2/n$, $\sum_{i=1}^{n} 1/x_i \geq n^2/k$ : 2° Solución y Generalización [in Spanish]. *Mathematicæ Notæ–Boletin del Instituto de Matemática (Rosario)* 2:197–199 (1942). (Translated into English, together with English translations of Bender [30], Frucht [71] and Levi [123], as Appendix A of Watson, Alpargu and Styan [266].)

[230] Norbert Schappacher and René Schoof. Beppo Levi and the arithmetic of elliptic curves. *The Mathematical Intelligencer* 18(1):57–69 (1996) [Zbl 849.01036].

[231] Andreas H. Schopf. On the Kantorovich inequality. *Numerische Mathematik* 2:344–346 (1960) [Zbl 095.25001]. (Paper written by A. S. Householder after the death of Andreas H. Schopf on September 29, 1959.)

[232] Peter Schreiber. The Cauchy–Bunyakovsky–Schwarz inequality [in English]. *Hermann Graßmann: Werk und Wirkung* (Peter Schreiber, ed.), Internationale Fachtagung anläßlich des 150. Jahrestages des ersten Erscheinens der "linealen Ausdehnungslehre" (Lieschow/Rügen, 23.–28.5.1994), Fachrichtungen Mathematik/Informatik, Ernst-Moritz-Arndt-Universität, Greifswald, pp. 64–70 (1995) [Zbl 841.01011].

[233] [Karl] H[ermann] A[mandus] Schwarz. Ueber ein die Flächen kleinsten Flächeninhalts betreffendes Problem der Variationsrechnung: Festschrift zum Jubelgeburtstage des Herrn Karl Weierstrass [in German]. *Acta Societatis Scientiarum Fennicæ(Helsinki)* 15:315–362 (1888). (Cf. pp. 343–345. Preface dated 31 October 1885. Reprinted in *Gesammelte Mathematische*

*Abhandlungen von H. A. Schwarz*, Julius Springer, Berlin, vol. 1, pp. 223–270 (1890); cf. pp. 251–253.)

[234] Pál Schweitzer. Egy egyenlőtlenség az aritmetikai középértékről [in Hungarian: "An inequality about the arithmetic mean"]. *Matematikai és Physikai Lapok* 23:257–261 (1914) [JFM 45:1245]. (Translated into English as Appendix A of Alpargu [2] and as Appendix B of Watson, Alpargu and Styan [266].)

[235] Alastair J. Scott and George P. H. Styan. On a separation theorem for generalised eigenvalues and a problem in the analysis of sample surveys. *Linear Algebra and Its Applications* 70:209–224 (1985) [MR 87i:62100, Zbl 587.62023].

[236] En-Wei Shi. An elementary proof of the Kantorovich inequality [in Chinese]. *Shuxue de Shijian yu Renshi / Mathematics in Practice and Theory (Beijing)* 1985(4):56–61 (1985) [MR 87g:15021].

[237] Oved Shisha and G. T. Cargo. On comparable means. *Pacific Journal of Mathematics* 14:1053–1058 (1964) [MR 30:4874, Zbl 141.24202].

[238] Oved Shisha and Bertram Mond. Differences of means. *Bulletin of the American Mathematical Society* 73:328–333 (1967) [MR 35:3024, Zbl 153.08101].

[239] Oved Shisha and Bertram Mond. Bounds on differences of means. In *Inequalities: Proceedings of a Symposium held at Wright-Patterson Air Force Base, Ohio, August 19–27, 1965* (Oved Shisha, ed.), Academic Press, New York, pp. 293–308 (1967) [MR 39:2930b].

[240] P. G. Spain. Operator versions of the Kantorovich inequality. *Proceedings of the American Mathematical Society* 124:2813–2819 (1996) [MR 96k:47034, Zbl 864.47006].

[241] James C. Spall. The Kantorovich inequality for error analysis of the Kalman filter with unknown noise distributions. *Automatica, A Journal of the International Federation of Automatic Control* 31:1513–1517 (1995) [MR 96e:93067, Zbl 836.93058].

[242] Wilhelm Specht. Zur Theorie der elementaren Mittel [in German]. *Mathematische Zeitschrift* 74:91–98 (1960) [MR 22:8090, Zbl 095:03801].

[243] W. Gilbert Strang. On the Kantorovich inequality. *Proceedings of the American Mathematical Society* 11:468 (1960) [MR 22:2904, Zbl 095.09601].

[244] George P. H. Styan. On some inequalities associated with ordinary least squares and the Kantorovich inequality. In *Festschrift for Eino Haikala on his Seventieth Birthday, Acta Universitatis Tamperensis, Series A*, vol. 153, pp. 158–166 (1983) [MR 85g:62120]. (Abstract: Styan and Zlobec [245].)

[245] George P. H. Styan and Sanjo Zlobec. An inequality connected with the efficiency of the least-squares estimator [abstract]. *The IMS Bulletin* 11:192 (1982). (Paper presented at the 181st IMS Meeting, San Diego, California, June 1982, and published as Styan [244].)

[246] Ji-guang Sun. Extensions of the Kantorovich inequality and the Bauer-Fike inequality. *Journal of Computational Mathematics* 9:360–368 (1991) [MR 93e:65074, Zbl 776.15014].

[247] Yves Thibaudeau and George P. H. Styan. Bounds for Chakrabarti's measure of imbalance in experimental design. In *Proceedings of the First International Tampere Seminar on Linear Statistical Models and their Applications, Tampere, Finland, August 30–September 2, 1983* (Tarmo Pukkila and Simo Puntanen, eds.), Dept. of Mathematical Sciences, University of Tampere, Tampere, Finland, pp. 323–347 (1985).

[248] George W. Thomson. Bounds for the ratio of range to standard deviation. *Biometrika* 42:268–269 (1955) [MR 16:841a, Zbl 064.13503].

[249] Makoto Tsukada and Sin-Ei Takahasi. On the Nakamura–Schweitzer–Kantorovich inequality [in Japanese]. In *Investigations on Applied Functional Analysis: Proceedings of a Symposium held at the Research Institute for Mathematical Sciences, Kyoto University, Kyoto, June 17–18, 1996*, Sūrikaisekikenkyūsho Kōkyūroku, no. 975, pp. 116–122 (1996) [MR 98e:00020].

[250] William N. Venables. Some implications of the union-intersection principle for tests of sphericity. *Journal of Multivariate Analysis* 6:185–190 (1976) [MR 54:3971, Zbl 332.62041].

[251] Chung-Lie Wang. Variants of the Hölder inequality and its inverses. *Canadian Mathematical Bulletin* 20:377–384 (1977) [MR 57:9929, Zbl 398.26018].

[252] Chung-Lie Wang. On development of inverses of the Cauchy and Hölder inequalities. *SIAM Review* 21:550–557 (1979) [MR 80g:26023, Zbl 414.26007].

[253] Jing-Long Wang and Dao-De Gao. Measures of inefficiency for generalised least square estimates of the Euclidean norms of matrices [in Chinese]. *Yingyong Gailü Tongji / Chinese Journal of Applied Probability and Statistics (Shanghai)* 7(4):361–366 (1991) [MR 93m:62160].

[254] Mingjin Wang. The mean inequalities of symmetric random variables [in Chinese with English summary]. *Journal of Jiangxi Normal University, Natural Sciences Edition* 19(4):301–303 (1995) [Zbl 863.60017].

[255] Song-Gui Wang and Shein-Chung Chow. *Advanced Linear Models: Theory and Applications.* Statistics: Textbooks and Monographs, vol. 141. Marcel Dekker, New York (1994) [MR 96c:62123, Zbl 822.62052]. (Cf. pp. 44–46, 212–215.)

[256] Song-Gui Wang and Zhongzhen Jia. *Inequalities in Matrix Theory* [in Chinese]. Anhui Education Press, Hefei (1994).

[257] Song-Gui Wang and Jun Shao. Constrained Kantorovich inequalities and relative efficiency of least squares. *Journal of Multivariate Analysis* 42:284–298 (1992) [MR 93j:62179, Zbl 752.62047].

[258] Song-Gui Wang and Hu Yang. Kantorovich-type inequalities and the measures of inefficiency of the GLSE [in English]. *Acta Mathematicae Applicatae Sinica, English Series* 5(4):372–381 (1989) [MR 91d:62104, Zbl 711.15016].

[259] Song-Gui Wang and Wai-Cheung Ye. A matrix version of the Wielandt inequality and its applications to statistics [in Chinese]. *Kexue Tongbao / Chinese Science Bulletin (Beijing)* 43(18):1930–1933 (1998). (Expanded English translation by Song-Gui Wang and Wai-Cheung Ip [Wai-Cheung Ye], Preprint, 16 pp., 1999.)

[260] Geoffrey Stuart Watson. *Serial Correlation in Regression Analysis.* PhD thesis, Dept. of Experimental Statistics, North Carolina State College, Raleigh (1951). (Includes an appendix by Cassels [45], pp. 138–139. Results presented at the Econometric Society Conference, Louvain, Belgium, September 1951. Reprinted as University of North Carolina, Mimeograph Series, no. 49 (1952); published as Watson [261] and Watson and Hannan [267].)

[261] Geoffrey S. Watson. Serial correlation in regression analysis, I. *Biometrika* 42:327–341 (1955) [MR 17:382a, Zbl 068.33201]. (Includes an appendix by Cassels [46]. Part of Watson [260].)

[262] Geoffrey S. Watson. Linear least squares regression. *The Annals of Mathematical Statistics* 38:1679–1699 (1967) [Zbl 155.26801].

[263] Geoffrey S. Watson. *Selected Topics in Statistical Theory.* Mimeo-graphed Lecture Notes. Mathematical Association of America, Washington, DC (1971). (Cf. pp. 87–92. "Notes on lectures given at the 1971 MAA Summer Seminar, Williams College, Williamstown, Massachusetts.")

[264] Geoffrey S. Watson. Prediction and efficiency of least squares. *Biometrika* 59:91–98 (1972) [Zbl 232.62032].

[265] Geoffrey S. Watson. A method for discovering Kantorovich-type inequalities and a probabilistic interpretation. *Linear Algebra and Its Applications* 97:211–217 (1987) [MR 89a:15018, Zbl 631.15010].

[266] Geoffrey S. Watson, Gülhan Alpargu and George P. H. Styan. Some comments on six inequalities associated with the inefficiency of ordinary least squares with one regressor. *Linear Algebra and Its Applications*, 264:13–53 (1997) [MR 98i:15023]. (Includes English translations of Bender [30], Frucht [71], Levi [123], Saleme [229] and Schweitzer [234].

[267] Geoffrey S. Watson and E[dward] J[ames] Hannan. Serial correlation in regression analysis, II. *Biometrika* 42:436–448 (1956) [MR 19:694h]. (Part of [260].)

[268] Helmut Wielandt. Inclusion theorems for eigenvalues. In *Simultaneous Linear Equations and the Determination of Eigenvalues* (Lowell J. Paige and Olga Taussky, eds.), National Bureau of Standards Applied Mathematics Series, U.S. Government Printing Office, Washington, D.C., vol. 29, pp. 75–78 (1953) [MR 15:496d]. (Reprinted, with commentaries by Chandler Davis and by Hans Schneider and with an editor's note, in *Helmut Wielandt: Mathematische Werke / Mathematical Works–Volume 2: Linear Algebra and Analysis* (Bertram Huppert and Hans Schneider, eds.), Walter de Gruyter, Berlin, pp. 156–162 (1996).)

[269] Helmut Wielandt. An extremum property of sums of eigenvalues. *Proceedings of the American Mathematical Society* 6:106–110 (1955) [Zbl 064.24703]. (Reprinted, with commentary, in *Helmut Wielandt: Mathematische Werke / Mathematical Works–Volume 2: Linear Algebra and Analysis* (Bertram Huppert and Hans Schneider, eds.), Walter de Gruyter, Berlin, pp. 172–181 (1996).)

[270] H. S. Witsenhausen. An inequality of Kantorovitch [*sic*] type for Stieltjes integrals. *Journal of Mathematical Analysis and Applications* 28:590–593 (1969) [MR 40:1839, Zbl 187.31601].

[271] Henry Wolkowicz. Bounds for the Kantorovich ratio. Research Report. Dept. of Mathematics, The University of Alberta, Edmonton, 15 pp. (1981).

[272] Henry Wolkowicz and George P. H. Styan. Bounds for eigenvalues using traces. *Linear Algebra and Its Applications* 29:471–506 (1980) [MR 81k:15015, Zbl 435.15015].

[273] Hu Yang. Extensions of the Kantorovich inequality and the error ratio efficiency of the mean square [in Chinese with English summary]. *Mathematica Applicata Yingyong Shuxue / Chinese Journal of Mathematics and its Applications (Wuhan)* 1(4):85–90 (1988) [MR 90a:62186, Zbl 665.62064].

[274] Hu Yang. The inefficiency of the least squares estimator and its bound [in English]. *Applied Mathematics and Mechanics of China, English Edition* 11(11):1087–1093 (1990) [Zbl 748.62031].

[275] Hu Yang. A brief proof on the ganeralizd [*sic*] variance bound of the relative efficiency in statistics. *Communications in Statistics–Theory and Methods* 19:4587–4590 (1990).

[276] Hu Yang. Two new classes of the generalised Kantorovich inequalities and their applications [in Chinese]. *Kexue Tongbao* 35:965–967 (1990) [MR 92m:15016]. (Translated into English as Yang [279].)

[277] Hu Yang. *Kantorovich Inequality and its Applications in Statistics.* MSc thesis, Dept. of Statistics, Yunnan University, Kunming (1990).

[278] Hu Yang. Some extensions on the upper bound of the Rayleigh quotient [in Chinese]. *Journal of Chongqing University* 9:16–23 (1990).

[279] Hu Yang. Two new classes of the generalised Kantorovich inequalities and their applications [in English]. *Chinese Science Bulletin* 36(22):1849–1851 (1991) [MR 92m:15016, Zbl 748.62032]. (Translation of Yang [276] from the Chinese.)

[280] Hu Yang. Matrix norm versions of the Kantorovich inequality and its applications [in English]. *Applied Mathematics: A Journal of Chinese Universities, Series B (English Edition)* 10(2):133–140 (1995) [MR 96e:15035, Zbl 837.15015]. (Chinese summary: *Gaoxiao Yingyong Shuxue Xuebao / A Journal of Chinese Universities, Series A (Hangzhou)* 10(2):233 (1995).)

[281] Hu Yang. Efficiency matrix and the partial ordering of estimate. *Communications in Statistics–Theory and Methods* 25:457–468 (1996) [MR 97b:62112, 875.62210].

[282] Hu Yang and Zhiqing Dong. A new kind of measure of information loss in parameter estimation [in Chinese with English summary]. *Journal of Chongqing Jiaotong Institute* 15(Suppl.):43–49 (1996).

[283] Hu Yang and Song-Gui Wang. Condition numbers, the spectral norm and precision of estimators [in Chinese]. *Yingyong Gailü Tongji / Chinese Journal of Applied Probability and Statistics (Shanghai)* 7:337–343 (1991).

[284] Lina Yeh. A note on Wielandt's inequality. *Applied Mathematics Letters* 8(3):29–31 (1995) [MR 96e:15029, Zbl 827.15015].

[285] Guang-Rong You. On Hölder's inequalities for convexity. *Journal of Mathematical Analysis and Applications* 143:448-458 (1989) [Zbl 688.26010].

[286] Jinlong Zhan and Jianbao Chen. The inefficiency of least squares in Gauss–Markov and variance component models [in English]. Unpublished paper, Dept. of Basic Sciences, Kunming Institute of Technology, 12 pp. (1994).

[287] Fuzhen Zhang. *Matrix Theory: Basic Results and Techniques.* Springer-Verlag, New York, in press (1999).

[288] Shi-Qiong Zhou. On the weighted mean inequality [in Chinese]. *Changsha Jiaotong Xueyuan Xuebao / Journal of the Changsha Communications Institute (Changsha)* 7(3):52–61 (1991) [MR 95g:26036].

[289] Shi-Qiong Zhou. The inverses of the Hölder inequality and their matrix forms [in Chinese]. *Changsha Jiaotong Xueyuan Xuebao / Journal of the Changsha Communications Institute (Changsha)* 9:10–16 (1993) [MR 95h:26033].

CHAPTER 2

# On Matrix Trace Kantorovich-type Inequalities

Shuangzhe LIU

## 2.1 Introduction

In this chapter, a variety of matrix trace versions of the Kantorovich-type inequalities are collected. Some relevant matrix and determinant inequalities are also included. Their mathematical and statistical applications are studied. In the mathematical applications, some known results are used to derive new inequalities involving the Hadamard product, and to propose conjectures. In the statistical applications, new criteria for efficiency comparisons in the linear model are introduced. Examples, illustrated by figures, are used to examine upper bounds for known and new criteria.

The Kantorovich inequality (KI), its extensions and applications have been studied since the 1940's. A relevant result, the Schweitzer inequality, which could be viewed as a Kantorovich-type inequality (KTI), was introduced as early as the 1910's. See Watson, Alpargu and Styan (1997) in which the KI (in the vector case) was named as the Frucht–Kantorovich inequality. The first matrix trace version of a KTI may have been given by Bloomfield and Watson (1975), and the first determinant version was given by Schopf (1960). Certainly, the first matrix version of the KI was presented by Marshall and Olkin (1990).

In this chapter, we proceed as follows. In Section 2, we compile several known trace, matrix and determinant results. In Section 3, we study mathematical applications. First, we derive inequalities for the Hadamard product case. Then, in view of the results obtained, we propose a conjecture on the Hadamard product. Also, for the positive-semidefinite matrix case, we derive trace inequalities. In Section 4, we discuss statistical applications. We introduce new criteria, with their upper bounds, for efficiency comparisons between the ordinary least squares (OLS) and generalised least squares (GLS) estimators. Then we examine them with numerical examples in the context of the non-singular linear model. We also comment on the alternative criteria for a singular linear model.

## 2.2 Basic Inequalities

For a general introduction to matrices, inequalities and linear models, see, for example, Magnus and Neudecker (1991). We compile a selection of known KTIs in Theorem 1.

**Theorem 1.** Let $A > 0$ be an $n \times n$ matrix with eigenvalues $\lambda_1 \geq \cdots \geq \lambda_n$, $X$ be an $n \times k$ matrix such that $X'X = I$ and $n \geq 2k$ (without loss of generality). Then we have

$$0 \leq \text{tr}\,[X'A^2X - (X'AX)^2] \leq \frac{1}{4}\sum_{j=1}^{k}(\lambda_j - \lambda_{n-j+1})^2, \tag{1}$$

$$1 \leq \frac{\text{tr}\,X'A^2X}{\text{tr}\,(X'AX)^2} \leq \frac{\sum_{j=1}^{k}(\lambda_j + \lambda_{n-j+1})^2}{4\sum_{j=1}^{k}\lambda_j\lambda_{n-j+1}}, \tag{2}$$

$$0 \leq \text{tr}\,[X'AX - (X'A^{-1}X)^{-1}] \leq \sum_{j=1}^{k}(\lambda_j^{1/2} - \lambda_{n-j+1}^{1/2})^2, \tag{3}$$

$$1 \leq \frac{\text{tr}\,X'AX}{\text{tr}\,(X'A^{-1}X)^{-1}} \leq \left[\frac{\sum_{j=1}^{k}(\lambda_j + \lambda_{n-j+1})}{2\sum_{j=1}^{k}\lambda_j^{1/2}\lambda_{n-j+1}^{1/2}}\right]^2, \tag{4}$$

$$k \leq \text{tr}\,X'AXX'A^{-1}X \leq \sum_{j=1}^{k}\frac{(\lambda_j + \lambda_{n-j+1})^2}{4\lambda_j\lambda_{n-j+1}}, \tag{5}$$

$$k \leq \text{tr}\,X'A^2X(X'AX)^{-2} \leq \sum_{j=1}^{k}\frac{(\lambda_j + \lambda_{n-j+1})^2}{4\lambda_j\lambda_{n-j+1}}, \tag{6}$$

$$0 \leq X'A^2X - (X'AX)^2 \leq \frac{1}{4}(\lambda_1 - \lambda_n)^2 I, \tag{7}$$

$$0 \leq X'A^2X \leq \frac{(\lambda_1 + \lambda_n)^2}{4\lambda_1\lambda_n}(X'AX)^2, \tag{8}$$

$$0 \leq (X'A^2X)^{1/2} - X'AX \leq \frac{(\lambda_1 - \lambda_n)^2}{4(\lambda_1 + \lambda_n)}I, \tag{9}$$

$$0 \leq (X'A^2X)^{1/2} \leq \frac{\lambda_1 + \lambda_n}{2\lambda_1^{1/2}\lambda_n^{1/2}}X'AX, \tag{10}$$

$$1 \leq |X'AXX'A^{-1}X| \leq \Pi_{j=1}^{k}\frac{(\lambda_j + \lambda_{n-j+1})^2}{4\lambda_j\lambda_{n-j+1}}. \tag{11}$$

**Proof.** For (1) see Bloomfield and Watson (1975). For (2) see Liu (1997). For (3) see Rao (1985). For (4) see Yang (1988). For (5) see Khatri and Rao (1981). For (6) see Khatri and Rao (1982). For (7) and (10) see Liu and Neudecker (1996). For (8) see, for example, Wang and Shao (1992) and Mond and Pečarić (1993). For (7) through (10) see also Liu (1995). For (11) see both Bloomfield and Watson (1975) and Knott (1975).

Note that $A$ in (1) and (7) can be relaxed to become symmetric, see, for example, Styan (1983). Inequalities like those in Theorem 1 for the case of $X'X = I$ cover several specific cases including the Hadamard product case, the upper-left matrix case, the Fan's matrix-sum case, the block-diagonal matrix case and the positive-semidefinite matrix case, see Liu (1995, 1996).

## 2.3 Mathematical Results

We study mathematical applications of Theorem 1 to the Hadamard product case. For statistical applications of the Hadamard product, see, for example, Styan (1973), Neudecker (1975), Neudecker, Liu and Polasek (1995) and Neudecker, Polasek and Liu (1995).

An equality connecting the Kronecker and Hadamard products is given below, see, for example, Browne (1974), Liu and Neudecker (1996) or Kollo and Neudecker (1997):

$$J'(B \otimes C)J = (B \odot C), \tag{12}$$

where $\otimes$ and $\odot$ indicate the Kronecker product and the Hadamard product respectively, and $J$ is the selection matrix with property $J'J = I$.

In fact, $B$ and $C$ in (12) can be relaxed to become $m \times p$ matrices $(m \neq p)$, see Liu (1995) for details. Using this equality, we can derive many Hadamard product results from Kronecker product results.

**Theorem 2.** Let $B > 0$ and $C > 0$ be $m \times m$ positive definite matrices, $\lambda_1 \geq \cdots \geq \lambda_n$ be the eigenvalues of $B \otimes C, n = m^2$, and $k = m$, then

$$0 \leq \text{tr}\ [B^2 \odot C^2 - (B \odot C)^2] \leq \frac{1}{4}\sum_{j=1}^{k}(\lambda_j - \lambda_{n-j+1})^2, \tag{13}$$

$$1 \leq \frac{\text{tr}\ B^2 \odot C^2}{\text{tr}\ (B \odot C)^2} \leq \frac{\sum_{j=1}^{k}(\lambda_j + \lambda_{n-j+1})^2}{4\sum_{j=1}^{k}\lambda_j\lambda_{n-j+1}}, \tag{14}$$

$$0 \leq \text{tr}\ [B \odot C - (B^{-1} \odot C^{-1})^{-1}] \leq \sum_{j=1}^{k}(\lambda_j^{1/2} - \lambda_{n-j+1}^{1/2})^2, \tag{15}$$

$$1 \leq \frac{\text{tr}\ B \odot C}{\text{tr}\ (B^{-1} \odot C^{-1})^{-1}} \leq \left[\frac{\sum_{j=1}^{k}(\lambda_j + \lambda_{n-j+1})}{2\sum_{j=1}^{k}\lambda_j^{1/2}\lambda_{n-j+1}^{1/2}}\right]^2, \tag{16}$$

$$k \leq \text{tr}\ (B \odot C)(B^{-1} \odot C^{-1}) \leq \sum_{j=1}^{k}\frac{(\lambda_j + \lambda_{n-j+1})^2}{4\lambda_j\lambda_{n-j+1}}, \tag{17}$$

$$k \leq \text{tr}\ (B^2 \odot C^2)(B \odot C)^{-2} \leq \sum_{j=1}^{k}\frac{(\lambda_j + \lambda_{n-j+1})^2}{4\lambda_j\lambda_{n-j+1}}. \tag{18}$$

**Proof.** Using (1) through (6) with $A = B \otimes C$ and $X = J$, and using (12), we get (13)–(18).

Note that $B$ and $C$ in (13) can be relaxed to become symmetric.

**Theorem 3.** Let matrix $B$ be of order $m \times m$, $\lambda_1 \geq \cdots \geq \lambda_n$ be the eigenvalues of the $n \times n$ matrix $B \otimes B^{-1} > 0, n = m^2$ and $k = m$, then

$$k \leq \text{tr } (B \odot B^{-1})^2 \leq \sum_{j=1}^{k} \frac{(\lambda_j + \lambda_{n-j+1})^2}{4\lambda_j \lambda_{n-j+1}}, \tag{19}$$

$$1 \leq |B \odot B^{-1}| \leq \Pi_{j=1}^{k} \frac{\lambda_j + \lambda_{n-j+1}}{2\lambda_j^{1/2} \lambda_{n-j+1}^{1/2}}. \tag{20}$$

**Proof.** Inserting $A = B \otimes B^{-1}$ and $X = J$ into (5) and (11), and using (12), we obtain

$$k \leq \text{tr } (B \odot B^{-1})(B^{-1} \odot B) \leq \sum_{j=1}^{k} \frac{(\lambda_j + \lambda_{n-j+1})^2}{4\lambda_j \lambda_{n-j+1}},$$

$$1 \leq |(B \odot B^{-1})(B^{-1} \odot B)| \leq \Pi_{j=1}^{k} \frac{(\lambda_j + \lambda_{n-j+1})^2}{4\lambda_j \lambda_{n-j+1}},$$

and hence (19) and (20).

Note that, in a similar way, we can get further results if we use the result of Drury and Styan (1997) which gives an extension of the Bloomfield–Watson–Knott inequality (11).

Also, we have

$$I \leq B \odot B^{-1} \leq \frac{\lambda_1 + \lambda_n}{2\lambda_1^{1/2} \lambda_n^{1/2}} I, \tag{21}$$

where $\lambda_1$ and $\lambda_n$ are the largest and smallest eigenvalues of the $n \times n$ matrix $B \otimes B^{-1} > 0$ respectively, see Liu (1995).

Thus, in view of Theorem 3 and (21), we propose the following conjecture:

$$\text{tr } B \odot B^{-1} \leq \sum_{j=1}^{k} \frac{\lambda_j + \lambda_{n-j+1}}{2\lambda_j^{1/2} \lambda_{n-j+1}^{1/2}}, \tag{22}$$

where $B$ is an $m \times m$ matrix, $\lambda_1 \geq \cdots \geq \lambda_n$ are the eigenvalues of $n \times n$ matrix $B \otimes B^{-1} > 0$, $n = m^2$ and $k = m$.

Furthermore, we propose three inequalities, relevant to (9) and (10), to hold as follows.

$$\text{tr } [(X'A^2X)^{1/2} - X'AX] \leq \sum_{j=1}^{k} \frac{(\lambda_j - \lambda_{n-j+1})^2}{4(\lambda_j + \lambda_{n-j+1})}, \tag{23}$$

$$\frac{\text{tr } (X'A^2X)^{1/2}}{\text{tr } X'AX} \leq \frac{\sum(\lambda_j + \lambda_{n-j+1})}{2\sum \lambda_j^{1/2} \lambda_{n-j+1}^{1/2}}, \tag{24}$$

$$\operatorname{tr}(X'A^2X)^{1/2}(X'AX)^{-1} \leq \sum_{j=1}^{k} \frac{\lambda_j + \lambda_{n-j+1}}{2\lambda_j^{1/2}\lambda_{n-j+1}^{1/2}}, \tag{25}$$

where $\lambda_1 \geq \cdots \geq \lambda_n$ are the eigenvalues of $A > 0$, and $X'X = I$.

Obviously, if $k = 1$, then the matrix case above becomes the vector case, and (22)–(25) are true. If $B$ is a diagonal matrix, then (22) holds.

Now we consider the positive-semidefinite matrix case where $D \geq 0$ is an $m \times m$ matrix and $Z$ is an $m \times q$ matrix such that $\Re(Z) \subset \Re(D)$. Write $D = TAT', ZZ^+ = SS'$ and $S = TH$, where $A > 0, T'T = I_n, S'S = I_k$, $\operatorname{rank}(D) = \operatorname{rank}(A) = n \leq m$, $\operatorname{rank}(Z) = k \leq \min(q, n), H'H = H'T'TH = S'S = I_k$, $D^2 = TA^2T', D^+ = TA^{-1}T'$, and the matrices $A, T, S$ and $H$ are of order $n \times n$, $m \times n$, $m \times k$ and $n \times k$ respectively. Hence

$$ZZ^+DZZ^+ = S(H'AH)S' \neq 0, \tag{26}$$
$$ZZ^+D^2ZZ^+ = S(H'A^2H)S', \tag{27}$$
$$(ZZ^+DZZ^+)^2 = S(H'AH)^2S', \tag{28}$$
$$(ZZ^+D^+ZZ^+)^+ = S(H'A^{-1}H)^{-1}S' \neq 0, \tag{29}$$
$$ZZ^+D^+ZZ^+ = S(H'A^{-1}H)S'. \tag{30}$$

We now have Theorem 4, which will be used in Section 4 for efficiency comparisons in a singular linear model.

**Theorem 4.** Let $D$ and $Z$ be defined as above, and let $D$ have nonzero eigenvalues $\lambda_1 \geq \cdots \geq \lambda_n$. Then

$$0 \leq \operatorname{tr}[ZZ^+D^2ZZ^+ - (ZZ^+DZZ^+)^2] \leq \frac{1}{4}\sum_{j=1}^{k}(\lambda_j - \lambda_{n-j+1})^2, \tag{31}$$

$$1 \leq \frac{\operatorname{tr} ZZ^+D^2ZZ^+}{\operatorname{tr}(ZZ^+DZZ^+)^2} \leq \frac{\sum_{j=1}^{k}(\lambda_j + \lambda_{n-j+1})^2}{4\sum_{j=1}^{k}\lambda_j\lambda_{n-j+1}}, \tag{32}$$

$$0 \leq \operatorname{tr}[ZZ^+DZZ^+ - (ZZ^+D^+ZZ^+)^+] \leq \sum_{j=1}^{k}(\lambda_j^{1/2} - \lambda_{n-j+1}^{1/2})^2, \tag{33}$$

$$1 \leq \frac{\operatorname{tr} ZZ^+DZZ^+}{\operatorname{tr}(ZZ^+D^+ZZ^+)^+} \leq \left[\frac{\sum_{j=1}^{k}(\lambda_j + \lambda_{n-j+1})}{2\sum_{j=1}^{k}\lambda_j^{1/2}\lambda_{n-j+1}^{1/2}}\right]^2, \tag{34}$$

$$k \leq \operatorname{tr} ZZ^+DZZ^+D^+ZZ^+ \leq \sum_{j=1}^{k}\frac{(\lambda_j + \lambda_{n-j+1})^2}{4\lambda_j\lambda_{n-j+1}}, \tag{35}$$

$$k \leq \operatorname{tr} ZZ^+D^2ZZ^+(ZZ^+DZZ^+)^{2+} \leq \sum_{j=1}^{k}\frac{(\lambda_j + \lambda_{n-j+1})^2}{4\lambda_j\lambda_{n-j+1}}. \tag{36}$$

**Proof.** Using (26)–(30) and (1)–(6), we get the results.

## 2.4 Statistical Applications

We shall apply the various inequalities only to efficiency comparisons for the OLS and GLS estimators. For applications to hypothesis tests and for other matters, see, for example, Hillier and King (1987).

Consider the following linear model

$$Y = X\beta + u, \tag{37}$$

where $Y$ is an $n \times 1$ observation vector, $X = (x_1, \ldots, x_n)'$ is an $n \times k$ model matrix, $\beta$ is an unknown $k \times 1$ parameter vector, $u$ is an $n \times 1$ error vector with mean E($u$)= 0 and variance matrix var ($u$)= $\Gamma$, and the eigenvalues of $\Gamma$ are $\lambda_1 \geq \cdots \geq \lambda_n$. Let $B_O$ and $B_G$ denote the OLS and GLS estimators of $\beta$ respectively. Without loss of generality, we shall suppose that $\Gamma > 0$ and that $X'X = I$. For simplicity, we shall use $n \geq 2k$. Clearly

$$\begin{aligned} B_O &= (X'X)^{-1}X'Y = X'Y, &(38)\\ \text{var}\,(B_O) &= X'\Gamma X, &(39)\\ B_G &= (X'\Gamma^{-1}X)^{-1}X'\Gamma^{-1}Y, &(40)\\ \text{var}\,(B_G) &= (X'\Gamma^{-1}X)^{-1}. &(41)\end{aligned}$$

Note that, in general,

$$\text{var}\,(B_O) \geq \text{var}\,(B_G), \tag{42}$$

and, in particular,

$$\text{var}\,(B_O) = \text{var}\,(B_G)$$

if and only if $\Gamma X = XC$, for some matrix $C$,

if and only if $\Gamma X = XX'\Gamma X$,

if and only if $X'\Gamma^2 X = (X'\Gamma X)^2$,

if and only if $X'\Gamma X = (X'\Gamma^{-1}X)^{-1}$.

Also

$$(X'\Gamma X)^2 \leq X'\Gamma^2 X \leq \frac{(\lambda_1+\lambda_n)^2}{4\lambda_1\lambda_n}(X'\Gamma X)^2, \tag{43}$$

$$(X'\Gamma X)^{-1} \leq X'\Gamma^{-1} X \leq \frac{(\lambda_1+\lambda_n)^2}{4\lambda_1\lambda_n}(X'\Gamma X)^{-1}. \tag{44}$$

For (42) see, for example, Puntanen and Styan (1989). For (43) and (44) see, for example, Liu (1995).

To study efficiency comparisons between $B_O$ and $B_G$, we see from (42) and (43) that it is natural to consider a new criterion

$$r_1 = \frac{\text{tr}\ X'\Gamma^2 X}{\text{tr}\ (X'\Gamma X)^2} \tag{45}$$

with $\Gamma > 0$ and $X'X = I$. If $B_O$ is equivalent to $B_G$, then $X'\Gamma^2 X = (X'\Gamma X)^2$, and $r_1 = 1$ reaches the lower bound. If $B_O$ is quite different from $B_G$, then $r_1$ could be very large. However, it can have an upper bound, as in Liu (1997):

$$\frac{\text{tr } X'\Gamma^2 X}{\text{tr } (X'\Gamma X)^2} \leq \frac{\sum_{j=1}^{k}(\lambda_j + \lambda_{n-j+1})^2}{4\sum_{j=1}^{k}\lambda_j\lambda_{n-j+1}}. \tag{46}$$

To find out how poor $B_O$ can be relative to $B_G$, we need to examine the upper bound of $r_1$. Bloomfield and Watson (1975) have also discussed this topic in detail (in a slightly different way), and they have introduced a criterion for comparing the efficiency of $B_O$ and $B_G$ which is based on the following inequality:

$$d_1 = \text{tr } [X'\Gamma^2 X - (X'\Gamma X)^2] \leq \frac{1}{4}\sum_{j=1}^{k}(\lambda_j - \lambda_{n-j+1})^2. \tag{47}$$

We list the following alternatives:

$$r_2 = \frac{\text{tr } X'\Gamma X}{\text{tr } (X'\Gamma^{-1}X)^{-1}} \leq \left[\frac{\sum_{j=1}^{k}(\lambda_j + \lambda_{n-j+1})}{2\sum_{j=1}^{k}\lambda_j^{1/2}\lambda_{n-j+1}^{1/2}}\right]^2, \tag{48}$$

$$d_2 = \text{tr } [X'\Gamma X - (X'\Gamma^{-1}X)^{-1}] \leq \sum_{j=1}^{k}(\lambda_j^{1/2} - \lambda_{n-j+1}^{1/2})^2, \tag{49}$$

$$l_1 = \text{tr } X'\Gamma^2 X(X'\Gamma X)^{-2} \leq \sum_{j=1}^{k}\frac{(\lambda_j + \lambda_{n-j+1})^2}{4\lambda_j\lambda_{n-j+1}}, \tag{50}$$

$$l_2 = \text{tr } X'\Gamma X X'\Gamma^{-1}X \leq \sum_{j=1}^{k}\frac{(\lambda_j + \lambda_{n-j+1})^2}{4\lambda_j\lambda_{n-j+1}}. \tag{51}$$

For $r_2$ in (48) and $d_2$ in (49), as two criteria for efficiency comparisons, see Yang (1988) and Rao (1985) respectively. The inequalities for $l_1$ and $l_2$ are new. One way of comparing these criteria is to examine their upper bounds. We consider four examples for the model (37) with the following $n \times n$ variance matrices respectively.

$$\Gamma_1 = \begin{pmatrix} 1 & p & \cdots & p \\ p & 1 & \ddots & \vdots \\ \vdots & \ddots & \ddots & p \\ p & \cdots & p & 1 \end{pmatrix}, \tag{52}$$

$$\Gamma_2 = \frac{1}{1-p^2}\begin{pmatrix} 1 & p & \cdots & p^{n-1} \\ p & 1 & \ddots & \vdots \\ \vdots & \ddots & \ddots & p \\ p^{n-1} & \cdots & p & 1 \end{pmatrix}, \tag{53}$$

$$\Gamma_3 = \begin{pmatrix} 1+p^2 & p & 0 & \cdots & 0 \\ p & 1+p^2 & p & \cdots & 0 \\ 0 & p & 1+p^2 & \ddots & \vdots \\ \vdots & \vdots & \ddots & \ddots & p \\ 0 & 0 & \cdots & p & 1+p^2 \end{pmatrix}, \tag{54}$$

$$\Gamma_4 = \begin{pmatrix} 1^p & 0 & \cdots & 0 \\ 0 & 2^p & \ddots & \vdots \\ \vdots & \ddots & \ddots & 0 \\ 0 & \cdots & 0 & n^p \end{pmatrix}. \tag{55}$$

The second example is of an AR(1) variance structure and the third has a MA(1) variance structure, see, for example, Fomby et. al (1984) for details. We have used the S-Plus package to generates numerical results and to graph them. For all examples with $n = 15$ observations and $k = 2$ parameters, we find the eigenvalues of $\Gamma_1, \Gamma_2, \Gamma_3$ and $\Gamma_4$, and then the upper bounds of $r_1, d_1, r_2, d_2$, and $l = l_1$ or $l = l_2$ (because of (50) and (51) the upper bound of $l_2$ is the same as that of $l_1$). We present four figures, for these four examples respectively, to illustrate comparisons among the upper bounds of the criteria. In each figure, we plot five curves of the upper bounds against $p$. (Although $|p| < 1$ always applies, for simplicity we focus only on $0 \leq p \leq 0.96$.) The behaviour of these upper bounds depends on specific circumstances. The upper bounds of $d_1$ are more sensitive to $p$ than those of the others in Figures 1, 2 and 4, but not in Figure 3. The upper bounds of $r_1, r_2$ and $l$ seem to change similarly in all four figures. The curves for the upper bounds of $r_1$ and $r_2$ almost coincide in Figure 4.

Now we shall comment on these criteria for a singular linear model with variance matrix a $\Sigma \geq 0$ and model matrix a $Z$ such that $\Re(Z) \subset \Re(\Sigma)$. The OLS and GLS estimators of $Z\beta$ and their variance matrices are

$$\begin{aligned} Zb_O &= Z(Z'Z)^+Z'Y, && (56) \\ \text{var}\,(Zb_O) &= ZZ^+\Sigma ZZ^+, && (57) \\ Zb_G &= Z(Z'\Sigma^+Z)^+Z'\Sigma^+Y, && (58) \\ \text{var}\,(Zb_G) &= Z(Z'\Sigma^+Z)^+Z' = (ZZ^+\Sigma^+ZZ^+)^+. && (59) \end{aligned}$$

For the second equality in (59), see Liu (1995). We define $r_1, d_1, r_2, d_2, l_1$ and $l_2$ (a discussion in Puntanen and Styan (1989) corresponding to (42) is used) in the positive-semidefinite matrix case:

$$r_1 = \frac{\text{tr}\; ZZ^+\Sigma^2 ZZ^+}{\text{tr}\;(ZZ^+\Sigma ZZ^+)^2}, \tag{60}$$

$$d_1 = \text{tr}\;[ZZ^+\Sigma^2 ZZ^+ - (ZZ^+\Sigma ZZ^+)^2], \tag{61}$$

$$r_2 = \frac{\text{tr}\; ZZ^+\Sigma ZZ^+}{\text{tr}\;(ZZ^+\Sigma^+ ZZ^+)^+}, \tag{62}$$

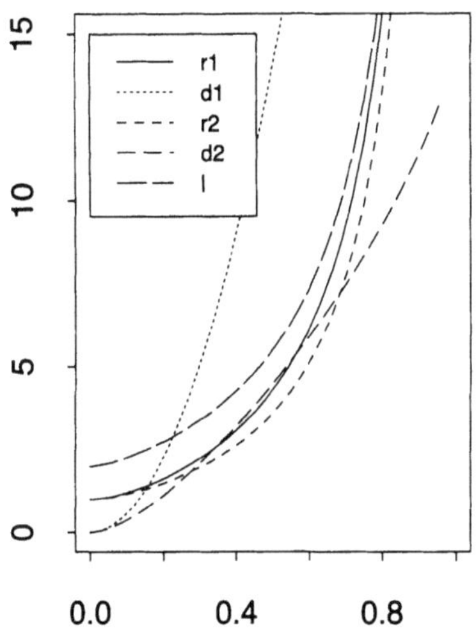


Figure 1

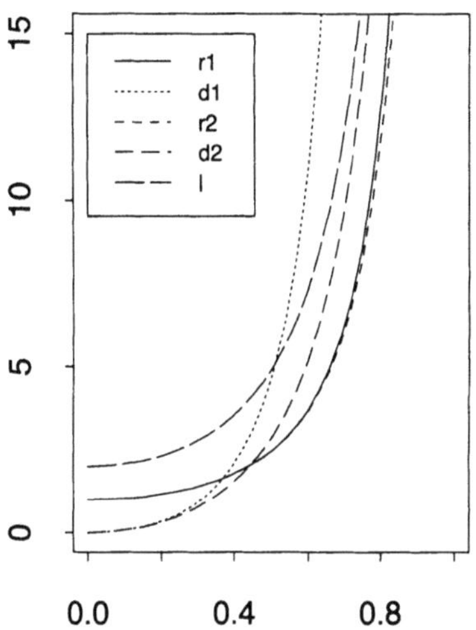


Figure 2

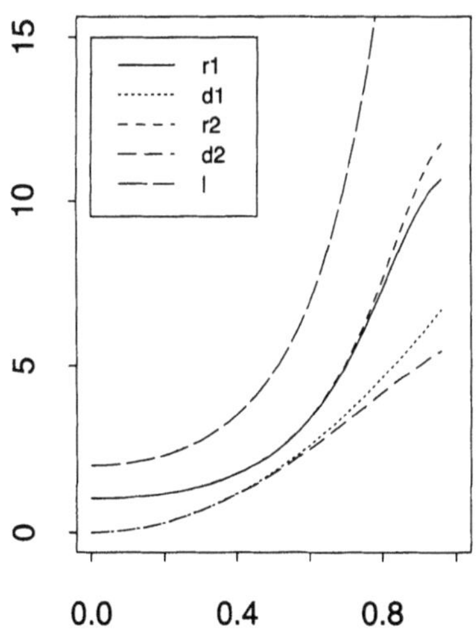


Figure 3

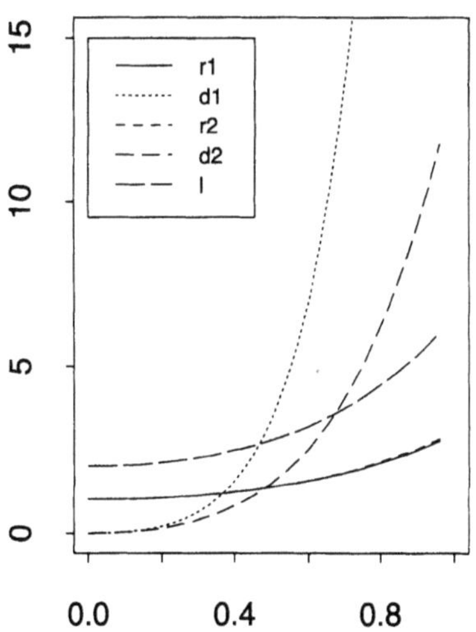


Figure 4

$$\begin{aligned} d_2 &= \operatorname{tr}\,[ZZ^+\Sigma ZZ^+ - (ZZ^+\Sigma^+ZZ^+)^+], &(63)\\ l_1 &= \operatorname{tr}\, ZZ^+\Sigma ZZ^+\Sigma^+ZZ^+, &(64)\\ l_2 &= \operatorname{tr}\, ZZ^+\Sigma^2ZZ^+(ZZ^+\Sigma ZZ^+)^{2+}. &(65)\end{aligned}$$

We can use Theorem 4 to give upper bounds for $r_1, d_1, r_2, d_2, l_1$ and $l_2$ in (60)–(65). This positive-semidefinite matrix case is more general than the one for (45) and (47)–(51). We see that this case for $r_2$ in (62) covers the case for the criterion introduced by Krämer (1980) in examining an AR(1) model. Also, (62) extends Yang's (1988) inequality. For a note relevant to (63), see Rao (1985).

## References

[1] Bloomfield, P., and G.S. Watson, (1975), The Inefficiency of Least Squares, *Biometrika,* 62, 121–128.

[2] Browne, M.W., (1974), Generalised Least Squares Estimators in the Analysis of Covariance Structures, *South African Statistical Journal,* 8, 1–24.

[3] Drury, S.W., and G.P.H. Styan, (1997), An Extension of an Inequality of Bloomfield–Watson and Knott, Presented by title in *6th International Workshop on Matrix Methods for Statistics, Istanbul, Turkey, Aug. 16–17, 1997.*

[4] Fomby, T.B., R.C. Hill and S.R. Johnson, (1984), *Advanced Econometric Methods,* Springer-Verlag, New York.

[5] Hillier, G.H., and M.L. King, (1987), Linear Regression with Correlated Errors: Bounds on Coefficient Estimates and *t*-values, pp. 74–80 in *Specification Analysis in the Linear Model,* M.L. King and D.E.A. Giles (eds.), Routledge and Kegan Paul, London.

[6] Khatri, C.G., and C.R. Rao, (1981), Some Extensions of the Kantotovitch Inequality and Statistical Applications, *Journal of Multivariate Analysis,* 11, 498–505.

[7] Khatri, C.G., and C.R. Rao, (1982), Some Generalisations of the Kantotovitch Inequality, *Sankhya A,* 44, 91–102.

[8] Knott, M., (1975), On the Minimum Efficiency of Least Squares, *Biometrika,* 62, 129–132.

[9] Kollo, T., and H. Neudecker, (1997), Asymptotics of Pearson–Hotelling Principal-Component Vectors of Sample Variance and Correlation Matrices, *Behaviormetrika,* 24(1), 51–69.

[10] Krämer, W., (1980), Finite Efficiency of Ordinary Least Squares in the Linear Regression Model with Autocorrelated Errors, *Journal of the American Statistical Association,* 75, 1005–1009.

[11] Liu, S., (1995), *Contributions to Matrix Calculus and Applications in Econometrics,* Thesis Publishers, Amsterdam.

[12] Liu, S., (1996), Reverse of a Convex Matrix Inequality, *Image,* 16, 32.

[13] Liu, S., (1997), Bounds for a Ratio of Matrix Traces, *Image,* 18, 32.

[14] Liu, S., and H. Neudecker, (1996), Several Matrix Kantorovich-type Inequalities, *J. Math. Anal. Appl.,* 197, 23–26.

[15] Magnus, J.R., and H. Neudecker, (1991), *Matrix Differential Calculus with Applications in Statistics and Econometrics,* Revised edition, John Wiley, Chichester.

[16] Marshall, A.W., and I. Olkin, (1990), Matrix Versions of the Cauchy and Kantorovich Inequalities, *Aequationes Math.,* 40, 89–93.

[17] Mond, M., and J.E. Pečarić, (1993), Matrix Versions of Some Means Inequalities, *Australian Math. Soc. Gaz.,* 20, 117–120.

[18] Neudecker, H., (1975), A Derivation of the Hessian of the (Concentrated) Likelihood Function of the Factor Model Employing the Schur Product, *British Journal of Math. Statist. Psychol.,* 28, 152–156.

[19] Neudecker, H., S. Liu and W. Polasek, (1995), The Hadamard Product and Some of its Applications in Statistics, *Statistics,* 26, 365–273.

[20] Neudecker, H., W. Polasek, and S. Liu, (1995), The Heteroskedastic Linear Model and the Hadamard Product—A note, *Journal of Econometrics,* 68, 361–366.

[21] Puntanen, S., and G.P.H. Styan, (1989), The Equality of the Ordinary Least-Squares Estimator and the Best Linear Unbiased Estimator, *The American Statistician,* 43 (3), 153–161.

[22] Rao, C. R., (1985), The Inefficiency of Least Squares: Extensions of the Kantorovich Inequality, *Linear Algebra and Its Applications,* 70, 249–255.

[23] Schopf, A.H., (1960), On the Kantorovich Inequality, *Numerische Mathematik,* 2, 344–346.

[24] Styan, G.P.H., (1973), Hadamard Products and Multivariate Analysis, *Linear Algebra and Its Applications,* 6, 217–240.

[25] Styan, G.P.H., (1983), On Some Inequalities Associated with Ordinary Least Squares and the Kantorovich Inequality, in *Festschrift for Eino Haikala on his Seventieth Birthday: Acta University Tampere,* Series A 153, 158–166.

[26] Wang, S.G., and J. Shao, (1992), Constrained Kantorovich Inequalities and Relative Efficiency of Least Squares, *Journal of Multivariate Analysis,* 42, 284–298.

[27] Watson, G.S., G. Alpargu, and G.P.H. Styan, (1997), Some Comments on Six Inequalities Associated with the Inefficiency of Ordinary Least Squares with One Regular Regressor, *Linear Algebra and Its Applications,* 264, 13–53.

[28] Yang, H., (1988), Extensions of the Kantorovitch Inequality and the Error Ratio of the Mean Square [in Chinese], *Mathematica Applicata-Yingyong Shuxue,* 1 (4), 85–90.

CHAPTER 3

# Matrix Inequality Applications in Econometrics

Paul BEKKER and Tom WANSBEEK

## 3.1 Introduction

Matrix inequalities play a large role in econometrics. Such inequalities essentially come in two kinds. One is concerned with inequalities of the scalar type involving scalar functions of matrices. A fine overview is offered by chapter 11 of the often-cited reference book of Magnus and Neudecker (1988). The title of this chapter refers to "Matrix inequalities" but the results given, though involving matrices, are in scalar form. Without distracting from the usefulness and the comprehensiveness of these results, an equally challenging type of inequalities are of the matrix type. Again, we may distinguish two types here. One involves element-by-element inequalities. These find many applications in mathematical economics. See e.g. Takayama (1974) for an overview. In econometrics (and in multivariate statistics in general) another type of matrix inequality occurs frequently. One is then concerned with so-called Löwner orderings, i.e., $A > B$ means that $A - B$ is a positive definite matrix, and $A \geq B$ means that $A - B$ is a positive semidefinite (or nonnegative definite) matrix.

In this essay we present a number of results on matrix inequalities relevant in econometrics. We do not attempt to give an overview of existing matrix inequalities in econometrics but rather present a number of new results stemming from a variety of problems. We discuss the following range of topics. In section 3.2 we consider the multinomial probit model where the covariance of the random utility components of the various alternatives is unrestricted. Since only the differences between the utilities of the alternatives play a role when spawning the choices, there is a certain freedom in varying a particular value of the covariance matrix without this having observational implications. Such varying should leave the covariance matrix positive definite, restricting the bounds, within which the variation can take place, in a surprisingly complicated way.

Section 3.3 presents an array of inequalities that play a role in regression analysis. Frequently, estimators can be ordered in terms of efficiency, and the results presented in this section offer a range of tools that can be useful in the process.

Often, in econometrics and multivariate statistics, the disturbance term or error term in a regression model is the sum of independent random elements with zero expectation and unknown variance. The MINQUE approach is often used to estimate these variances. A desirable property of the estimators of variances is of course nonnegativity. Section 3.4 discusses some matrix inequalities that play a role when establishing sufficient conditions under which the MINQUE is positive.

When matrices are subject to a Löwner ordering, there are also implied orderings on their eigenvalues. This is a field that has not yet been fully researched. Section 3.5 presents a number of basic results. The results have a number of potential applications in the context of measurement error models. Such models are more explicitly the focus of section 3.6. One question facing a researcher estimating a model in which some regressors are measured with error is whether to omit the noisy regressors or to leave them in. This choice problem is shown to involve some matrix inequality puzzles. Section 3.7 concludes the paper.

## 3.2 Equivalent Covariance Matrices in the Multinomial Probit Model

The multiple probit model is frequently used to model choices from multiple alternatives. One of the attractive features of this model is the simple way in which covariances between the random elements in the random utility equations can be modelled. Omitting subscripts pertaining to individual observational units, the multiple probit model then is

$$u \sim N_M(X\beta, \Omega),$$

where $u$ is an $M$-vector containing the (unobserved) utility attached to each of $M$ alternatives; $X$ is an $M \times k$ matrix containing $k$ characteristics for each of the alternatives; $\beta$ contains the regression coefficients; and $e \equiv u - X\beta$ is a disturbance vector, independent of $X$, with zero expectation and covariance matrix $\Omega$, of order $M \times M$. Random utility theory postulates that individuals choose the alternative yielding maximal utility. Hence the observations are only informative about utility differences.

The model formulated in this way is not identified. There are basically two sources of underidentification. One is a matter of scale invariance. That is, a parameter configuration $(\beta, \Omega)$ is observationally equivalent to $(c\beta, c^2\Omega)$ for any scalar $c$. This is simple to see since a change of scale does not affect the ordering of the latent utilities.

Another source of underidentification, relevant here, has to do with $\Omega$ alone. Assume that we have been offered an estimate of $\Omega$, and we wish to use this estimate to say something about the structure of preferences. When doing so, we have to take into account that $\Omega$ is also not identified in the following sense.

Let $\alpha$ be an as yet arbitrary vector of order $M$. Then we may freely add the matrix $\iota_M\alpha' + \alpha\iota_M'$ of order $M$ and rank 2 to $\Omega$ since $\Omega$ and $\Omega + \iota_M\alpha' + \alpha\iota_M'$ are observationally equivalent since the data are only informative on the differences in utility between alternatives. The only limitation to the choice of $\alpha$ is that the resulting matrix should still be positive definite.

This motivates the following problem. Given $\Omega$, with $\Omega > 0$, describe the set of $M$-vectors $\alpha$ such that

$$\Omega + \iota_M\alpha' + \alpha\iota_M' > 0. \tag{1}$$

This short question appears to be in fact quite complicated to answer. A simple case is when $\alpha$ is taken to be proportional to $\iota_M$. Then (1) holds when the factor of proportionality is larger than $-\frac{1}{2}\iota_M'\Omega^{-1}\iota_M$.

When handling the general case, it is convenient to work with a transformation of $\Omega$:

$$A \equiv \left(\Omega + \iota_M\iota_M'\right)^{-1}.$$

Using this definition, (1) is equivalent with

$$\Omega + \iota_M\iota_M' + \alpha\alpha' > \iota_M\iota_M' - \iota_M\alpha' - \alpha\iota_M' + \alpha\alpha'$$

or

$$A^{-1} + \alpha\alpha' > (\iota_M - \alpha)(\iota_M - \alpha)'.$$

Corollary 1 in Bekker, Kapteyn, and Wansbeek (1984) implies the equivalence of this with

$$(\iota_M - \alpha)'\left(A^{-1} + \alpha\alpha\right)^{-1}(\iota_M - \alpha) < 1.$$

We can develop a string of equivalences from here:

$$\begin{array}{rcl}
(\iota_M - \alpha)'\left\{A - \dfrac{A\alpha\alpha' A}{1 + \alpha' A\alpha}\right\}(\iota_M - \alpha) & < & 1 \\
& \Longleftrightarrow & \\
(\iota_M - \alpha)'A(\iota_M - \alpha)\cdot(1 + \alpha' A\alpha) - \left\{(\iota_M - \alpha)'A\alpha\right\}^2 & < & 1 + \alpha' A\alpha \\
& \Longleftrightarrow & \\
\left\{\iota_M' A\iota_M + \alpha' A\alpha - 2\iota_M' A\alpha\right\}\cdot(1 + \alpha' A\alpha) & & \\
-\left(\iota_M' A\alpha\right)^2 - \left(\alpha' A\alpha\right)^2 + 2\iota_M' A\alpha\alpha' A\alpha & < & 1 + \alpha' A\alpha \\
& \Longleftrightarrow & \\
\iota_M' A\iota_M - 2\iota_M' A\alpha + \alpha' A\alpha\iota_M' A\iota_M - \alpha' A\iota_M\iota_M' A\alpha & < & 1 \\
& \Longleftrightarrow & \\
(1 - \iota_M' A\iota_M) + 2\iota_M' A\alpha - \alpha'\left(\iota_M' A\iota_M\cdot A - A\iota_M\iota_M' A\right)\alpha & > & 0 \\
& \Longleftrightarrow & \\
(1, \alpha')\begin{bmatrix} 1 - \iota_M' A\iota_M & \iota_M' A \\ A\iota_M & -\left(\iota_M' A\iota_M\cdot A - A\iota_M\iota_M' A\right)\end{bmatrix}\begin{bmatrix} 1 \\ \alpha\end{bmatrix} & > & 0 \\
& \Longleftrightarrow & \\
(1, \alpha')\left\{\begin{bmatrix} 1 & \iota_M' A \\ A\iota_M & A\iota_M\iota_M' A\end{bmatrix} - \iota_M' A\iota_M\begin{bmatrix} 1 & 0 \\ 0 & A\end{bmatrix}\right\}\begin{bmatrix} 1 \\ \alpha\end{bmatrix} & > & 0
\end{array}$$

or

$$(1,\alpha')\begin{bmatrix} 1 & 0 \\ 0 & A \end{bmatrix}\begin{bmatrix} 1 \\ \alpha \end{bmatrix} < (1,\alpha')\frac{\begin{bmatrix} 1 \\ A\iota_M \end{bmatrix}\begin{bmatrix} 1 \\ A\iota_M \end{bmatrix}'}{\iota_M' A\iota_M}\begin{bmatrix} 1 \\ \alpha \end{bmatrix}. \tag{2}$$

Now let $x$ denote a vector proportional to $(1,\alpha')'$, i.e.

$$x \equiv \lambda \begin{bmatrix} 1 \\ \alpha \end{bmatrix} \tag{3}$$

for arbitrary scalar $\lambda$. Let $\phi$ be a scalar such that $\phi^2 = \iota_M' A\iota_M$. Then (2) is equivalent with

$$\frac{x'\begin{bmatrix} 1 \\ A\iota_M \end{bmatrix}\begin{bmatrix} 1 \\ A\iota_M \end{bmatrix}'x}{x'\begin{bmatrix} 1 & 0 \\ 0 & A \end{bmatrix}x} > \phi^2. \tag{4}$$

When

$$y \equiv \begin{bmatrix} 1 & 0 \\ 0 & A^{1/2} \end{bmatrix}x,$$

then (4) can be written as

$$\frac{y'\begin{bmatrix} 1 & 0 \\ 0 & A^{-1/2} \end{bmatrix}\begin{bmatrix} 1 \\ A\iota_M \end{bmatrix}\begin{bmatrix} 1 \\ A\iota_M \end{bmatrix}'\begin{bmatrix} 1 & 0 \\ 0 & A^{-1/2} \end{bmatrix}y}{y'y} > \phi^2$$

or

$$y'\begin{bmatrix} 1 \\ A^{1/2}\iota_M \end{bmatrix}\begin{bmatrix} 1 \\ A^{1/2}\iota_M \end{bmatrix}'y > \phi^2 y'y. \tag{5}$$

Let

$$r \equiv \frac{1}{\phi}\begin{bmatrix} 1 \\ A^{1/2}\iota_M \end{bmatrix},$$

so $r'r = (1+\phi^2)/\phi^2 > 1$, then (5) is equivalent with

$$y'rr'y > y'y. \tag{6}$$

Let the regression of $y$ on $r$ be given by $y = r\beta + u$, where $r'u = 0$ so $r'y = r'r\beta$. Then (6) is equivalent with

$$\beta^2(r'r)^2 > \beta^2 r'r + u'u$$

or

$$\beta^2 > \frac{u'u}{r'r(r'r-1)}. \tag{7}$$

In terms of $x$, we now have

$$\begin{aligned} x &= \begin{bmatrix} 1 & 0 \\ 0 & A^{-1/2} \end{bmatrix} r\beta + \begin{bmatrix} 1 & 0 \\ 0 & A^{-1/2} \end{bmatrix} u \\ &= \frac{1}{\phi}\begin{bmatrix} 1 & 0 \\ 0 & A^{-1/2} \end{bmatrix}\begin{bmatrix} 1 \\ A^{1/2}\iota_M \end{bmatrix}\beta + \begin{bmatrix} 1 & 0 \\ 0 & A^{-1/2} \end{bmatrix} u \\ &= \frac{1}{\phi}\iota_{M+1}\beta + \begin{bmatrix} 1 & 0 \\ 0 & A^{-1/2} \end{bmatrix} u. \end{aligned} \tag{8}$$

Let $u$ be partitioned as

$$u = \begin{bmatrix} u_1 \\ \tilde{u} \end{bmatrix}.$$

Since $r'u = 0$ or $(1, \iota_M' A^{1/2})u = 0$ there holds

$$u_1 + \tilde{u}' A^{1/2}\iota_M = 0, \tag{9}$$

and (8) becomes

$$\begin{aligned} x &= \frac{1}{\phi}\iota_{M+1}\beta + \begin{bmatrix} u_1 \\ A^{-1/2}\tilde{u} \end{bmatrix} \\ &= \frac{\beta}{\phi}\iota_{M+1} + \begin{bmatrix} -\iota_M' A^{1/2} \\ A^{-1/2} \end{bmatrix}\tilde{u}. \end{aligned}$$

Taking the first element gives, because of (3),

$$\lambda = \frac{\beta}{\phi} - \iota_M' A^{1/2}\tilde{u},$$

and hence the other elements imply

$$\begin{aligned} \alpha &= \frac{\dfrac{\beta}{\phi}\iota_M + A^{-1/2}\tilde{u}}{\dfrac{\beta}{\phi} - \iota_M' A^{1/2}\tilde{u}} \\ &= \frac{\iota_M + \dfrac{\phi}{\beta}A^{-1/2}\tilde{u}}{1 - \dfrac{\phi}{\beta}\iota_M' A^{1/2}\tilde{u}} \\ &= \frac{\iota_M + z}{1 - \iota_M' A z}, \end{aligned} \tag{10}$$

with

$$z \equiv \frac{\phi}{\beta}A^{-1/2}\tilde{u}.$$

Notice that $z$ is unrestricted, since we have employed the orthogonality condition to which $u$ was subject in the derivation.

So we have now expressed $\alpha$ in terms of an unrestricted vector $z$ and a scalar $\beta$ that was restricted due to (7). Elaborating the latter gives

$$\begin{aligned}
1 &> \frac{1}{\beta^2}\frac{u'u}{r'r(r'r-1)} \\
&= \frac{1}{\beta^2}\frac{\phi^4}{1+\phi^2}(u_1^2+\tilde{u}'\tilde{u}) \\
&= \frac{1}{\beta^2}\frac{\phi^4}{1+\phi^2}\tilde{u}'(I+A^{1/2}\iota_M\iota_M'A^{1/2})\tilde{u} \\
&= \frac{\phi^2}{1+\phi^2}\left(\frac{\phi}{\beta}\tilde{u}'A^{-1/2}\right)(A+A\iota_M\iota_M'A)\left(\frac{\phi}{\beta}A^{-1/2}\tilde{u}\right) \\
&= \frac{\iota_M'A\iota_M}{1+\iota_M'A\iota_M}z'(A+A\iota_M\iota_M'A)z,
\end{aligned}$$

or

$$z'(A+A\iota_M\iota_M'A)z < 1+\frac{1}{\iota_M'A\iota_M}. \tag{11}$$

So the set of $\alpha$'s such that (1) holds is given by (10) where $z$ is restricted to lie within the ellipsoid given by (11).

## 3.3 Matrix Inequalities in Regression Analysis

Streibel and Harvey (1993) consider the estimation of a single equation, within a system of simultaneous equations, that contains a stochastic trend component. This makes it implicitly possible to allow for latent factors such as technical progress and tastes. They consider, in particular, two estimators for this model, a LIML estimator and a generalised 2SLS estimator. They show that the latter has a lower asymptotic variance. In order to prove this, they employ a result due to Harvey, Neudecker and Streibel (1991) on a matrix inequality involving partitioned matrices. In this section, we provide an alternative proof of this inequality. We do so by putting the problem in a wider perspective and consider an array of matrix inequalities.

Let

$$A=\begin{bmatrix} A_{11} & A_{12} \\ A_{21} & A_{22}\end{bmatrix} \text{ and } B=\begin{bmatrix} B_{11} & B_{12} \\ B_{21} & B_{22}\end{bmatrix}$$

be two positive definite matrices, where $A_{ij}$ and $B_{ij}$ $(i,j=1,2)$ are of the same order. Then there holds

$$\left(A_{11}^{-1}+B_{11}^{-1}\right)^{-1} \geq \left\{\left(A^{-1}+B^{-1}\right)^{-1}\right\}_{11}.$$

It is worthwhile to put the problem in perspective and to consider the following array of matrix inequalities (which is to be read from left to right or from top to bottom) which hold for any set of positive definite matrices $Q_i, i = 1, \ldots, n$ (so $Q_i > 0$) and for any $X$ such that $X'X = I$:

$$
\begin{array}{ccc}
\sum_{i=1}^{n} X'Q_iX & \overset{\text{(i8)}}{\geq} & \{X'(\sum_{i=1}^{n} Q_i)^{-1}X\}^{-1} \\
 & & \text{(i7)} \\
\text{(i1)} & & \geq \\
\geq & & \sum_{i=1}^{n}(X'Q_i^{-1}X)^{-1} \\
 & & \text{(i6)} \\
 & \text{(i9)} & \geq \\
X'Q_iX & \geq & (X'Q_iX)^{-1} \\
\text{(i2)} & & \\
\geq & & \text{(i5)} \\
\{\sum_{i=1}^{n}(X'Q_iX)^{-1}\}^{-1} & & \geq \\
\text{(i3)} & & \\
\geq & \text{(i4)} & \\
X'(\sum_{i=1}^{n} Q_i^{-1})^{-1}X & \geq & \{X'(\sum_{i=1}^{n} Q_i^{-1})X\}^{-1}.
\end{array}
$$

To prove this, first notice that, when (i1)–(i4) have been proven, (i5)–(i8) follow after replacing $Q_i$ by $Q_i^{-1}$ and inverting the expressions; (i1) is trivial, and since (i6) is equally trivial (i2) follows. To prove (i3), notice that for any $Z$ of full column rank there holds $I \geq Z(Z'Z)^{-1}Z'$. Substitute $Q_i^{1/2}$ for $Z$ and pre- and postmultiply both sides by $Q_i^{-1/2}X$ to obtain

$$Q_i^{-1} \geq X(X'Q_iX)^{-1}X'. \tag{12}$$

Summation over $i$ gives

$$\sum_{i=1}^{n} Q_i^{-1} \geq X \left\{ \sum_{i=1}^{n} (X'Q_iX)^{-1} \right\} X'. \tag{13}$$

This implies (i3) due to the following argument, whose proof is adapted from Bekker and Neudecker (1989). Let $A, B, C$ be matrices such that $A > 0, C > 0$ and $A - BC^{-1}B' \geq 0$, then

$$
\begin{aligned}
0 &\leq \begin{bmatrix} I & BC^{-1} \\ 0 & I \end{bmatrix} \begin{bmatrix} A - BC^{-1}B' & 0 \\ 0 & C \end{bmatrix} \begin{bmatrix} I & 0 \\ C^{-1}B' & I \end{bmatrix} \\
&= \begin{bmatrix} A & B \\ B' & C \end{bmatrix} \\
&= \begin{bmatrix} I & 0 \\ B'A^{-1} & I \end{bmatrix} \begin{bmatrix} A & 0 \\ 0 & C - B'A^{-1}B \end{bmatrix} \begin{bmatrix} I & A^{-1}B \\ 0 & I \end{bmatrix},
\end{aligned}
$$

so $C - B'A^{-1}B \geq 0$. Now (i3) follows from (13) if we choose

$$A = \sum_{i=1}^{n} Q_i^{-1},$$

$$\begin{aligned} B &= X, \\ C &= \left\{ \sum_{i=1}^{n} (X'Q_iX)^{-1} \right\}^{-1}. \end{aligned}$$

The proof of (i4) is as follows: apply (12) with $\sum_{i=1}^{n} Q_i^{-1}$ substituted for $Q_i$ to obtain

$$\left\{ \sum_{i=1}^{n} Q_i^{-1} \right\}^{-1} \geq X \left\{ X' \left( \sum_{i=1}^{n} Q_i^{-1} \right) X \right\}^{-1} X'.$$

Pre- and postmultiply both sides by $X'$ and $X$. Finally, (i9) follows from (12) in the same way after substituting $Q_i^{-1}$ for $Q_i$.

Now choose $n = 2, Q_1 = A, Q_2 = B, X = (I, 0)'$, i.e. $X'$ and $X$ together select the upper left block of a matrix: $X'AX = A_{11}, X'BX = B_{11}$. Then the inequalities become

$$\begin{array}{ccc} A_{11} + B_{11} & \geq & \left[ \left\{ (A+B)^{-1} \right\}_{11} \right]^{-1} \\ & & \geq \\ \geq & & \left\{ (A^{-1})_{11} \right\}^{-1} + \left\{ (B^{-1})_{11} \right\}^{-1} \\ & & \geq \\ A_{11} \ \ (\text{or } B_{11}) & \geq & \left\{ (A^{-1})_{11} \right\}^{-1} \ \ (\text{or} \left\{ (B^{-1})_{11} \right\}^{-1}) \\ \geq & & \\ \left\{ (A_{11})^{-1} + (B_{11})^{-1} \right\}^{-1} & & \geq \\ \geq & & \\ \left\{ (A^{-1} + B^{-1})^{-1} \right\}_{11} & \geq & \left\{ (A^{-1})_{11} + (B^{-1})_{11} \right\}^{-1}. \end{array}$$

The lower left inequality is the one considered by Harvey, Neudecker and Streibel (1991).

## 3.4 A Condition for the Positivity of the MINQUE

In this section we are concerned with the linear regression model with a non-scalar covariance matrix of the disturbances. This matrix contains unknown parameters. The elements of the covariance matrix are linear functions of these parameters. In particular, we have the following situation. Let

$$y = X\beta + e,$$

with $y$ $(n \times 1)$ and $X$ $(n \times k)$ observed, $\beta$ $(k \times 1)$ a vector of unknown parameters, and $e$ $(n \times 1)$ an unobservable vector with zero expectation and variance

$$D(e) = \sum_{i=1}^{s} \sigma_i^2 V_i,$$

where the matrices $V_i, i = 1, \ldots, s$, are known positive (semi)definite matrices, with

$$V \equiv \sum_{i=1}^{s} V_i > 0.$$

and the $\sigma_i^2, i = 1, \ldots, s$, called the variance components, are unknown parameters.

Various ways of estimating these parameters have been put forward in the literature; see e.g. Rao and Kleffe (1988) excellent overview. One well-known estimator is the Minimum Norm Quadratic Unbiased Estimator (MINQUE).

There are various ways to motivate and describe the MINQUE. Here we just describe succinctly one such way. Consider estimation of a linear function

$$\sum_{i=1}^{s} p_i \sigma_i^2$$

of the variance components by a quadratic function $y'Ay$, where $A$ is a symmetric matrix to be determined. Unbiasedness of the estimator implies

$$\begin{aligned} X'AX &= 0, \\ \text{tr}AV_i &= p_i, i = 1, \ldots, s, \end{aligned} \tag{14}$$

and the MINQUE is obtained if, under the unbiasedness conditions,

$$\text{tr}AVAV + 2r^2 \text{tr}AXX'AV$$

is minimised, where $r^2$ is a constant to be specified by the researcher.

The solution to this problem has been given by Rao (1971), and a succinct alternative derivation has been given by Neudecker and Wansbeek (1991). The solution is

$$A = V^{-1} \sum_{i=1}^{s} \lambda_i (V_i - P_V V_i P_V') V^{-1}$$

with

$$P_V \equiv X(X'V^{-1}X)^{-1}X'V^{-1},$$

and the constants $\lambda_i$ are determined by the unbiasedness condition (14).

A problem with estimators of this kind, when estimating a positive quantity, is that they are not guaranteed to be positive. Evidently, a sufficient condition for positivity (or rather, nonnegativity) is that all matrices of the form $V_i - P_V V_i P_V'$ are p.s.d. This suggests the following problem. Let $W \geq 0$ and let $P$ be a symmetric projection matrix. Under which condition does it hold that $W - PWP \geq 0$? The answer to this question is

$$W - PWP \geq 0 \iff PWP = WP.$$

The proof of this is as follows.

First, $W \geq 0$ implies

$$\begin{bmatrix} P \\ I \end{bmatrix} W(P, I) \geq 0,$$

or

$$\begin{bmatrix} PWP & PW \\ WP & W \end{bmatrix} \geq 0.$$

If $PWP = WP$, then

$$\begin{bmatrix} PWP & PWP \\ PWP & W \end{bmatrix} \geq 0.$$

Hence for the Schur complement there holds

$$W \geq PWP(PWP)^{+}PWP \text{ or } W \geq PWP.$$

Second, $P(W - PWP)P = 0$. If $W - PWP \geq 0$, then $(W - PWP)P = 0$ or $WP = PWP$ follow directly.

We now elaborate the solution. The condition $PWP = WP$ is equivalent with $PW = WP$, i.e. $P$ and $W$ are commuting matrices. Hence they have the same eigenvectors and can be diagonalised simultaneously. In other words, we can write $P = QQ'$, with $Q$ the matrix containing the eigenvectors of $W$, so $WQ = Q\Psi$, with $\Psi$ the diagonal matrix of eigenvalues of $W$. Summarising, the condition $W - PWP \geq 0$ is equivalent to the condition that $P$ projects onto the space spanned by the eigenvectors of $W$.

## 3.5 Eigenvalues and Eigenvectors of a Bounded Matrix

Let $A < B$ be given matrices, and let $\Sigma$ be a symmetric matrix whose only known property is

$$A \leq \Sigma \leq B, \tag{15}$$

i.e., $\Sigma$ is bounded in the Löwner sense by $A$ and $B$. This implies bounds on the eigenvalues of $\Sigma$. The purpose of this section is to derive such bounds.

Let $C$ and $D$ be defined by $A + B = 2C$ and $B - A = 2D$. Substitution in (15) gives

$$C - D \leq \Sigma \leq C + D,$$

with $D > 0$. This is equivalent to

$$\begin{bmatrix} \Sigma - (C - D) & 0 \\ 0 & (C + D) - \Sigma \end{bmatrix} \geq 0,$$

or

$$\begin{bmatrix} I & I \\ I & -I \end{bmatrix} \begin{bmatrix} \Sigma - (C - D) & 0 \\ 0 & (C + D) - \Sigma \end{bmatrix} \begin{bmatrix} I & I \\ I & -I \end{bmatrix} \geq 0,$$

or again

$$\begin{bmatrix} D & \Sigma - C \\ \Sigma - C & D \end{bmatrix} \geq 0.$$

According to corollary 4.3 in Bekker (1988), the set

$$\mathcal{X} \equiv \{x \mid x = \Sigma f, C - D \leq \Sigma \leq C + D\}$$

is given by the set of vectors $x$ satisfying

$$\begin{bmatrix} f'Df & x' - f'C \\ x - Cf & D \end{bmatrix} \geq 0.$$

In particular, the set $\{x \mid x \in \mathcal{X}, x = \lambda f\}$ is given by those vectors $\lambda f$ where $\lambda$ satisfies

$$\begin{bmatrix} f'Df & \lambda f' - f'C \\ \lambda f - Cf & D \end{bmatrix} \geq 0.$$

Since $D > 0$, Albert's theorem (cf. Bekker and Neudecker (1989)) implies that this set is given by those $\lambda$'s that satisfy

$$\begin{aligned} f'Df &\geq f'(\lambda I - C)D^{-1}(\lambda I - C)f \\ &= \lambda^2 f'D^{-1}f - 2\lambda f'CD^{-1}f + f'CD^{-1}Cf. \end{aligned}$$

This is satisfied by vectors $f$ satisfying

$$(f'CD^{-1}f)^2 - f'D^{-1}f(f'CD^{-1}Cf - f'Df) \geq 0,$$

or

$$f'Df \geq f'C\left(D^{-1} - \frac{D^{-1}ff'D^{-1}}{f'D^{-1}f}\right)Cf. \tag{16}$$

All values from the interval $[\lambda_1, \lambda_2]$ are then possible eigenvalues:

$$\lambda_{1,2} = \frac{f'CD^{-1}f}{f'D^{-1}f} \pm \sqrt{\frac{f'Df - f'C\left(D^{-1} - \frac{D^{-1}ff'D^{-1}}{f'D^{-1}f}\right)Cf}{f'D^{-1}f}}. \tag{17}$$

In particular, if $D = \alpha C$, with $C > 0$, (16) and (17) are given by

$$\frac{(f'f)^2}{f'Cff'C^{-1}f} \geq 1 - \alpha^2 \tag{18}$$

$$\lambda_{1,2} = \frac{f'f \pm \sqrt{f'f - (1 - \alpha^2)f'Cff'C^{-1}f}}{f'C^{-1}f}. \tag{19}$$

For example, an interesting question is how large $\alpha$ must at least be such that

$$(1 - \alpha)C \leq \Sigma \leq (1 + \alpha)C$$

and

$$\Sigma = \begin{bmatrix} \Sigma_{11} & 0 & \Sigma_{12} \\ 0 & \sigma_i^2 & 0 \\ \Sigma_{21} & 0 & \Sigma_{22} \end{bmatrix}.$$

The solution is as follows: let $e_i$ be the $i$th unit vector. Then

$$\Sigma e_i = \sigma_i^2 e_i = \lambda e_i.$$

Hence, according to (18) with $f = e_i$,

$$\frac{1}{c_{ii}c^{ii}} \geq 1 - \alpha^2,$$

with $c_{ii} \equiv e_i' C e_i$ and $c^{ii} \equiv e_i' C^{-1} e_i$. So

$$\alpha^2 \geq 1 - \frac{1}{c_{ii}c^{ii}}.$$

Take

$$\alpha = \sqrt{1 - \frac{1}{c_{ii}c^{ii}}}.$$

Then, cf. (19): $\sigma_i^2 \in [\lambda_1, \lambda_2]$, where $\lambda_1 = \lambda_2 = 1/c^{ii}$, the Schur complement.

An open question is the following: how large must $\alpha$ be at least such that $\Sigma$ can be diagonal?

## 3.6 Proxies and Measurement Error

The standard linear multiple regression model can be written as

$$y = \Xi\beta + \varepsilon, \tag{20}$$

where $y$ is an observable $N$-vector and $\varepsilon$ an unobservable $N$-vector of random variables. The latter are independently normally distributed with zero expectation and variance $\sigma_\varepsilon^2 > 0$. The $k$-vector $\beta$ is fixed but unknown. The $N \times k$-matrix $\Xi$ contains the regressors. Its rows are assumed to be independently normally distributed with zero expectation and nonsingular covariance matrix, and uncorrelated with $\varepsilon$.

The first $g$ regressors are measured with error. So instead of $\Xi$ we observe the matrix $X$ of the same order as $\Xi$,

$$X = \Xi + V,$$

where $V$ $(N \times k)$ is the matrix of measurement errors. Its rows are assumed to be independently normally distributed with zero expectation and covariance matrix $\Omega$ $(k \times k)$ and uncorrelated with $\Xi$ and $\varepsilon$. The last $k - g$ columns of $V$ are identically equal to zero so $\Omega$ has structure

$$\Omega = \begin{bmatrix} \Omega_{11} & 0 \\ 0 & 0 \end{bmatrix} \geq 0. \tag{21}$$

Let

$$\begin{aligned} \Sigma &= \begin{bmatrix} \Sigma_{11} & \Sigma_{12} \\ \Sigma_{21} & \Sigma_{22} \end{bmatrix} \\ &\equiv \operatorname{plim} \frac{1}{N}(y, X)'(y, X) = \begin{bmatrix} \gamma' A \gamma + \omega & \gamma' A \\ A\gamma & A \end{bmatrix}. \end{aligned} \tag{22}$$

So $\gamma$ is the plim of the OLSE from regressing $y$ on $X$, and $\omega$ is the plim of the resulting usual estimator of $\sigma_\varepsilon^2$. Since we are concerned with asymptotic properties only in this note we disregard the distinction between finite-sample properties and asymptotic properties, and so consider $\Sigma$ (hence $A$, $\gamma$, and $\omega$) known. Due to the assumed normality of all random variables in the model, $\Sigma$ also contains *all* sample knowledge about the parameters (cf. Bekker (1986)); when one or more of the distributions would be nonnormal we would in addition have knowledge from higher-order moments but since this type of inference has found little application in practice we do not consider such cases here. Given the assumptions made there holds

$$\text{plim}\frac{1}{N}(y, \Xi)'(y, \Xi) = \Sigma - \begin{bmatrix} 0 & 0 \\ 0 & \Omega \end{bmatrix} > 0; \tag{23}$$

this constitutes, for a given value of $\Sigma$, a restriction on $\Omega$ in addition to (21). Apart from (21) and (23) we have no information on $\Omega$ and we in particular do not assume the existence of exogenously given bounds on $\Omega$.

The relationship between $\beta$ and $\gamma$ is

$$(A - \Omega)\beta = A\gamma. \tag{24}$$

The set of all possible $\beta$'s given the value of $\Sigma$ is described by

$$\mathcal{B} \equiv \{\beta \mid \beta = (A - \Omega)^{-1}A\gamma, \quad \Omega \text{ satisfies (21) and (23)}\}.$$

Let $\beta$ be partitioned into two subvectors, corresponding with the "noisy" and the correctly measured variables, respectively:

$$\beta = \begin{bmatrix} \beta_1 \\ \beta_2 \end{bmatrix} \begin{matrix} \} & g \\ \} & k - g. \end{matrix}$$

The set of all possible $\beta_2$'s compatible with the sample information is described by

$$\mathcal{B}_2 \equiv \{\beta_2 \mid \beta \in \mathcal{B}\}.$$

Bekker and Wansbeek (1996) characterised the sets $\mathcal{B}$ and $\mathcal{B}_2$, so answering the question where the OLSE of $\beta$ and $\beta_2$ might be located, given measurement error on the first $g$ regressors.

Here we consider a complementary or 'dual' problem. Assume that we have observed the true data, so we know the true value of $\beta$, $\sigma^2$ and $K \equiv \text{plim}\Xi\Xi'/N$, at least in the asymptotic sense as used before. What is the set of (plims of) OLSE's of $\beta$ if we allow for measurement error on the first $g$ variables?

To pose the problem formally, we know

$$\begin{bmatrix} \beta' K\beta + \sigma^2 & \beta' K \\ K\beta & K \end{bmatrix}$$

and we want to describe the set

$$\mathcal{B}^* = \left\{ f \middle| f = (K + \Omega)^{-1}K\beta, \quad \Omega = \begin{bmatrix} \Omega_{11} & 0 \\ 0 & 0 \end{bmatrix} \geq 0 \right\}.$$

The solution is given by

$$\mathcal{B}^* = \left\{ f \middle| f = \beta - \begin{bmatrix} K^{11} \\ K^{21} \end{bmatrix} H\beta_1, \quad 0 \leq H < \left(K^{11}\right)^{-1} \right\},$$

where superscripts denote blocks of inverse matrices. The proof of this is as follows. Let

$$H \equiv \Omega_{11} - \Omega_{11} \left\{ \Omega_{11} + \left(K^{11}\right)^{-1} \right\}^{-1} \Omega_{11}.$$

Then

$$\begin{aligned} f &= (K + \Omega)^{-1} K\beta \\ &= \left\{ K + \begin{bmatrix} I_g \\ 0 \end{bmatrix} \Omega_{11}(I_g, 0) \right\}^{-1} K\beta \\ &= \left\{ K^{-1} - K^{-1} \begin{bmatrix} I_g \\ 0 \end{bmatrix} H(I_g, 0) K^{-1} \right\} K\beta \\ &= \beta - \begin{bmatrix} K^{11} \\ K^{21} \end{bmatrix} H\beta_1. \end{aligned}$$

Lemma 2 from

Bekker and Wansbeek (1996) implies that $0 \leq H < \left(K^{11}\right)^{-1}$. We now characterise the set $\mathcal{B}_2^*$ pertaining to the subvector $\beta_2$ corresponding with the correctly measured variables. Let

$$M \equiv K^{21} \left(K^{11}\right)^{-1} K^{12},$$

then

$$\mathcal{B}_2^* = \{\beta_2\} \cup \mathcal{A}$$

where $\mathcal{A}$ is given by the set of $f_2$'s ($f_2$ denoting the second subvector of $f$) satisfying

$$\left\{ f_2 - \beta_2 - \frac{1}{2}(\zeta_2 - \beta) \right\}' M^+ \left\{ f_2 - \beta_2 - \frac{1}{2}(\zeta_2 - \beta) \right\} < \frac{1}{4}\beta_1' \left\{K^{11}\right\}^{-1} \beta_1$$

and

$$MM^+ (f_2 - \beta_2) = (f_2 - \beta_2),$$

where

$$\zeta_2 \equiv -K^{21} \left(K^{11}\right)^{-1} \beta_1 + \beta_2.$$

Before we prove this, we notice the following. If $\beta_1 = 0$, then $\beta_2 = \zeta_2 = f_2$ because of the definition of $\zeta_2$. If $\beta_1 \to 0$, then both $f_2$ and $\zeta_2 \to \beta_2$. If $\beta_1 \in \mathcal{R}\left(K^{21}\right)$ is an element of the range spanned by the columns of $K^{21}$, then

$$(\zeta_2 - \beta_2)' M^+ (\zeta_2 - \beta_2) = \beta_1' \left(K^{11}\right)^{-1} \beta_1,$$

hence $\beta_2$ and $\zeta_2$ lie on the boundary of the ellipsoid. So, now also $\beta_2$ is an element of the set, i.e. a point on the boundary. If $\beta_1 \notin \mathcal{R}\left(K^{21}\right)$, then $\beta_2$ and $\zeta_2$ lie inside the ellipsoid.

We now turn to a proof of the characterisation of $\mathcal{B}_2^*$. For $\varepsilon > 0$ there holds

$$0 \leq H \leq (1-\varepsilon)\left(K^{11}\right)^{-1}$$

is equivalent to

$$\begin{bmatrix} \left(K^{11}\right)^{-1} & \frac{2H}{1-\varepsilon} - \left(K^{11}\right)^{-1} \\ \frac{2H}{1-\varepsilon} - \left(K^{11}\right)^{-1} & \left(K^{11}\right)^{-1} \end{bmatrix} \geq 0.$$

The set of possible vectors $-K^{21}H\beta_1$ is then given by

$$\begin{bmatrix} \beta'\left(K^{11}\right)^{-1}\beta & 2\frac{(f_2-\beta)'}{1-\varepsilon} - \beta'\left(K^{11}\right)^{-1}K^{12} \\ 2\frac{(f_2-\beta)}{1-\varepsilon} - K^{21}\left(K^{11}\right)^{-1}\beta & K^{21}\left(K^{11}\right)^{-1}K^{12} \end{bmatrix} \geq 0.$$

This can be written as

$$\left\{\frac{f_2-\beta_2}{1-\varepsilon} - \frac{1}{2}(\zeta_2-\beta)\right\}' M^+ \left\{\frac{f_2-\beta_2}{1-\varepsilon} - \frac{1}{2}(\zeta_2-\beta)\right\} \leq \frac{1}{4}\beta_1'\left(K^{11}\right)^{-1}\beta_1$$

and

$$MM^+\left\{\frac{f_2-\beta_2}{1-\varepsilon} - \frac{1}{2}(\zeta_2-\beta)\right\} = \frac{f_2-\beta_2}{1-\varepsilon} - \frac{1}{2}(\zeta_2-\beta).$$

There moreover holds

$$MM^+(\zeta_2-\beta) = \zeta_2-\beta,$$

which concludes the proof.

## 3.7 Conclusion

In this essay we presented an array of econometric situations offering problems in the field of matrix inequalities. The problems considered are all related to a Löwner ordering of matrices, which can frequently be reformulated in terms of the positive definiteness of partitioned matrices, cf. Bekker (1988). Such reformulations allow for applications of general results on Schur complements. A nice survey of such results is given, in a historical perspective, by Ouellette (1981). This paper also gives an exhaustive bibliography of books and articles related to Schur complements. The inequalities considered here form an illustration of the wide applicability of matrix properties related to Schur complements.

Many people are intrigued by matrix inequalities. A new website on matrix inequalities in science and engineering has recently been started and can be found under

`http://www.wm.edu/CAS/MINEQ/matrix.html`.

## References


[1] Bekker, P.A., (1986), Comment on Identification in the Linear Errors in Variables Model, *Econometrica,* 54, 215–217.

[2] Bekker, P.A., (1988), The Positive Semidefiniteness of Partitioned Matrices, *Linear Algebra and its Applications,* 111, 261–278.

[3] Bekker, P.A., and A. Kapteyn and T.J. Wansbeek, (1984), Measurement Error and Endogeneity in Regression: Bounds for ML and IV Estimates, in T.K. Dijkstra (ed.), *Misspecification Analysis,* Springer Verlag, Berlin.

[4] Bekker, P.A., and H. Neudecker, (1989), Albert's Theorem Applied to Problems of Efficiency and MSE Superiority, *Statistica Neerlandica,* 43, 157–167.

[5] Bekker, P.A., and T.J. Wansbeek, (1996), Proxies versus Omitted Variables in Regression Analysis, *Linear Algebra and its Applications,* 237/238, 301–312.

[6] Harvey, A.C., H. Neudecker and M. Streibel, (1991), Problem 91.4.1: An Inequality for the Block-partitioned Inverse, *Econometric Theory,* 7, 543.

[7] Magnus, J.R., and H. Neudecker, (1988), *Matrix Differential Calculus,* John Wiley, Chichester.

[8] Neudecker, H., and T.J. Wansbeek, (1991), Rao's MINQUE-without-invariance Revisited, *Journal of Quantitative Economics,* 7, 239–246.

[9] Ouellette, D.V., (1981), Schur Complements and Statistics, *Linear Algebra and its Applications,* 36, 187–295.

[10] Rao, C.R., (1971), Estimation of Variance and Covariance Components-MINQUE Theory, *Journal of Multivariate Analysis,* 1, 257–275.

[11] Rao, C.R., and J. Kleffe, (1988), *Estimation of Variance Components and Applications,* North-Holland Publishing Co., Amsterdam.

[12] Streibel, M., and A.C. Harvey, (1993), Estimation of Simultaneous Equation Models with Stochastic Trend Components, *Journal of Economic Dynamics and Control,*17, 263–287.

[13] Takayama, A., (1974), *Mathematical Economics,* The Dryden Press, Hinsdale, Illinois.

CHAPTER 4

# On a Generalisation of the Covariance Matrix of the Multinomial Distribution

Götz TRENKLER[1]

## 4.1 Introduction

A few years ago, Heinz Neudecker sent me a complimentary copy of his paper entitled "Mathematical properties of the variance of the multinomial distribution". On the first page was the handwritten message: "Lieber Herr Trenkler, vielleicht interessiert Sie diese Arbeit! Ihr HN." In fact, I found this paper, Neudecker (1995), so impressive that it motivated me to work on the same topic, namely the covariance of the multinomial distribution and related matrix theory. Of course, the reader will recognise the use of some of Neudecker's favourite tools and tricks in the results below.

Consider the $n \times 1$ vector $p$ with elements $p_i > 0$ and the $n \times n$ diagonal matrix $P = \text{diag}(p_i)$. Tanabe and Sagae (1992) derived the square–root free Cholesky decomposition of the matrix

$$R = P - pp', \tag{1.1}$$

where they assumed that $s'p \leq 1$, with $s$ being the $n \times 1$ vector of ones. Furthermore, they were interested in the Moore–Penrose inverse and the product of the non–zero eigenvalues of $R$. Neudecker (1995) resumed this analysis, provided more elegant proofs and offered some new results. Watson (1996) was successful in identifying the eigenvalues of $R$ when $s'p = 1$.

In the following, we shall generalise the cited work by investigating the matrix

$$S = D + ab', \tag{1.2}$$

where $D = \text{diag}(d_i)$ is a nonsingular $n \times n$ diagonal matrix and $a = (a_i)$ and $b = (b_i)$ are $n \times 1$–vectors. In particular, we are interested in the eigenvalues and the Moore–Penrose inverse of $S$.

[1]Support by Deutsche Forschungsgemeinschaft, Grant No. TR 253/2-3 is gratefully acknowledged.

Obviously, (1.1) is a special case of (1.2), but there are still other examples in the literature that can be treated under the set–up (1.2). For instance, Vermeulen (1967) and Klamkin (1970) were interested in the determination of eigenvalues of the matrix

$$\begin{pmatrix} a_1+e_1 & a_1 & \dots & a_1 \\ a_2 & a_2+e_2 & \dots & a_2 \\ \vdots & \vdots & \ddots & \vdots \\ a_n & a_n & \dots & a_n+e_n \end{pmatrix}, \tag{1.3}$$

where the $a_i$ and $e_i$ are all strictly positive. This problem arose in connection with a physical investigation on the electronic properties of particle–counting diamonds. When choosing $D = \text{diag}(e_i)$ and $b = s$, we see that (1.3) can be written as $D + as'$. We may note that Graybill (1983, Section 8.3) provides inverse, eigenvalues and determinant of $S$ in case of $S$ being nonsingular. Subsequently, however, we shall dispense with this assumption.

## 4.2 Moore–Penrose Inverse

For $S = D + ab'$, two cases may occur. $S$ is either singular or not. In the second case, the inverse of $S$ is readily seen to be

$$S^{-1} = D^{-1} - \frac{D^{-1}ab'D^{-1}}{1 + b'D^{-1}a}. \tag{2.1}$$

Actually, this result holds when $D$ is not necessarily diagonal (cf. Rao and Bhimasankaram, 1992, p.145). When $S$ is singular, then only the Moore–Penrose inverse exists. For this situation, the following relatively general result is useful.

**Theorem 1.** Let $T = A + ab'$, where $A$ is a nonsingular symmetric matrix. Then the following holds.

(i) $T$ is singular if and only if $1 + b'A^{-1}a = 0$.

(ii) If $T$ is singular, then the Moore–Penrose inverse of $T$ is

$$T^{+} = (I_n - kk^{+})A^{-1}(I_n - hh^{+}),$$

where $h = A^{-1}b$ and $k = A^{-1}a$.

(iii) If $T$ is singular, then $A^{-1}$ is a $g$–inverse of $T$.

**Proof.**

(i) $\det(T) = \det(A + ab') = \det(A)(1 + b'A^{-1}a)$ by a well–known formula (cf. Rao and Bhimasankaram, 1992, p.246). Since $\det(A) \neq 0$, the assertion is shown.

(ii) Let $T$ be singular, i.e. $1+b'A^{-1}a=0$. Then $(A+ab')k=a+a(b'A^{-1}a)=0$ which implies $(A+ab')kk^+=0$ and, consequently,

$$(A+ab')(I_n-kk^+)=A+ab'. \tag{2.2}$$

Using the symmetry of $A$, we may show, likewise, that

$$(I_n-hh^+)(A+ab')=A+ab'. \tag{2.3}$$

Hence it follows that

$$\begin{aligned}(I_n-kk^+)A^{-1}(I_n-hh^+)(A+ab') &= (I_n-kk^+)A^{-1}(A+ab')\\ &= I_n-kk^+ \end{aligned} \tag{2.4}$$

and

$$\begin{aligned}(A+ab')(I_n-kk^+)A^{-1}(I_n-hh^+) &= (A+ab')A^{-1}(I_n-hh^+)\\ &= I_n-hh^+. \end{aligned} \tag{2.5}$$

From (2.2), (2.3), (2.4) and (2.5), we finally get the four properties which the Moore–Penrose inverse has to satisfy.

(iii) Since $b'A^{-1}a+1=0$ we get

$$\begin{aligned}(A+ab')A^{-1}(A+ab') &= (A+ab')(I_n+A^{-1}ab')\\ &= A+ab'+ab'+ab'A^{-1}ab'\\ &= A+ab'.\end{aligned}$$

Taking $A=P, a=p$ and $b=-p$, and assuming $s'p=1$, from Theorem 1(ii) we have the result derived in Tanabe and Sagae (1992), Neudecker (1995) and Watson (1995), namely

$$\begin{aligned}R^+ &= (P-pp)^+\\ &= NP^{-1}N,\end{aligned} \tag{2.6}$$

where $N=I_n-n^{-1}ss'$.

Theorem 1(iii) admits an elegant characterisation of all solutions to the equation $Tx=c$, where $c$ is a given vector and $T$ is singular. This equation is consistent if and only if $TT^-c=c$, and then the general solution is

$$x=T^-c+(I-T^-T)z \tag{2.7}$$

with $z$ being an arbitrary vector.

As seen above, one choice of $T^-$ is $A^{-1}$. Thus we obtain: $Tx=c$ is consistent if and only if $a=0$ or $b'A^{-1}c=0$, in which case the general solution is

$$x=A^{-1}[c-(b'z)a], \tag{2.8}$$

where $z$ is an arbitrary vector.

## 4.3 Eigenvalues

Let us now investigate the eigenvalues of $T = A + ab'$. Since $A$ is symmetric, there exists an orthogonal matrix $Q$ such that $A = QDQ'$, where $D$ is a nonsingular diagonal matrix. Since $A+ab'$ and $D+Q'ab'Q$ have the same eigenvalues, without loss of generality we may investigate the matrix

$$S = D + ab', \tag{3.1}$$

again, where $D = \text{diag}(d_i)$ with $d_i \neq 0$. To determine the eigenvalues of $S$, we have to solve the determinantal equation

$$\det(D + ab' - \lambda I_n) = 0. \tag{3.2}$$

Since

$$D + ab' - \lambda I_n = D(I_n + D^{-1}ab' - \lambda D^{-1}), \tag{3.3}$$

equation (3.2) is equivalent to

$$\det\{D(\lambda) + eb'\} = 0, \tag{3.4}$$

where $e = (e_j) = D^{-1}a$ and $D(\lambda) = I_n - \lambda D^{-1}$. It is readily seen that $\lambda = 0$ is an eigenvalue of $S$ if and only if $b'e = -1$, or, equivalently, $b'D^{-1}a+1 = 0$ which is the condition for $S$ to be singular. From (2.8), it follows that the eigenvectors corresponding to $\lambda = 0$ are scalar multiples of $D^{-1}a = e$.

**Theorem 2.**

(i) If all $d_i$ are different and all $a_ib_i \neq 0$, $i = 1, \ldots, n$, then none of the $d_i$ is an eigenvalue of $S = D + ab'$.

(ii) If all $d_i$ are different, but, for index $j$, we have $a_jb_j = 0$, then $d_j$ is an eigenvalue of $S = D + ab'$.

**Proof.**

(i) Assume that one of the $d_i$, $d_j$ say, is an eigenvalue of $D + ab'$. Then we have

$$\det\{D(d_j) + eb'\} = 0. \tag{3.5}$$

Hence $D(d_j) = I_n - d_jD^{-1}$ has rank $n-1$ with 0 as a simple eigenvalue. By Lemma 1 of the appendix,

$$D(d_j)^{\#} = \mu\{D(d_j)\}e_je_j', \tag{3.6}$$

where $D(d_j)^{\#}$ is the adjoint matrix corresponding to $D(d_j)$, $\mu\{D(d_j)\} = \prod_{i\neq j}(1-d_j/d_i)$ and $e_j$ is the $j$th unit vector (see Magnus and Neudecker, 1988, p.41). From Lemma 2 of the appendix, we obtain

$$\begin{aligned}\det\{D(d_j) + eb'\} &= \det\{D(d_j)\} + b'D(d_j)^{\#}e \\ &= \mu\{D(d_j)\}b_je_j \\ &= \mu\{D(d_j)\}a_jb_j/d_j,\end{aligned} \tag{3.7}$$

which, by assumption, is different from zero. Thus we have obtained a contradiction.

(ii) As in (i) we derive

$$\det\{D(d_j) + eb'\} = \mu\{D(d_j)\}a_j b_j / d_j.$$

Since $a_j b_j = 0$, we see that $d_j$ is an eigenvalue of $S$.

For any $d_j$, define

$$\begin{aligned} N(d_j) &= \{d_i | d_i = d_j\} \\ \nu(d_j) &= |N(d_j)|, \end{aligned}$$

where $|\cdot|$ denotes the number of elements of a finite set.

**Theorem 3.** If $\nu(d_j) > 1$, then $d_j$ is an eigenvalue of $S = D + ab'$.

**Proof.** If $\nu(d_j) > 1$, then there exists at least one index $i$ such that $i \neq j$ and $d_i = d_j$. Then $r\{D(d_j)\} = r(I_n - d_j D^{-1}) \leq n - 2$, and, consequently, by Lemma 3 of the appendix, we have $D(d_j)^{\#} = 0$. Hence $\det\{D(d_j) + eb'\} = \det\{D(d_j)\} + b'D(d_j)^{\#}e = 0$ by virtue of Lemma 2 of the appendix.

Let us now characterise all eigenvalues of $T$ that are different from the $d_i$.

**Theorem 4.** Let $\lambda \neq d_i$ for all $i$. Then $\lambda$ is an eigenvalue of $S = D + ab'$ if and only if

$$\sum_{i=1}^{n} \frac{a_i b_i}{d_i - \lambda} = -1. \tag{3.8}$$

**Proof.**

$$\det\{D(\lambda) + eb'\} = \det\{D(\lambda)\}\{1 + b'D(\lambda)^{-1}e\}.$$

Therefore we get $\det\{D(\lambda) + eb'\} = 0$ if and only if $1 + b'D(\lambda)^{-1}e = 0$, i.e. $\sum_{i=1}^{n}(a_i b_i)/(d_i - \lambda) = -1$ by using $e_i d_i = a_i$ for all $i$.

When we look for eigenvalues different from the $d_i$, by Theorem 4 we have to solve equation (3.8). It is readily seen that the solutions are the roots of a polynomial of degree $n$, provided $\nu(d_i) = 1$ and $a_i b_i \neq 0$ for all $i$. However, if some $d_i$ coincide or some $a_i b_i = 0$, the degree of the polynomial decreases, in accordance with Theorems 2 and 3. Clearly $\lambda = 0$ is a solution to (3.8) if and only if $b'D^{-1}a + 1 = 0$, i.e. $S$ is singular.

Furthermore, if we arrange the $d_i$ in ascending order with some possible blocks of equal elements, then the eigenvalues can be located among the $d_i$. For instance, suppose $n = 8$ such that $d_1 = d_2 = d_3 < d_4 < 0 < d_5 < d_6 = d_7 < d_8$, $a_i b_i > 0$, $i \neq 5$, $a_5 b_5 = 0$ and $b'D^{-1}a + 1 = 0$. Then four eigenvalues are

$d_1, d_1, d_5, d_6$. The other four eigenvalues (including 0) can be found as zeros of the rational function

$$\begin{aligned} Q(\lambda) &= \frac{1}{d_1 - \lambda}(a_1 b_1 + a_2 b_2 + a_3 b_3) + \frac{a_4 b_4}{d_4 - \lambda} \\ &+ \frac{1}{d_6 - \lambda}(a_6 b_6 + a_7 b_7) + \frac{a_8 b_8}{d_8 - \lambda} + 1. \end{aligned} \tag{3.9}$$

The three nonzero roots of (3.9) fall into the open intervals $(-\infty, d_1), (d_1, d_4)$ and $(d_6, d_8)$.

This information is helpful if one tries to calculate the roots numerically by the Newton–Raphson method.

It should be noted that, according to the remark at the beginning of Section 3, we have shown the surprising fact that the matrix $T = A + ab'$ with $A$ being symmetric and nonsingular, has always real eigenvalues.

In conclusion let us consider $R = P - pp'$ from Section 1. When setting $D = P$, $a = -p$ and $b = p$, we get $R = D + ab'$. Since $p_i > 0$, it follows that $a_i b_i \neq 0$ for all $i$, and condition (ii) of Theorem 2 does not apply.

**Case 1.** $R$ singular
Then 0 is an eigenvalue of $R$. If $\nu(p_j) > 1$, then $p_j$ is also an eigenvalue of $R$. The eigenvalues different from the $p_i$ can be found from the equation

$$\sum_i \frac{p_i^2}{p_i - \lambda} = 1. \tag{3.10}$$

**Case 2.** $R$ nonsingular
Then 0 is not an eigenvalue. The other eigenvalues can be identified as in Case 1.

For example, if $p_1 < p_2 < \cdots < p_n$ and $R$ is nonsingular, then the eigenvalues $\lambda_i$ (in ascending order) satisfy

$$0 < \lambda_1 < p_1 < \lambda_2 < p_2 < \cdots < p_{n-1} < \lambda_n < p_n.$$

For a similar discussion, we refer to Watson (1995). Further statistical implications can be found in Johnson, Kotz and Balakrishnan (1997, Ch. 35), Ronning (1982) and Huschens (1990).

## Appendix

**Lemma 1.** (Magnus and Neudecker, 1988, p.41):
Let $B$ be a real $n \times n$ matrix ($n \geq 2$) possessing a simple eigenvalue 0, and $x$ and $y$ are $n \times 1$ vectors. Then $r(B) = n - 1$ and

$$B^{\#} = \mu(B) xy' / x'y,$$

where $\mu(B)$ is the product of the $n-1$ non–zero eigenvalues of $B$, and $x$ and $y$ satisfy $Bx = B'y = 0$.

**Lemma 2.** (Rao and Bhimasankaram, 1992, p.246)
Let $B$ be a real $n \times n$ matrix ($n \geq 2$) and $x$ and $y$ are $n \times 1$ vectors. Then

$$\det(B + xy') = \det(B) + y'B^{\#}x$$

**Lemma 3.** (Magnus and Neudecker, 1988, p.41)
Let $B$ be a real $n \times n$ matrix ($n \geq 2$). If $r(B) \leq n - 2$, then $B^{\#} = 0$.

## References


[1] Graybill, F.A., (1983), *Matrices with Applications in Statistics,* Wadsworth Publishing Company, Belmont, USA.

[2] Huschens, S., (1990), Necessary Sample Sizes for Categorical Data, *Statistical Papers,* 31, 47–53.

[3] Johnson, N.L., S. Kotz and N. Balakrishnan, (1997), *Discrete Multivariate Distributions,* John Wiley, New York.

[4] Klamkin, M.S., (1970), On the Roots of a Certain Determinantal Equation, *The Mathematical Gazette,* 54, 57–58.

[5] Magnus, J.R., and H. Neudecker, (1988), *Matrix Differential Calculus with Applications in Statistics and Econometrics,* John Wiley, New York.

[6] Neudecker, H., (1995), Mathematical Properties of the Variance of the Multinomial Distribution, *Journal of Mathematical Analysis and Applications,* 189, 757–762.

[7] Rao, A.R., and P. Bhimasankaram, (1992), *Linear Algebra,* McGraw-Hill, New Delhi.

[8] Ronning, G., (1982), Characteristic Values and Triangular Factorisation of the Covariance Matrix for Multinomial, Dirichlet and Multivariate Hypergeometric Distributions and some Related Results. *Statistische Hefte,* 23, 151–176.

[9] Tanabe, K., and M. Sagae, (1992), An Exact Cholesky Decomposition and the Generalised Inverse of the Variance–Covariance Matrix of the Multinomial Distribution, with Applications, *Journal of the Royal Statistical Society,* B 54, 211–219.

[10] Vermeulen, L.A., (1967), The Solution of a Certain Polynomial Equation. *The Mathematical Gazette,* 51, 308–309.

[11] Watson, G.S., (1996), Spectral Decomposition of the Covariance Matrix of a Multinomial, *Journal of the Royal Statistical Society,* B 58, 289–291.

CHAPTER 5

# A General Method of Testing for Random Parameter Variation in Statistical Models

B.P.M. McCABE and S.J. LEYBOURNE

## 5.1 Introduction

Testing the adequacy of an estimated statistical model is a perennial problem for data analysts. If some specific alternative construction of the model is entertained—for example, if allowance should be made for autocorrelation—then it is fairly straightforward to write down specific test statistics tailored for that situation. On the other hand, in many circumstances what is required is some indication of the overall suitability of the model rather looking in some specific direction. In cases where the model can be formulated in such a way that maximum likelihood (or pseudo maximum likelihood) methods may be used, an elegant technique was introduced by White (1982) which exploits the well known information-matrix equality that holds for properly specified models. If the specified model is not consistent with the data, this equality will not hold; and this fact may be used as the basis for a general specification test.

Subsequently, Chesher (1984) showed that this so-called information-matrix test could be interpreted as a test that the parameters of the model were constants (as is usually considered in a non Bayesian framework) against the alternative that they were random variables. Treating parameters as random variables has a long tradition, even in classical statistics, as the literature on random effects ANOVA attests. By considering random effects models, we are able to extend our inferences to a broader class of problems than is possible in the fixed effects case. Thus, the insight of the Chesher approach allows us to view the information-matrix test in an enhanced light. We also provide examples on the use of the results which include testing for multivariate random walks, nonlinear regressions and generalised linear models.

The purpose of this chapter is to reformulate the approach of Chesher (1994) and Cox (1993) so that the techniques may be applied to much more general settings than considered earlier. In particular, we wish to allow for quite arbitrary dependence and non-stationarity in the data generating mechanism. In addition, we consider a multivariate situation which relies heavily on the matrix derivative approach of Magnus and Neudecker (1988).

## 5.2 Derivation of the Test

Let $y_1, \ldots, y_T$ be a sequence of $T$ vectors of order $p \times 1$ containing observations, and let $\alpha_1, \ldots, \alpha_N$ be a sequence of $N$ vectors of order $k \times 1$ containing parameters which are considered to be random. Other parameters in the model are not separately distinguished.

Let the joint density of $(y_1, \ldots, y_T, \alpha_1, \ldots, \alpha_N)$ be denoted by $f(y, \alpha)$, where $y = [y_1', \ldots, y_T']'$ and $\alpha = [\alpha_1', \ldots, \alpha_N']'$ are $Tp \times 1$ and $Nk \times 1$ vectors, respectively. By definition,

$$f(y, \alpha) = f(y \mid \alpha) g(\alpha), \tag{1}$$

where $f(y \mid \alpha)$ denotes the conditional distribution of $y$ and $g(\alpha)$ is the marginal distribution of $\alpha$. From (1), the marginal distribution of $y$, $h(y)$, say, is given by

$$\begin{aligned} h(y) &= \int \cdots \int f(y \mid \alpha) g(\alpha) d\alpha_1 \cdots d\alpha_N \\ &= E_\alpha \{ f(y \mid \alpha) \}, \end{aligned}$$

by definition. Next, suppose $E(\alpha) = \mu$ and $V(\alpha) = \Omega$ and we assume the required moments are finite. Then, a Taylor series expansion of $f(y \mid \alpha)$ about $\mu$ yields

$$\begin{aligned} E_\alpha\{f(y \mid \alpha)\} &= E_\alpha \left\{ f(y \mid \alpha)|_{\alpha=\mu} + (\alpha - \mu)' \left. \frac{\partial f(y \mid \alpha)}{\partial \alpha} \right|_{\alpha=\mu} \right. \\ &\quad \left. + \frac{1}{2} (\alpha - \mu)' \left. \frac{\partial^2 f(y \mid \alpha)}{\partial \alpha \partial \alpha'} \right|_{\alpha=\mu} (\alpha - \mu) + \mathrm{O}(\|\alpha - \mu\|^3) \right\} \\ &= f(y \mid \alpha)|_{\alpha=\mu} + \frac{1}{2} E_\alpha \left\{ (\alpha - \mu)' \left. \frac{\partial^2 f(y \mid \alpha)}{\partial \alpha \partial \alpha'} \right|_{\alpha=\mu} (\alpha - \mu) \right\} \\ &\quad + \mathrm{O}\left\{ E_\alpha \left( \|\alpha - \mu\|^3 \right) \right\} \\ &= f(y \mid \alpha)|_{\alpha=\mu} + \frac{1}{2} \mathrm{tr} \left\{ \left. \frac{\partial^2 f(y \mid \alpha)}{\partial \alpha \partial \alpha'} \right|_{\alpha=\mu} \Omega \right\} + \mathrm{O}\left\{ E_\alpha \left( \|\alpha - \mu\|^3 \right) \right\}. \end{aligned}$$

Hence, an approximation to the marginal density of $y$ is given by

$$h^*(y) = f(y \mid \alpha)|_{\alpha=\mu} + \frac{1}{2} \mathrm{tr} \left\{ \left. \frac{\partial^2 f(y \mid \alpha)}{\partial \alpha \partial \alpha'} \right|_{\alpha=\mu} \Omega \right\} \tag{2}$$

which only depends on the moments of $\alpha$. Thus there is no need to specify a (marginal) density for $\alpha$ or, indeed, to assume that the $y_t$ are independent or stationary and so on.

We now wish to obtain a test that $\alpha = \mu$, i.e. that $\Omega = 0$, based on the approximate density (2). It proves convenient to express the approximate density, $h^*(y)$, in terms of the likelihood and log-likelihood of $y$; that is $L(\alpha \mid y) = f(y \mid \alpha)$ and $l(\alpha \mid y) = \log\{L(\alpha \mid y)\}$. This is accomplished by means of the well-known equality

$$\frac{\partial^2 L}{\partial\alpha\partial\alpha'} = \left(\frac{\partial l}{\partial\alpha}\frac{\partial l}{\partial\alpha'} + \frac{\partial^2 l}{\partial\alpha\partial\alpha'}\right) L(\alpha \mid y),$$

using which, $h^*(y)$ may be written as

$$h^*(y) = L(\alpha \mid y)|_{\alpha=\mu}\left[1 + \frac{1}{2}\mathrm{tr}\left\{\frac{\partial l}{\partial\alpha}\frac{\partial l}{\partial\alpha'} + \left.\frac{\partial^2 l}{\partial\alpha\partial\alpha'}\right|_{\alpha=\mu}\Omega\right\}\right].$$

A local test is then obtained by differentiating $h^*$ with respect to the appropriate elements of $\Omega$ and then dividing by $L$. Assuming that it is possible to parameterise $\Omega$ as $\Omega(\pi)$, where $\pi$ is a non-negative scalar such that $\Omega = 0$ for $\pi = 0$ and $\Omega$ is positive definite for $\pi > 0$, the foregoing development can be summarised in the following Lemma.

**Lemma 1.** *One Parameter Problem.*

Let $l(\alpha \mid y)$ be the log-likelihood of a vector of observations, $y$, given a vector of parameters, $\alpha$. A locally most powerful test of the null of constancy of the parameters against the alternative that they are random with mean $\mu$ and variance $\Omega(\pi)$, that is, a test of $H_{01} : \pi = 0$ against $H_{11} : \pi > 0$, is given by

$$S_1 = \mathrm{tr}\left(M\frac{\partial\Omega}{\partial\pi}\right),$$

where $M = (\partial l/\partial\alpha \cdot \partial l/\partial\alpha' + \partial^2 l/\partial\alpha\partial\alpha')\big|_{\alpha=\mu}$ and $\partial\Omega/\partial\pi = \partial\Omega/\partial\pi|_{\pi=0}$.

In the more general case where $\Omega$ depends on an $m \times 1$ vector of parameters $\Pi = [\pi_1, \ldots, \pi_m]'$, we appeal to King and Wu (1990) who show the locally mean most powerful test in the one sided case is given by the sum of the individual tests constructed for members of $\Pi$.

**Lemma 2.** *Multivariate Parameter Problem.*

In this multivariate context, a test of $H_{0m} : \Pi = 0$ against $H_{1m}$ : at least one $\pi_i > 0, i = 1, \ldots, m$ is given by

$$S_m = \mathrm{tr}\left(M\sum_{i=1}^{m}\frac{\partial\Omega}{\partial\pi_i}\right),$$

where $M$ is as in Lemma 1 and $\partial\Omega/\partial\pi_i$ is evaluated at $\Pi = 0$ i.e. $\partial\Omega/\partial\pi_i|_{\Pi=0}$.

**Corollary 1.** A computationally neater version of $S_m$ can be obtained by using result (A1) in the Appendix. This yields

$$\begin{aligned} S_m &= \text{vec}'(M)\text{vec}\left(\sum_{i=1}^{m} \frac{\partial \Omega}{\partial \pi_i}\right) \\ &= \text{vec}'(M)\frac{\partial \text{vec}(\Omega)}{\partial \Pi'}e, \end{aligned} \tag{3}$$

where $e$ is an $m \times 1$ vector of ones and the derivative expression is evaluated at $\Pi = 0$.

## 5.3 Examples of the Test Procedure

In this section, we present a number of specific examples which demonstrate the simplicity and flexibility of the test procedure.

**Example 1.** *Testing for vector white noise against the local level model.*
Let the $p \times 1$ vector $y_t, t = 1, \ldots, T$ be generated as

$$y_t = \alpha_t + \varepsilon_t, \qquad \varepsilon_t \sim IIN(0, \Sigma),$$

where $\alpha_t$ follows the $p$-variate random walk process

$$\alpha_t = \alpha_{t-1} + \eta_t, \qquad \alpha_0 = 0,$$

with $E(\eta_t) = 0$, $V(\eta_t) = \Lambda(\Pi) = \text{diag}(\Pi)$. Here, then, $N = T$ and $k = m = p$. We also suppose that the $\eta_t$ are independent of each other and the $\varepsilon_t$. In the context of this model, we might wish to test the null $H_{0p} : \Pi = 0$ i.e. $\alpha_t = 0$, in which case $y_t$ is a $p$-variate white noise process against the alternative $H_{1p}$ : at least one $\pi_i > 0$, i.e. at least one element of $\alpha_t$ follows a random walk, so at least one element of $y_t$ is a local level model (random walk plus noise process).

Writing the above model in vector form (for all $t$), we obtain

$$y = \alpha + \varepsilon,$$

where $y = [y_1', \ldots, y_T']'$, $\alpha = [\alpha_1', \ldots, \alpha_T']'$ and $\varepsilon = [\varepsilon_1', \ldots, \varepsilon_T']'$ are all $Tp \times 1$ vectors. Further, since by back substitution $\alpha_t = \sum_{j=1}^{t} \eta_j$, it follows from the properties of $\eta_t$ that $E(\alpha) = \mu = 0$, $V(\alpha) = \Omega = V \otimes \Lambda$, where $V = LL'$ and $L$ is a $T \times T$ lower triangular matrix of ones.

The conditional distribution $y \mid \alpha$ is $N(\alpha, I_T \otimes \Sigma)$ and so,

$$l(\alpha \mid y) = -\frac{1}{2}(y - \alpha)'(I_T \otimes \Sigma)^{-1}(y - \alpha) + k,$$

where $k$ does not involve $\alpha$. Thus, the derivatives we need to construct $M$ are

$$\begin{aligned} \frac{\partial l}{\partial \alpha} &= -(I_T \otimes \Sigma)^{-1}(y - \alpha), \\ \frac{\partial^2 l}{\partial \alpha \partial \alpha'} &= (I_T \otimes \Sigma)^{-1}. \end{aligned}$$

So,

$$\begin{aligned} M|_{\alpha=\mu} &= (I_T \otimes \Sigma^{-1})(y-\alpha)(y-\alpha)'(I_T \otimes \Sigma^{-1}) + (I_T \otimes \Sigma^{-1})\big|_{\alpha=\mu} \\ &= (I_T \otimes \Sigma^{-1})yy'(I_T \otimes \Sigma^{-1}) + (I_T \otimes \Sigma^{-1}), \end{aligned}$$

where we have used (A2). To compute $S_p$ we also require $\partial \text{vec}(V \otimes \Lambda)/\partial \Pi' e|_{\Pi=0}$. It follows from Lemma A.1, proved in the Appendix, that

$$\left.\frac{\partial \text{vec}(V \otimes \Lambda)}{\partial \Pi' e}\right|_{\Pi=0} = \text{vec}(V \otimes I_p).$$

Hence, on ignoring those terms not involving $y$, we have

$$S_p = \text{vec}'\left\{(I_T \otimes \Sigma^{-1})yy'(I_T \otimes \Sigma^{-1})\right\} \text{vec}(V \otimes I_p).$$

Using (A1) and (A3), $S_p$ can be rearranged as

$$\begin{aligned} S_p &= \text{tr}\left\{(I_T \otimes \Sigma^{-1})yy'(I_T \otimes \Sigma^{-1})(V \otimes I_p)\right\} \\ &= y'(I_T \otimes \Sigma^{-1})(V \otimes I_p)(I_T \otimes \Sigma^{-1})y \\ &= y'(V \otimes \Sigma^{-2})y. \end{aligned}$$

A simpler form for this last expression is obtained if we define the $T \times p$ matrix $Y = [y_1, \ldots, y_T]'$ from which it follows that $y = \text{vec}(Y)$. So,

$$\begin{aligned} S_p &= \text{vec}'(Y')(V \otimes \Sigma^{-2})\text{vec}(Y') \\ &= \text{vec}'(Y')\text{vec}(\Sigma^{-2}Y'V) \\ &= \text{tr}(\Sigma^{-2}Y'VY), \end{aligned} \tag{4}$$

using (A1) and (A4). In the univariate case when $p = 1$, (so that $Y = y$) and $V(\varepsilon_t) = \sigma^2$, then (4) simplifies to

$$S_1 = \frac{y'Vy}{\sigma^4},$$

which is the basic form of the stationarity tests proposed by Kwiatkowski, Phillips, Schmidt and Shin (1992) and Leybourne and McCabe (1994).

**Example 2.** *Testing for a vector random walk against a heteroscedastic alternative.*

Let the $p \times 1$ vector $y_t, t = 1, \ldots, T$ be generated as

$$\begin{aligned} y_t &= Ay_{t-1} + \varepsilon_t, \qquad \varepsilon_t \sim IIN(0, \Sigma), \qquad y_0 = 0, \\ A_t &= I_p + \Gamma_t, \end{aligned}$$

where the $p \times p$ random matrix $\Gamma_t$ satisfies $E\{\text{vec}(\Gamma_t)\} = 0$ and $V\{\text{vec}(\Gamma_t)\} = \Lambda(\Pi) = \text{diag}(\Pi)$ with $\Pi$ of dimension $p^2 \times 1$. We suppose that the $\Gamma_t$ are independent of each other and the $\varepsilon_t$. A little calculation then shows

$$\begin{aligned} E_{t-1}(y_t) &= y_{t-1}, \\ V_{t-1}(y_t) &= (y_{t-1}' \otimes I_p)\Lambda(\Pi)(y_{t-1} \otimes I_p) + \Sigma, \\ V(y_t) &= (I_p + A^*)V(y_{t-1})(I_p + A^*)' + \Sigma, \end{aligned}$$

where $E_{t-1}$ is the conditional expectation up to time $t-1$ and $A^*$ is a symmetric positive definite matrix which is a function of $\Lambda(\Pi)$. Thus, by virtue of the recursion, the unconditional variance of $y_t$ is heteroscedastic with an increasing trend. We wish to test $H_{0pp} : \Pi = 0$ i.e. $A_t = I$, so that $y_t$ is a $p$-variate random walk process, against $H_{1pp}$ : at least one $\pi_i > 0$ , i.e. the process for $y_t$ has at least one member with a heteroscedastic variance. Next, make the definition that $\alpha_t = \text{vec}(A_t)$ and $\alpha$ is the vector of the $\alpha_t$. Then, $E(\alpha) = [\text{vec}'(I_p), \ldots, \text{vec}'(I_p)]'$ and $V(\alpha) = \Omega = I_T \otimes \Lambda$. The conditional distribution $y_t \mid \alpha_t, y_{t-1}$ is $N(A_t y_{t-1}, \Sigma)$ so,

$$\begin{aligned} l(\alpha \mid y) &= -\frac{1}{2}\sum_{t=1}^{T}(y_t - A_t y_{t-1})'\Sigma^{-1}(y_t - A_t y_{t-1}) + k \\ &= -\frac{1}{2}\sum_{t=1}^{T}(y_t\Sigma^{-1}y_t - 2y_{t-1}'A_t'\Sigma^{-1}y_t + y_{t-1}'A_t'\Sigma^{-1}A_t y_{t-1}) + k \\ &= -\frac{1}{2}\sum_{t=1}^{T}\{-2\text{vec}'(\Sigma^{-1}y_t)\text{vec}(A_t y_{t-1}) \\ &\qquad + \text{vec}'(\Sigma^{-1}A_t y_{t-1})\text{vec}(A_t y_{t-1}) + y_t'\Sigma^{-1}y_t + k\}, \end{aligned}$$

using (A1). Further, in view of (A4) and (A3), we can write

$$\begin{aligned} l(\alpha \mid y) &= -\frac{1}{2}\sum_{t=1}^{T}\{-2\text{vec}'(\Sigma^{-1}y_t)(y_{t-1}' \otimes I_p)\alpha_t \\ &\qquad + \alpha_t'(y_{t-1}y_{t-1}' \otimes \Sigma^{-1})\alpha_t + y_t'\Sigma^{-1}y_t\}. \end{aligned}$$

From standard results for differentiating quadratic forms, we obtain

$$\begin{aligned} \frac{\partial l}{\partial \alpha_t} &= (y_{t-1} \otimes I_p)\text{vec}(\Sigma^{-1}y_t) - (y_{t-1}y_{t-1}' \otimes \Sigma^{-1})\alpha_t \\ &= \text{vec}\left\{\Sigma^{-1}(y_t - A_t y_{t-1})y_{t-1}'\right\}, \end{aligned}$$

using (A4) and

$$\frac{\partial^2 l}{\partial \alpha_t \partial \alpha_t'} = -(y_{t-1}y_{t-1}' \otimes \Sigma^{-1}).$$

The cross second derivative terms are all zero. From Lemma A.1, we have

$$\left.\frac{\partial \text{vec}(I_T \otimes \Lambda)}{\partial \Pi'}\right|_{\Pi=0} \cdot e = \text{vec}(I_T \otimes I_{pp}) = \text{vec}(I_{Tpp}),$$

and so

$$\begin{aligned} S_{pp} &= \text{vec}'(M|_{\alpha=\mu}) \left.\frac{\partial \text{vec}(\Omega)}{\partial \Pi'}\right|_{\Pi=0} \cdot e \\ &= \text{vec}'(M|_{\alpha=\mu})\text{vec}(I_{Tpp}) \end{aligned}$$

$$
\begin{aligned}
&= \operatorname{tr}(M|_{\alpha=\mu}) \\
&= \sum_{t=1}^{T} \operatorname{tr}\left( \frac{\partial l}{\partial \alpha_t}\frac{\partial l}{\partial \alpha_t'} + \left.\frac{\partial^2 l}{\partial \alpha_t \partial \alpha_t'}\right|_{\alpha=\mu} \right) \\
&= \sum_{t=1}^{T} \left[ \operatorname{tr}\left\{ \operatorname{vec}'(\Sigma^{-1}\Delta y_t y_{t-1}')\operatorname{vec}(\Sigma^{-1}\Delta y_t y_{t-1}') \right\} - \operatorname{tr}(y_{t-1}y_{t-1}' \otimes \Sigma^{-1}) \right] \\
&= \sum_{t=1}^{T} \left\{ \operatorname{tr}(y_{t-1}\Delta y_t'\Sigma^{-2}\Delta y_t y_{t-1}') - \operatorname{tr}(y_{t-1}y_{t-1}')\cdot \operatorname{tr}(\Sigma^{-1}) \right\} \\
&= \sum_{t=1}^{T} y_{t-1}'y_{t-1}\left\{ \Delta y_t'\Sigma^{-2}\Delta y_t - \operatorname{tr}(\Sigma^{-1}) \right\}, \qquad (5)
\end{aligned}
$$

using (A1) and (A5). In the case where $p = 1$ and $V(\varepsilon_t) = \sigma^2$, then (5) simplifies to

$$
S_1 = \frac{\sum_{t=1}^{T} y_{t-1}^2(\Delta y_t^2 - \sigma^2)}{\sigma^4},
$$

which is the est for a stochastic unit root suggested in McCabe and Tremayne (1995) and Leybourne, McCabe and Tremayne (1996).

**Example 3.** *Testing for a vector of fixed coefficient nonlinear relationships against a random coefficient model.*

Let the $p \times 1$ vector $y_t$ be generated as

$$
y_t = f_t(\alpha_t) + \varepsilon_t, \qquad \varepsilon_t \sim IIN(0, \Sigma),
$$

where the $m \times 1$ vector $\alpha_t$ is such that $E(\alpha_t) = \gamma$, $V(\alpha_t) = \Lambda(\Pi) = \operatorname{diag}(\Pi)$. We suppose that the $p \times 1$ vector $f_t$ has continuous partial derivatives up to the second order. We also assume that the $\alpha_t$ are independent of each other and the $\varepsilon_t$. Here we wish to test $H_{0m} : \Pi = 0$ i.e. $\alpha_t = \gamma$, so that $y_t$ is a $p$-variate nonlinear regression model with fixed coefficient vector $\gamma$ against $H_{1m}$ : at least one $\pi_i > 0$, i.e. that all of the nonlinear regressions contain some random coefficients.

Here $E(\alpha) = \mu = [\gamma', \ldots, \gamma']'$ and $V(\alpha) = \Omega = I_T \otimes \Lambda$. The conditional distribution $y_t \mid \alpha_t$ is $N(f_t(\alpha_t), \Sigma)$, hence

$$
l(\alpha \mid y) = -\frac{1}{2}\sum_{t=1}^{T} \{y_t - f_t(\alpha_t)\}' \Sigma^{-1} \{y_t - f_t(\alpha_t)\} + k
$$

and

$$
\begin{aligned}
\frac{\partial l}{\partial \alpha_t} &= \frac{\partial f_t'}{\partial \alpha_t}\Sigma^{-1}\{y_t - f_t(\alpha_t)\} \\
&= \left[\{y_t - f_t(\alpha_t)\}'\Sigma^{-1} \otimes I_m\right] \operatorname{vec}\left(\frac{\partial f_t'}{\partial \alpha_t}\right), \\
\frac{\partial^2 l}{\partial \alpha_t \partial \alpha_t'} &= \left[\{y_t - f_t(\alpha_t)\}'\Sigma^{-1} \otimes I_m\right] \frac{\partial}{\partial \alpha_t'}\operatorname{vec}\left(\frac{\partial f_t'}{\partial \alpha_t}\right) - \frac{\partial f_t'}{\partial \alpha_t}\Sigma^{-1}\frac{\partial f_t}{\partial \alpha_t'}.
\end{aligned}
$$

The cross second derivative terms are zero. From Lemma A.1, we have

$$\left.\frac{\partial \text{vec}(I_T \otimes \Lambda)}{\partial \Pi'}\right|_{\Pi=0} \cdot e = \text{vec}(I_T \otimes I_m) = \text{vec}(I_{Tm}),$$

so

$$\begin{aligned} S_m &= \text{vec}'(M|_{\alpha=\mu})\text{vec}(I_{Tm}) \\ &= \text{tr}(M|_{\alpha=\mu}) \\ &= \sum_{t=1}^{T}\left(\frac{\partial l}{\partial \alpha_t}\frac{\partial l}{\partial \alpha_t'} + \left.\frac{\partial^2 l}{\partial \alpha_t \partial \alpha_t'}\right|_{\alpha=\mu}\right) \\ &= \sum_{t=1}^{T}\text{tr}\left(\frac{\partial f_t'}{\partial \alpha_t}\Sigma^{-1}w_t w_t'\Sigma^{-1}\frac{\partial f_t}{\partial \alpha_t'}\right) \\ &\quad + \sum_{t=1}^{T}\text{tr}\left[(w_t'\Sigma^{-1}\otimes I_m)\frac{\partial}{\partial \alpha_t'}\text{vec}\left(\frac{\partial f_t'}{\partial \alpha_t}\right) - \frac{\partial f_t'}{\partial \alpha_t}\Sigma^{-1}\frac{\partial f_t}{\partial \alpha_t'}\right]\Bigg|_{\alpha=\mu}, \end{aligned}$$

where $w_t = y_t - f_t(\mu)$. In the case where $p = 1$ and $V(\varepsilon_t) = \sigma^2$, $S_m$ simplifies to

$$\begin{aligned} S_m &= \frac{1}{\sigma^4}\sum_{t=1}^{T} w_t^2 \frac{\partial f_t}{\partial \alpha_t'}\frac{\partial f_t}{\partial \alpha_t} + \frac{1}{\sigma^2}\sum_{t=1}^{T} w_t \,\text{tr}\left(\frac{\partial^2 f_t}{\partial \alpha_t \partial \alpha_t}\right) - \frac{1}{\sigma^2}\sum_{t=1}^{T}\left.\frac{\partial f_t}{\partial \alpha_t}\frac{\partial f_t}{\partial \alpha_t'}\right|_{\alpha=\mu} \\ &= \frac{1}{\sigma^2}\sum_{t=1}^{T}\left(\frac{w_t^2}{\sigma^2} - 1\right)\frac{\partial f_t}{\partial \alpha_t'}\frac{\partial f_t}{\partial \alpha_t} + w_t \,\text{tr}\left.\left(\frac{\partial^2 f_t}{\partial \alpha_t \partial \alpha_t}\right)\right|_{\alpha=\mu}, \end{aligned}$$

which is a multivariate version of the test for random coefficients in a nonlinear regression derived in McCabe and Leybourne (1993).

**Example 4.** *Testing for overdispersion in Generalised Linear Models.*

Let $y_i$, a scalar random variable, be independently distributed with an exponential family density given by

$$f(y_i \mid \theta_i, \lambda) = \exp\left[\frac{y_i\theta_i - b(\theta_i)}{\lambda} + c(y_i, \lambda)\right],$$

where $\theta_i$ is a scalar parameter and $b(\cdot)$ is a twice differentiable function. The function $c(\cdot)$ is such that the density integrates to unity. Included in this family are the normal, binomial, Poisson, gamma and inverse Gaussian distributions. It is well known that

$$E(y_i) = \frac{\partial b(\theta_i)}{\partial \theta_i} \equiv \mu(\theta_i), \qquad V(y_i) = \frac{\partial^2 b(\theta_i)}{\partial \theta_i^2} \equiv \Sigma(\theta_i),$$

and further details on generalised linear models may be found in Fahrmeir and Tutz (1994) or McCullagh and Nelder (1989). The variable $y$ is to be explained linearly by a vector of covariates, $x$ $(k \times 1)$, i.e. by $\eta_i = x_i'\beta$ and so, we relate the *mean* of $y_i$, $\mu(\theta_i)$, to the linear predictor, $\eta_i = x_i'\beta$, by the link equation $\mu(\theta_i) = h(\eta_i)$ where $h(\cdot)$ is such that

$$\eta_i = h^{-1}\{\mu(\theta_i)\}$$

and $g(\cdot) = h^{-1}(\cdot)$ is called the link function. A very commonly occurring case is the so-called natural link function where

$$\eta_i = \theta_i, \qquad [h^{-1}\{\mu(\theta_i)\} = \theta_i],$$

so that $g(\cdot) = \mu^{-1}(\cdot)$.

In exponential families, there is a connection between the mean and variance of the random variables in question. However, in practice, it is often found that such a mean variance relationship does not hold for real data; in particular, the variance of $y_i$ often exceeds the nominal variance and $y_i$ is termed *overdispersed.* Thus it would seem to be important to test for overdispersion and model it in the event it was found to be present. There are many reasons why overdispersion might be present, but one obvious source is that the responses, $\beta$, associated with the covariates, $x_i$, are not constant. One way to model this is to assume that $\beta$ is random i.e. $\beta_i = \beta + \xi_i$, where $\xi_i$ are independently distributed random vectors $(k \times 1)$ with mean zero and covariance matrix $\Omega$ $(k \times k)$. For more detail on estimation of these models, consult Fahrmeir and Tutz (1994) (chapter 7).

Lemma 1 provides an easy way of testing if $\beta_i$ is constant. For simplicity, we will use the natural link functions and assume that $\Omega = \Omega(\pi)$ is diagonal based on the scalar parameter $\pi$. Now the log likelihood of $(y_1, \ldots, y_n)$, apart from constants, is

$$\begin{aligned} l(\beta) &= \sum_{i=1}^{n} l_i(\beta), \\ l_i(\beta) &= \frac{y_i\theta_i - b(\theta_i)}{\lambda}, \\ \theta_i &= x_i'\beta. \end{aligned}$$

Thus, taking $\lambda$ to the left hand side,

$$\lambda\frac{\partial l_i(\beta)}{\partial\beta} = \{y_i - \mu_i(\beta)\}\, x_i,$$

where $\mu_i(\beta) \equiv \partial b(\eta_i)/\partial\eta_i|(\eta_i = x_i'\beta)$ and the second derivative is given by

$$\lambda\frac{\partial^2 l_i(\beta)}{\partial\beta\partial\beta'} = -\Sigma(\beta)\cdot x_i x_i',$$

where $-\Sigma(\beta) \equiv \partial^2 b(\eta_i)/\partial\eta_i^2|(\eta_i = x_i'\beta)$. We may construct the test statistic by summation and use of the trace operator viz.

$$\begin{aligned} S_1 &= \frac{1}{\lambda}\mathrm{tr}\left[\sum_{i=1}^{n}\left\{\frac{\{y_i - \mu_i(\beta)\}^2}{\lambda} - \Sigma_i(b)\right\} x_i x_i'\right] \\ &= \frac{1}{\lambda}\sum_{i=1}^{n}\left[\frac{\{y_i - \mu_i(\beta)\}^2}{\lambda} - \Sigma_i(b)\right] x_i x_i'. \end{aligned}$$

To implement the test statistic for any member of the exponential family, one simply substitutes the appropriate expressions for the mean and variance.

## 5.4 Summary

This paper provides a general method of deriving tests of the hypothesis that the parameters of a model are constants versus the alternative that they are random. We have illustrated its use in several cases of practical interest and shown that easily constructed test statistics arise.

## Appendix

We use the following results, often without comment, throughout the text.

(A1) $\mathrm{tr}(AB) = \mathrm{vec}'(A')\mathrm{vec}(B)$,

(A2) $(A \otimes B)^{-1} = A^{-1} \otimes B^{-1}$,

(A3) $(A \otimes B)(C \otimes D) = (AC \otimes BD)$,

(A4) $\mathrm{vec}(ABC) = (C' \otimes A)\mathrm{vec}(B)$,

(A5) $\mathrm{tr}(A \otimes B) = \mathrm{tr}(A) \cdot \mathrm{tr}(B)$.

**Lemma A.1.** Let $X$ be an $n \times q$ matrix, $Z(\Pi)$ a $p \times r$ matrix which is a function of an $m \times 1$ vector $\Pi$ and $e$ an $m \times 1$ vector of ones. Then

$$\frac{\partial \mathrm{vec}(X \otimes Z)}{\partial \Pi'} e = \mathrm{vec}(X \otimes z),$$

where $z$ is a $p \times r$ matrix with $ij$'th element $\partial z_{ij}/\partial \Pi' e$.

**Proof.** From Magnus and Neudecker (1988), we have

$$\mathrm{vec}(X \otimes Z) = (I_q \otimes K_{rn} \otimes I_p)\{\mathrm{vec}(X) \otimes I_{pr}\}\mathrm{vec}(Z),$$

where $K_{rn}$ is the $rn \times rn$ commutation matrix. So,

$$\frac{\partial \mathrm{vec}(X \otimes Z)}{\partial \Pi'} e = (I_q \otimes K_{rn} \otimes I_p)\{\mathrm{vec}(X) \otimes I_{pr}\} \frac{\partial \mathrm{vec}(Z)}{\partial \Pi'} e.$$

Now, $\partial \mathrm{vec}(Z)/\partial \Pi' e = \mathrm{vec}(z)$ where $z$ is a $p \times r$ matrix with $ij$'th element $\partial z_{ij}/\partial \Pi' e$. Therefore we may write

$$\begin{aligned} \frac{\partial \mathrm{vec}(X \otimes Z)}{\partial \Pi'} e &= (I_q \otimes K_{rn} \otimes I_p)\{\mathrm{vec}(X) \otimes I_{pr}\}\mathrm{vec}(z) \\ &= \mathrm{vec}(X \otimes z). \end{aligned}$$

In the special case where $Z(\Pi) = \mathrm{diag}(\Pi)$ then $r = p = m$ and $z = I_m$.

## References


[1] Cox, D.R., (1983), Some Remarks on Overdispersion, *Biometrika,* 70, 269–274.

[2] Chesher, A.D., (1984), Testing for Neglected Heterogeneity, *Econometrica,* 52, 865–872.

[3] Fahrmeir, L., and G. Tutz, (1994), *Multivariate Statistical Modelling Based on Generalised Linear Models,* Springer Verlag, New York.

[4] King, M.L., and P.X. Wu, (1990), Locally Optimal One-Sided Tests for Multiparameter Hypotheses, Paper Presented at the Sixth World Congress of the Econometric Society, Barcelona.

[5] Kwaitowski, D., P.C.B Phillips, P. Schmidt and Y. Shin, (1992), Testing the Null Hypothesis of Stationarity Against the Alternative of a Unit Root: How Sure are we that Economic Time Series have a Unit Root? *Journal of Econometrics,* 54, 159–178.

[6] Leybourne, S.J., and B.P.M. McCabe, (1994), A Consistent Test for a Unit Root, *Journal of Business and Economic Statistics,* 12, 157–166.

[7] Leybourne, S.J., B.P.M. McCabe and A.R. Tremayne, (1996), Can Economic Time Series be Differenced to Stationarity? *Journal of Business and Economic Statistics,* 14, 435–446.

[8] Magnus, J.R., and H. Neudecker, (1988), *Matrix Differential Calculus with Applications in Statistics and Econometrics,* John Wiley and Sons, New York.

[9] McCabe, B.P.M., and A.R. Tremayne, (1995), Testing if a Time Series is Difference Stationary, *Annals of Statistics,* 23, 1015–1028.

[10] McCullagh, P., and J.A. Nelder, (1989), *Generalised Linear Models, 2nd Edition,* Chapman and Hall, London.

[11] McCabe, B.P.M., and S.J. Leybourne, (1993), Testing for Parameter Variation in Nonlinear Regression Models, *Journal of the Royal Statistical Society, Series B,* 55, 133–144.

[12] White, H., (1982), Maximum Likelihood Estimation of Misspecified Models, *Econometrica,* 50, 1–26.

CHAPTER 6

# Dual Scaling and Correspondence Analysis of Rank Order Data

Michel van de VELDEN

## 6.1 Introduction

Dual scaling and correspondence analysis are closely related, mathematically equivalent, data analysis methods; see, for example, Tenenhaus and Young (1984), Greenacre (1984). Typically, the data one subjects to either correspondence analysis or dual scaling are in the format of a so-called contingency matrix, i.e. a matrix consisting of frequencies of co-occurrences. There exist however modifications of the methods that make it possible to analyse also other types of data. One of these modifications involves the treatment of preference data. Both Greenacre (1984, 1993) and Nishisato (1980, 1994) treat the analysis of such data, however, partly due to differences in the format and exact nature of the data, the methods they propose appear to be different. Because of this it would be, as Greenacre pointed out in his (1996) book review, "interesting to compare the two approaches more closely".

In this chapter, we clarify the differences between dual scaling and correspondence analysis applied to rank order data using an alternative approach for dual scaling analysis of rank order data. This new approach, which involves a doubling of the data, makes it possible to perform Nishisato's dual scaling of rank order data using conventional correspondence analysis software. Moreover, the approach presented here permits the analysis of paired comparison data by correspondence analysis.

One reason for the differences between the analysis of preference data by dual scaling or correspondence analysis lies in the format of the data. Nishisato (1980, 1994) for example, describes the analysis of so-called paired comparison data, i.e. observations on comparisons between pairs of objects. As special cases of such paired comparison data he then presents the analysis of rank order data (data consisting of rankings of objects according to some criterion) and rating data (data where objects are assigned a rating).

In the case of rank order data Nishisato transforms the original data to paired comparison data by comparing the rank numbers between all pairs of objects. He then subjects them to the same analysis as ordinary paired comparison data. One could propose a similar procedure in the analysis of rating data, i.e. compare the ratings between all pairs of objects and analyse the resulting paired comparisons data. Clearly such a procedure can lead to a significant loss of information since the sizes of the differences between ratings are completely ignored. As an alternative Nishisato therefore proposes a coding scheme which he refers to as "successive categories", see Nishisato (1994, p 221). Once again, the thus obtained transformed data are subjected to paired comparison analysis.

Greenacre (1984, 1993) does not treat paired comparison data. Instead he focuses on the analysis of rating data. As mentioned before these rating data can be transformed into paired comparison data. The opposite however is not typically true: Paired comparison data cannot be transformed to rating data. This "singularity" makes an algebraic comparison between the two approaches cumbersome. The difference between them is apparently the result of a different treatment of differently coded data.

In the case of rank order data this singularity problem does not occur due to the restrictive nature of the data; it is possible to retrieve the original rank numbers from the paired comparison data. Moreover, if one interprets the rank numbers as ratings, one may directly apply Greenacre's analysis. Because of this "non-singularity" the relationship between the two approaches can be described mathematically.

To facilitate a clear exposition of the relationship between dual scaling and correspondence analysis applied to rank order data, we will start with a summary of several equations essential in both methods. The origins of these equations will not be discussed here and we refer the interested reader to standard texts on the subjects such as Greenacre (1984, 1993), Lebart, Morineau and Warwick (1984), Benzécri (1992) (correspondence analysis) or Nishisato (1980, 1994) (dual scaling). Then, following a brief introduction to rank order data and some essential notation, we shall give an alternative approach of dual scaling of rank order data.[1] This new approach has as advantage over Nishisato's approach that it does not require the use of specific dual scaling software when analysing rank order data. Instead ordinary correspondence analysis software can be used to analyse rank order data. Moreover, the new approach permits the analysis of paired comparison data by correspondence analysis.

## 6.2 Correspondence Analysis and Dual Scaling

As mentioned before, correspondence analysis and dual scaling are mathematically equivalent methods, see, for example, Tenenhaus and Young (1984) and Greenacre (1984). The most important differences between the methods concern the underlying rationales. For a comprehensive treatment of the methods, we refer the reader to any of the standard texts mentioned in the first section.

---

[1] A similar approach has been proposed, independently, by Anna Torres of the Universitat Pompeu Fabra, Barcelona.

Here we will restrict ourselves to the introduction of some essential equations. We shall do so by using Greenacre's correspondence analysis framework. Important differences with Nishisato's formulation will be shortly highlighted when necessary.

### 6.2.1 General Formulation

Let $N$ denote an $n \times p$ nonnegative data matrix. Then, in accordance with Greenacre (1984), the correspondence matrix $P$ is defined as

$$P \equiv \frac{1}{1_n' N 1_p} N, \tag{1}$$

so that

$$1_n' P 1_p = 1.$$

In addition

$$r \equiv P 1_p, \qquad D_r \equiv \mathrm{diag}\,(r)\,, \tag{2}$$

and

$$c \equiv P' 1_n, \qquad D_c \equiv \mathrm{diag}(c). \tag{3}$$

In dual scaling there is no correspondence matrix $P$. The scaling matrices $D_r$ and $D_c$ are then defined as in (2) and (3) with $N$ inserted for $P$. This difference in scaling is clearly of no importance.

Correspondence analysis and dual scaling amounts to considering the following singular value decomposition:

$$D_r^{-1/2} \left(P - rc'\right) D_c^{-1/2} = U \Lambda^{1/2} V', \tag{4}$$

where

$$U'U = V'V = I_k,$$

and $k$ is the rank of $P - rc'$. Note that, by selecting the $\kappa(\kappa < k)$ largest singular values and corresponding singular vectors, we obtain a least-squares approximation of rank $\kappa$ of $D_r^{-1/2} \left(P - rc'\right) D_c^{-1/2}$. In correspondence analysis, one is usually interested in this approximation rather than in the full decomposition since it permits to interpret the multidimensional data in a space of smaller dimension, i.e. $\kappa$.

Matrices $A$ and $B$ are defined as

$$A \equiv D_r^{1/2} U \tag{5}$$

and

$$B \equiv D_c^{1/2} V. \tag{6}$$

The columns of $A$ and $B$ are called the principal axes for the rows and columns of $P$. They are standardised as

$$A' D_r^{-1} A = B' D_c^{-1} B = I_k.$$

Moreover, from (4) and (5) it immediately follows that

$$1'_n A = 1'_n D_r^{1/2} U = 1'_n (P - rc') D_c^{-1/2} V \Lambda^{-1/2} = (c' - c') D_c^{-1/2} V \Lambda^{-1/2} = 0.$$

A similar argument leads to $1'_p B = 0$.

In addition, Greenacre (1984) defines so-called standard coordinate matrices $G$ and $H$ as

$$G \equiv D_r^{-1/2} U \tag{7}$$

and

$$H \equiv D_c^{-1/2} V, \tag{8}$$

so that

$$G' D_r G = H' D_c H = I_k. \tag{9}$$

These standard coordinate matrices are known to be identical to the so-called row and column weight matrices $Y$ and $X$ obtained in dual scaling, see, for example, Greenacre (1984).

It is not difficult to see that the standard coordinate matrices in (7) and (8) can be obtained from the eigenequations

$$D_r^{-1} (P - rc') D_c^{-1} (P - rc')' G = G\Lambda \tag{10}$$

and

$$D_c^{-1} (P - rc')' D_r^{-1} (P - rc') H = H\Lambda, \tag{11}$$

both of which follow from (4).

Finally, we have transition formulae linking $H$ and $G$, viz

$$G = D_r^{-1} P H \Lambda^{-1/2} \tag{12}$$

and

$$H = D_c^{-1} P' G \Lambda^{-1/2}. \tag{13}$$

To obtain (12) rewrite (4) as

$$U = D_r^{-1/2} (P - rc') D_c^{-1/2} V \Lambda^{-1/2}.$$

Then, using (7) and (8), we get

$$D_r^{1/2} G = D_r^{-1/2} (P - rc') H \Lambda^{-1/2}$$

$$\rightarrow G = D_r^{-1} (P - rc') H \Lambda^{-1/2},$$

and, because

$$c' H = 1'_p B = 0,$$

where we used (6) and (8), transition formula (12) immediately follows. In a similar fashion, we obtain (13).

## 6.3 Rank Order Data

Suppose $n$ individuals are asked to rank $q$ objects with respect to a criterion such that the $q$ objects ($q \leq n$) receive $q$ different ranks, i.e. a strict ranking is required. The resulting data are usually referred to as rank order data. The data can be coded in the following way. For each individual assign rank numbers ranging from 0 (for the lowest ranked object) to $q-1$ (for the highest ranked object), and collect these in an $n \times q$ matrix $T$. Thus, the rows of $T$ correspond to the observations (the individuals) and the columns correspond to the $q$ objects. In addition, we define a matrix $S$ in a similar way, however we reverse the scale, i.e. rank numbers are assigned such that 0 indicates the highest ranked object and $q-1$ the lowest ranked object. The reason for the introduction of the matrix $S$ is that the direction of the scale should not influence our results.

Clearly

$$T + S = (q-1)1_n 1_q' \tag{14}$$

and

$$T1_q = S1_q = \frac{1}{2}q(q-1)1_n. \tag{15}$$

We can collect the matrices $T$ and $S$ in one matrix as

$$F_c \equiv \begin{pmatrix} T, & S \end{pmatrix}, \tag{16}$$

or also as

$$F_r \equiv \begin{pmatrix} T \\ S \end{pmatrix}. \tag{17}$$

The first representation suggests the availability of data on $2q$ objects whereas the second representation suggests the availability of data on $2n$ individuals. Using (14) and (15), we immediately derive

$$F_c 1_{2q} = q(q-1)1_n, \tag{18}$$

$$F_r 1_q = \frac{1}{2}q(q-1)1_{2n} \tag{19}$$

and

$$F_r' 1_{2n} = n(q-1)1_q. \tag{20}$$

In his analysis of rank order data, Nishisato (1994) introduces a so-called dominance matrix $E$ defined as

$$E \equiv T - S, \tag{21}$$

which he subsequently submits to dual scaling. From (15) it immediately follows that

$$E1_q = 0_n. \tag{22}$$

**Remark 1.** If we allow subjects to give the same rank number to more than one object, i.e. if we allow ties to occur, then, by assigning the average rank number to the tied objects properties, (14) and (15) still hold. Thus, (18) through (22) are also still valid.

## 6.4 Dual Scaling of Rank Order Data

Due to the mathematical equivalence between correspondence analysis and dual scaling, eigenequations (10) and (11) given in section 6.2.1 can normally be used to obtain the dual scaling solution. In the analysis of rank order data, however, the methods as proposed by Nishisato (1980, 1994) and Greenacre (1984, 1993) appear to differ significantly. Without going into much detail, we will present the main results as obtained by Nishisato (1994). For a complete treatment, including a more extensive argumentation of some of the equations, we refer the reader to Nishisato (1994).

### 6.4.1 Nishisato's Dual Scaling of Rank Order Data

Nishisato introduces the dominance matrix $E$ as defined in (21) and he subjects this matrix to dual scaling. Clearly this dominance matrix $E$ is, as opposed to the usual data matrices for dual scaling analysis, *not* nonnegative and the usual procedure breaks down because from (22) it follows that the scaling matrix $D_r$ is a zero matrix. Therefore alternative definitions are required.

Nishisato proposes the following definitions of $D_r$ and $D_c$:

$$D_r \equiv q(q-1)I_n \qquad \text{and} \qquad D_c \equiv n(q-1)I_q. \tag{23}$$

These definitions are justified in the following way: In the dual scaling analysis of a contingency matrix $N$ the diagonal matrices $D_r$ and $D_c$ have as elements the total number of observations in a row or a column. In the analysis of rank order data this analogy should still hold. Now, since there are $q$ columns and each column element is the result of $q-1$ comparisons, we have $q(q-1)$ as total number of "observations" in each row. Hence the definition of $D_r$. Similarly, $n(q-1)$ denotes the total number of "observations" in each column; $n$ rows of which each element is the result of $q-1$ comparisons. Hence the definition of $D_c$.

Using these alternative scaling matrices Nishisato obtains the dual scaling solutions for the rows and columns of $E$ from the following eigenequations:

$$\gamma EE'Y = Y\Lambda \tag{24}$$

and

$$\gamma E'EX = X\Lambda, \tag{25}$$

where we used

$$D_r^{-1}ED_c^{-1}E' = \gamma EE',$$
$$D_c^{-1}E'D_r^{-1}E = \gamma E'E$$

and

$$\gamma \equiv \frac{1}{nq(q-1)^2}, \tag{26}$$

where $\Lambda$ is a $k \times k$ ($k$ denoting the rank of $E$) diagonal matrix with as elements the nontrivial eigenvalues of $\gamma E'E$, and $X$ and $Y$ are matrices of corresponding eigenvectors. Furthermore, we assume that there are exactly $k$ distinct eigenvalues. The columns of the $q \times k$ matrix $X$ in (25) are usually referred to as vectors of object weights, whereas the columns of the $n \times k$ matrix $Y$ in (24) denote the so-called vectors of subject weights. Usually, one is not interested in all $k$ object or subject weight vectors but only in $\kappa < k$ weight vectors. In that case, one selects $\kappa$ eigenvectors corresponding to the $\kappa$ largest eigenvalues. Once again, we refer the reader to Nishisato (1980, 1994) for a more elaborate discussion concerning the argumentation and interpretation of these results.

Note that the equations (25) and (24) are the same as the eigenequations for the standard coordinates (10) and (11) where $E$ is substituted for $P - rc'$.

Nishisato (1994) standardises $X$ and $Y$ in such a way that

$$X'D_cX = Y'D_rY = nq(q-1)I_k,$$

or, equivalently,

$$X'X = qI_k \qquad \text{and} \qquad Y'Y = nI_k. \tag{27}$$

In accordance with formulae (12) and (13) obtained at the end of section 6.2.1, we have transition formulae

$$Y = D_r^{-1}EX\Lambda^{-1/2} = \frac{1}{q(q-1)}EX\Lambda^{-1/2} \tag{28}$$

and

$$X = D_c^{-1}E'Y\Lambda^{-1/2} = \frac{1}{n(q-1)}E'Y\Lambda^{-1/2}. \tag{29}$$

These formulae can also be obtained from the eigenequations (24) and (25), e.g. premultiplying (25) by $D_r^{-1}E$ yields the expression for $Y$.

Nishisato (1980, 1994) remarks that the approach as described above successfully analyses a data matrix with both negative and positive entries. Therefore, he argues, it is an important extension of dual scaling. The definitions of the scaling matrices $D_r$ and $D_c$ in (23), however, are rather cumbersome. In fact, the choice of scaling matrices appears to be somewhat arbitrary. One could, for example, also argue that after assigning a rank number to the first object there only remain $q - 2$ comparisons for the second object, and so on. Or, one may propose $D_r = I_n$, since each row consists in fact of exactly one observation.

In addition, the claim that a data matrix consisting of both negative and positive elements can be analysed by the approach described above is in fact based on the a-priori subtraction of the so-called trivial solution, i.e. the analysis of $E$ can be seen as the analysis of $P - rc'$ To see this, we will show in section 6.4.2 that Nishisato's dual scaling of rank order data is equivalent to simple correspondence analysis of the doubled data matrix $F_r$ as defined in (17).

### 6.4.2 An Alternative Approach to Dual Scaling of Rank Order Data

Instead of analysing the dominance matrix $E$, we can analyse the so-called doubled data matrix $F_r$ as defined in (17). As will be shown below, the results of such an analysis prove to be equivalent to those obtained in Nishisato's analysis of the dominance matrix $E$. In fact, the only difference between the two approaches is due to the different order of $F_r$ as compared to $E$. An important advantage of the method to be proposed here is that it does not require any alternative and contrived definitions of scaling matrices $D_r$ and $D_c$. Furthermore, it will help to clarify the difference between the method proposed by Greenacre and the analysis of Nishisato.

Following the procedure as described in section 6.2.1, we obtain, after substituting $F_r$ for $N$, the following equations:

$$P_r = \frac{1}{s} F_r, \tag{30}$$

$$r = \frac{1}{s} F_r 1_q \tag{31}$$

and

$$c = \frac{1}{s} F_r' 1_{2n}, \tag{32}$$

where

$$s \equiv 1_{2n}' F_r 1_q = nq(q-1).$$

Note that

$$\frac{1}{s} = (q-1)\gamma. \tag{33}$$

Using (19) and (20), we can further simplify (31) and ( 32) to

$$r = \frac{1}{2n} 1_{2n} \qquad \text{and} \qquad c = \frac{1}{q} 1_q, \tag{34}$$

so that

$$D_r = \frac{1}{2n} I_{2n} \qquad \text{and} \qquad D_c = \frac{1}{q} I_q. \tag{35}$$

As mentioned in section 6.2.1 the matrix of so-called object weights obtained in dual scaling is identical to the matrix of standard column coordinates as obtained in the correspondence analysis of $P$. Hence, to obtain the object weight matrix $H$, we employ the following eigenequation

$$2nq\,(P_r - rc')'\,(P_r - rc')\,H = H\Lambda, \tag{36}$$

where we used

$$D_c^{-1}\,(P_r - rc')'\,D_r^{-1}\,(P_r - rc') = 2nq\,(P_r - rc')'\,(P_r - rc')\,.$$

**Proposition 1.** The object weight matrix $H$ as obtained in the dual scaling analysis of $F_r$ is identical to the object weight matrix $X$ obtained in Nishisato's analysis of $E$.

**Proof.** Define a matrix L as

$$L \equiv (\ I_n,\ \ I_n\ )\,.$$

Then

$$P_r - rc' = \frac{1}{s}F_r - \frac{1}{2s}(q-1)1_{2n}1_q' = \frac{1}{s}\left(I_{2n} - \frac{1}{2}L'L\right)F_r.$$

Clearly, as $LL' = 2I_n$, the matrix $I_{2n} - \frac{1}{2}L'L$ is idempotent, i.e.

$$\left(I_{2n} - \frac{1}{2}L'L\right)^2 = I_{2n} - \frac{1}{2}L'L.$$

Thus, eigenequation (36) can be rewritten as

$$2\gamma F_r'\left(I_{2n} - \frac{1}{2}L'L\right)F_rH = H\Lambda, \tag{37}$$

where we used (26) and (33). The matrix $H$ of objects weights is standardised as in (9), i.e.

$$H'D_cH = I_k \rightarrow H'H = qI_k.$$

Note that this standardisation is the same as the one for $X$ in (27).

Now, since the matrices $H$ and $X$, are already standardised in the same way and there are no nonzero eigenvalues of multiplicity greater than one, the identity

$$\gamma E'E = 2\gamma F_r'\left(I_{2n} - \frac{1}{2}L'L\right)F_r \tag{38}$$

implies that $H = X$.

To show that (38) holds, we define a matrix $J$ as

$$J \equiv (\ I_n, \quad -I_n\ ),$$

so that

$$E = JF_r.$$

Insertion of this expression in (38) immediately yields the result because

$$J'J = 2I_{2n} - L'L = 2\left(I_{2n} - \frac{1}{2}L'L\right).$$

Thus, the object weight matrix $H$ obtained from dual scaling applied to the data matrix $F_r = (T', S')'$, is identical to the object weight matrix $X$ obtained from Nishisato's analysis of the dominance matrix $E = T - S$.

From the orders of $F_r$ and $E$, it immediately follows that no identity exists for the row coordinate matrix $G$ and the subject weight matrix $Y$. However, using transition formulae (28) and (12), we can show that the two are closely related.

Insertion of (21) in (28) yields, after applying (14),

$$\begin{aligned} Y &= \frac{1}{q(q-1)}(T-S)X\Lambda^{-1/2} \\ &= \frac{1}{q(q-1)}\{2T - (q-1)1_n1_q'\}X\Lambda^{-1/2} \\ &= \frac{2}{q(q-1)}TX\Lambda^{-1/2}, \end{aligned} \tag{39}$$

because

$$X'1_q = 0.$$

Now, inserting the definitions for $P_r$ and $D_r$ into transition formula (12), we can calculate standard coordinates for the rows, i.e.

$$G = D_r^{-1} P_r H \Lambda^{-1/2} = \frac{2}{q(q-1)} \begin{pmatrix} T \\ S \end{pmatrix} H\Lambda^{-1/2} = \frac{2}{q(q-1)} \begin{pmatrix} TH\Lambda^{-1/2} \\ SH\Lambda^{-1/2} \end{pmatrix}.$$

Hence

$$G = \frac{2}{q(q-1)} \begin{pmatrix} TX\Lambda^{-1/2} \\ -TX\Lambda^{-1/2} \end{pmatrix} = \begin{pmatrix} Y \\ -Y \end{pmatrix},$$

where we have used equations (40) and (14) together with the identity of $X$ and $H$.

From the foregoing it follows that Nishisato's dual scaling of the dominance matrix $E$ is equivalent to dual scaling, or correspondence analysis, applied to the doubled data matrix $F_r$ where the scaling matrices $D_r$ and $D_c$ are defined in the usual way. Moreover, the fact that dual scaling can be applied to the dominance matrix $E$ with both positive and negative elements can be explained by noting that it is equivalent to the analysis of $P_r - rc'$, i.e. the correspondence matrix with the trivial solution subtracted.

Thus far, we have restricted ourselves to rank order data. As mentioned in the introduction, Nishisato (1980, 1994) treats the analysis of rank order data as a special case of paired comparison data. Let us therefore briefly discuss this more general type of preference data with respect to our new approach.

**Paired Comparison Data.** If there are $q$ objects, one can form $\frac{1}{2}q(q-1)$ pairs. When $n$ individuals are asked to indicate for each pair which object they prefer, we obtain paired comparison data. Rank order data can be seen as a special case of paired comparison data where the comparisons must be transitive. Following Guttman's (1946) approach of quantifying paired comparisons, we define matrices $T_g$ and $S_g$ as:

- $T_g$ : an $n \times q$ matrix with elements $t_{ij}$ the number of times individual $i$ prefers object $j$ over other objects.
- $S_g$ : an $n \times q$ matrix with elements $s_{ij}$ the number of times individual $i$ prefers other objects over object $j$.

(Once again, we do not allow ties to occur. Using a similar coding as described in remark 6.3 one could, however, easily introduce the possibility of ties.) Dual scaling of paired comparison data is defined in the same way as before (see section 6.4.1) with dominance matrix $E_g = T_g - S_g$.

If we have rank order data, the number of times that an object is preferred is equal to its rank number, where zero denotes the lowest rank and $q-1$ the highest. Hence, the definitions of $T_g$ and $S_g$ do not conflict with our previous definitions of $T$ and $S$. Furthermore, all properties described in section 6.3 still hold, i.e. the row sums of $T_g$ and $S_g$ are $\frac{1}{2}q(q-1)$ and $T_g + S_g = (q-1)1_n1_q'$. Hence, the approach described here can also be applied to paired comparison data. One could therefore refer to this method as correspondence analysis of paired comparison data.

## 6.5 Correspondence Analysis of Rank Order Data

In the previous section, we showed that dual scaling of the dominance matrix E is equivalent to correspondence analysis of the doubled data matrix $F_r$. Given the usual equivalence between the two methods, one would therefore expect that correspondence analysis of rank order data is defined accordingly. This, however, is not the case. Careful inspection of the texts on the analysis of preference data using correspondence analysis, i.e. Greenacre (1984, Chapter 8) and Greenacre (1993, Module 19) leads to the conclusion that Greenacre proposes a doubling with respect to the objects. In the case of rank order data this implies a doubling of the columns, i.e. the analysis of the matrix $F_c$ as defined in (16), rather than a doubling of the rows, i.e. the analysis of $F_r$ as defined in (17). Hence, in Greenacre's approach, we have for each observation (row) a rank number corresponding to the original scale and one to the reversed scale.

Obviously correspondence analysis of $F_c$ differs from correspondence analysis of $F_r$ (simply insert $F_c$ in the equations of section 6.2.1 to see that the equations thus obtained differ from the equations obtained in sections 6.4.1 and 6.4.2). Hence, correspondence analysis of rank order data differs from dual scaling of rank order data.

## 6.6 Concluding Remarks

We have shown that in the case of paired comparison data, of which rank order data are a special case, dual scaling and correspondence analysis are closely related. In fact, when a particular doubling of the data- different from the one common in the literature- is employed, the two methods are equivalent up to some row order effects. This relationship is of great practical and theoretical interest. It clarifies the different results obtained when analysing rank order data using dual scaling as introduced by Nishisato (1980, 1994) and correspondence analysis as described by Greenacre (1984, 1993). The approach introduced in this chapter enables researchers to apply dual scaling of paired comparison data using standard correspondence analysis software and it permits the analysis of paired comparison data by correspondence analysis.

Finally, it should be noted that we have constrained ourselves here to the mathematical properties of dual scaling and correspondence analysis. We have found a way to describe the methods in such a way that they are mathematically equivalent. One should, however, be careful in interpreting the results using typical correspondence analysis or dual scaling concepts. For example, correspondence analysis can be seen as the simultaneous low-rank approximation of row and column profiles where one minimises the chi-squared distance. The interpretation of chi-squared distance, however, is far from obvious in the analysis of paired comparison data. Moreover, the duality that is of such importance in both correspondence analysis and dual scaling becomes rather a cumbersome concept in the treatment of paired comparison data since there is in fact only one variable. Perhaps an optimal scaling justification like the one provided by Guttman (1946) is more appropriate when dealing with the kind of data treated in this chapter.

## References

[1] Benzécri, J.-P., (1992), *Correspondence Analysis Handbook,* Marcel Dekker, Inc., New York.

[2] Gifi, A., (1990), *Nonlinear Multivariate Analysis,* John Wiley and Sons, New York.

[3] Greenacre, M.J., (1984), *Theory and Applications of Correspondence Analysis,* Academic Press, London.

[4] Greenacre, M.J., (1993), *Correspondence Analysis in Practice,* Academic Press, London.

[5] Greenacre, M.J., (1996), Book Review: Shizuhiko Nishisato. *Elements of Dual Scaling: an Introduction to Practical Data Analysis,* Lawrence Erlbaum, Hillsdale, N.J., *Psychometrika,* 61, 177–179.

[6] Guttman, L., (1946), An Approach for Quantifying Paired Comparisons and Rank Order, *Annals of Mathematical Statistics,* 17, 144–163.

[7] Lebart, L., A. Morineau, and K.M. Warwick, (1984), *Multivariate Descriptive Statistical Analysis: Correspondence Analysis and Related Techniques for Large Matrices,* John Wiley and Sons, New York.

[8] Mardia, K.V., J.T. Bibby, and J.M. Kent, (1994), *Multivariate Analysis,* Academic Press, London.

[9] Nishisato, S., (1978), Optimal Scaling of Paired Comparison and Rank Order Data: An alternative to Guttman's Formulation, *Psychometrika,* 43, 263–271.

[10] Nishisato, S., (1980), *Analysis of Categorical Data: Dual Scaling and its Applications,* University of Toronto Press, Toronto.

[11] Nishisato, S., (1994), *Elements of Dual Scaling: an Introduction to Practical Data Analysis,* Lawrence Erlbaum, Hillsdale, N.J.

[12] Tenenhaus, M., and F.W. Young, (1985), An Analysis and Synthesis of Multiple Correspondence Analysis, Optimal Scaling, Dual Scaling, Homogeneity Analysis and other Methods for Quantifying Categorical Multivariate Data, *Psychometrika*, 50, 91–119.

CHAPTER 7

# Continuous Extensions of Matrix Formulations in Correspondence Analysis, with Applications to the FGM Family of Distributions

C.M. CUADRAS, J. FORTIANA
and M.J. GREENACRE

## 7.1 Introduction

Correspondence analysis (CA) is a method designed to give a graphical representation of a contingency table $N$ and thus to interpret the association between rows and columns. To be specific, correspondence analysis visualises the so-called *correspondence matrix* $P$, which is the discrete bivariate density obtained by dividing $N$ by its grand total $n$: $P = (1/n)N$. A continuous extension of CA can be obtained by replacing $P$ with a bivariate probability density $h(x, y)$. The marginal densities $f(x), g(y)$ are the continuous counterparts of the row and column margins of $P$.

This extension was studied by Naouri [23], although in a rather formal presentation. Some aspects are also briefly commented in Benzécri [2], who illustrated the continuous version with an example taken from [23]. Greenacre [14] also used the same example to illustrate the Guttman effect.

The aim of this chapter is to present a continuous extension of CA in a more accessible form, following a terminology closer to more recent works, such as Greenacre [14]. Section 2 describes the saturated canonical correlation model, see Goodman [13], equivalent to the CA model to represent categorical data. Section 3 relates this model to the chi-square distance, showing the method as a weighted metric scaling. Sections 4 and 5 are the continuous versions of Sections 2 and 3. Section 6 studies the Pearson contingency coefficient from the perspective of the geometric variability of a distance. Sections 7 and 8 deal with the application of this continuous extension to the construction of probability distributions given the marginals. The Farlie–Gumbel–Morgenstern

(FGM) distribution is presented from a continuous CA perspective, and generalised in terms of diagonal expansions in Lancaster [20], also called Lancaster probabilities, see Koudou [19].

This chapter seeks to be of pedagogical interest. The matrix formulas in CA are labelled from (1) to (10), while the continuous extensions are respectively labelled from (11) to (20). The formulas starting in (21) deal with the construction of probabilities. The proof is given of most formulas in the continuous extensions of CA.

Other developments connected with this extension can be found in several papers by Cuadras and Fortiana [6], [7], [9], and especially [8], all of which deal with some continuous extension of multidimensional scaling.

## 7.2 Discrete Correspondence Analysis

Let $N = (n_{ij})$ be a $I \times J$ contingency table crosstabulating the categories of two discrete variables. Let $P = (1/n)N$, where $n = \sum_i \sum_j n_{ij}$ is the grand total of table $N$. Let

$$r = P1, \qquad c = P'1,$$
$$D_r = \mathrm{diag}(r), \qquad D_c = \mathrm{diag}(c),$$

be the vectors of row and column marginal frequencies of $P$ and diagonal matrices of these vectors, respectively.

Suppose that we wish to assign scale values, contained in the vectors $a$ and $b$, to the rows and columns of $N$ giving maximum correlation between them. It is well-known that canonical correlation analysis with singular covariance matrices leads to the singular value decomposition

$$D_r^{-1/2}(P - rc')D_c^{-1/2} = UD_\lambda V', \tag{1}$$

where the columns of $U$ and $V$ are left and right singular vectors respectively, each set of vectors being orthonormal: $U'U = V'V = I$, and $D_\lambda$ is the diagonal matrix of singular values in descending order (see, for example, Meyer [22]). Other equivalent approaches are discriminant analysis (see Lebart, Morineau and Warwick [21]) and dual or optimal scaling (see Nishisato [24]).

The scale values in the above canonical correlation approach are given by the column vectors of

$$\begin{aligned} A &= D_r^{-1/2}UD_\lambda, \\ B &= D_c^{-1/2}VD_\lambda. \end{aligned} \tag{2}$$

The first singular value $\lambda_1$ is the highest canonical correlation, corresponding to the scale values $a$ and $b$ in the first columns of $A$ and $B$ respectively. Subsequent singular values $\lambda_2, \lambda_3, \ldots$, are canonical correlations associated with scale values in the corresponding columns of $A$ and $B$, in each case maximised under the constraint of being uncorrelated to higher-order vectors.

It is common to use the first two column vectors of $A$ and $B$ as Cartesian coordinates to represent the $I$ rows and $J$ columns of $N$, respectively, in a planar representation. The elements of matrices $A$ and $B$ are called *principal coordinates* of the rows and columns of $N$. This terminology will be justified in Section 3. Matrices $A$ and $B$ are related by

$$\begin{aligned} A &= D_r^{-1}(P - rc')BD_\lambda^{-1}, \\ B &= D_c^{-1}(P' - cr')AD_\lambda^{-1}. \end{aligned} \tag{3}$$

From (1) and (2), $c'B = 0$ and $r'A = 0$, so that these relations reduce to

$$\begin{aligned} A &= D_r^{-1}PBD_\lambda^{-1}, \\ B &= D_c^{-1}P'AD_\lambda^{-1}. \end{aligned} \tag{4}$$

Thus, except for the dilating factor $D_\lambda^{-1}$ (the $\lambda$'s are all less than 1, being correlations), the row principal coordinates in $A$ are weighted means of the column principal coordinates in $B$, and vice versa, where the weights are the row conditional frequencies $p_{ij}/r_i$, $j = 1, \ldots, J$, or column conditional frequencies $p_{ij}/c_j$, $i = 1, \ldots, I$, respectively, in the two formulas of (4). This is the basis of the joint symmetric representation of rows and columns.

The *standard coordinates* (see Greenacre and Hastie [15])

$$\begin{aligned} A_0 &= D_r^{-1/2}U, \\ B_0 &= D_c^{-1/2}V, \end{aligned} \tag{5}$$

are also used in a variant of the graphical display called the asymmetric representation. For example, using $A$ and $B_0$, the rows are displayed in principal coordinates and the columns in standard coordinates and the row points are exactly at weighted means of the column points.

These representations are related to the Pearson contingency coefficient

$$\begin{aligned} \phi^2 &= \sum_{i=1}^{I}\sum_{j=1}^{J}(p_{ij} - r_i c_j)^2/r_i c_j \\ &= \sum_{i=1}^{I}\sum_{j=1}^{J} r_i c_j \left(\frac{p_{ij}}{r_i c_j} - 1\right)^2, \end{aligned} \tag{6}$$

which in turn is equal to $\chi^2/n$, where $\chi^2$ is the chi-squared statistic for testing independence between rows and columns of $N$. This coefficient gives an overall measure of dependence and is decomposed as the sum of squared canonical correlations:

$$\phi^2 = \sum_{k \geq 1} \lambda_k^2. \tag{7}$$

## 7.3 The Chi-square Distance

From (1), we have

$$D_r^{-1/2}(P - rc')D_c^{-1}(P' - cr')D_r^{-1/2} = UD_\lambda^2 U'. \tag{8}$$

The matrix of row conditional frequencies, or row profile matrix, is

$$Q = D_r^{-1}P = (p_{ij}/r_i) = (q_{ij}),$$

so that the vectors of marginal frequencies $r$ and $c$ of $P$ are related by

$$c = Q'r.$$

Rewriting (8) in terms of $Q$, we obtain

$$D_r^{1/2}(I - 1r')QD_c^{-1}Q'(I - r1')D_r^{1/2} = UD_\lambda^2U'. \tag{9}$$

The CA of $P$ can be defined equivalently as the geometric study of the "average" variation of the ratios

$$r(i,j) = p_{ij}/r_ic_j,$$

see (6). Thus, to represent rows using $A$, the usual *chi-square distance* between the profiles of rows $i, i'$ can be written as

$$\begin{aligned}\delta_{ii'}^2 &= \sum_{j=1}^{J}(p_{ij}/r_i - p_{i'j}/r_{i'})^2/c_j \\ &= \sum_{j=1}^{J}\{r(i,j) - r(i',j)\}^2c_j.\end{aligned} \tag{10}$$

The inner product matrix related to this distance is $K = QD_c^{-1}Q'$, so that the matrix of squared distances $\Delta^{(2)} = (\delta_{ii'}^2)$ satisfies

$$\Delta^{(2)} = q1' + 1q' - 2QD_c^{-1}Q',$$

where $q$ is the vector with the diagonal entries of $QD_c^{-1}Q'$. Hence (9) is equivalent to

$$D_r^{1/2}(I - 1r')(-1/2\Delta^{(2)})(I - r1')D_r^{1/2} = UD_\lambda^2U'.$$

Consequently, $A = D_r^{-1/2}UD_\lambda$ is a *weighted metric scaling* solution for $\Delta^{(2)}$ with weight vector $r$. This means that each row $i$ is represented with a weight $r_i$ (see Cuadras and Fortiana [8] and Greenacre [14]). The chi-squared distances between profiles are equal to the Euclidean distances between the row points represented by their full set of principal coordinates $A$:

$$\delta_{ii'}^2 = \sum_{k=1}^{K}(a_{ik} - a_{i'k})^2,$$

where $K = \min(I, J) - 1$ is the total number of dimensions in the CA solution.

Rao [25] proposed an alternative to CA based on Hellinger distance between profiles and discussed some advantages. The same distance function had been considered in a CA context by Domenges and Volle [12].

## 7.4 Continuous Random Variable Extension

Let $(X, Y)$ be a continuous random vector with probability density $h(x, y)$, marginal densities $f(x), g(y)$ and supports $[a, b]$, $[c, d]$, respectively. Suppose that the measure $h(x, y)\, dx\, dy$ is absolutely continuous with respect to the product $f(x)\, g(y)\, dx\, dy$.

Let us denote by $\circ$ the integral operation. For example, the integral operator $\Phi$ with symmetric kernel $\varphi(s, t)$, $t \in [\alpha, \beta]$, is written as:

$$\Phi\{u(s)\} = \varphi(s,t) \circ u(t) = \int_{\alpha}^{\beta} \varphi(s,t)\, u(t)\, dt.$$

In general, $\circ$ stands for the integration with respect to the repeated variable, for example, the variable $t$ above, and is the continuous counterpart of the row×column vector product in a matrix multiplication.

An eigenvalue and associated eigenfunction of $\Phi$ is the pair $(\mu, u)$ satisfying the equation $\Phi u = \mu u$. If $\{\mu_k, u_k(x), k \geq 1\}$ is a countable set of eigenvalues and orthonormal eigenfunctions of $\Phi$, the spectral decomposition of $\varphi$ is

$$\varphi(x, x') = \sum_{k \geq 1} \mu_k u_k(x) u_k(x').$$

We can now introduce the continuous version of (1) as the singular value decomposition

$$z(x,y) = f(x)^{-1/2} \{h(x,y) - f(x)g(y)\}\, g(y)^{-1/2} = \sum_{k\geq 1} \lambda_k u_k(x) v_k(y), \qquad (11)$$

where $\{u_k(x), k \geq 1\}$ and $\{v_k(y), k \geq 1\}$ are complete orthonormal countable sets of functions on the supports in $L^2([a, b])$ and $L^2([c, d])$. This means that $\{\lambda_k^2, u_k(x)\}$ are eigenvalues and eigenvectors of the symmetric kernel

$$\varphi(x, x') = z(x, y) \circ z(x', y) = \sum_{k \geq 1} \lambda_k^2 u_k(x) u_k(x').$$

This expansion will also appear in (19). A similar expansion concerning $\{\lambda_k^2, v_k(y)\}$ holds.

Let us suppose that $z$, see (11), satisfies

$$\int_a^b \int_c^d z^2(x,y) dx dy < \infty.$$

We then have

$$\sum_{k \geq 1} \lambda_k^2 = \int_a^b \int_c^d z^2(x,y) dx dy < \infty,$$

which implies that $\lambda_k \to 0$ as $k \to \infty$.

The continuous version of (3) are the scores $A(x) = \{a_1(x), a_2(x), \ldots\}$ and $B(y) = \{b_1(y), b_2(y), \ldots\}$, which can be interpreted as functions giving the uncorrelated sequences $\{a_k(X), k \geq 1\}$ and $\{b_k(Y), k \geq 1\}$ such that the correlations $\mathrm{corr}\{a_k(X), b_k(Y)\}$, $k \geq 1$, are maximal. $A(x)$ and $B(y)$ are given by

$$\begin{aligned} A(x) &= f(x)^{-1/2}\{\lambda_1 u_1(x), \lambda_2 u_2(x), \ldots\}, \\ B(y) &= g(y)^{-1/2}\{\lambda_1 v_1(y), \lambda_2 v_2(y), \ldots\}, \end{aligned}$$

i.e., using compact notation $D_\lambda = \mathrm{diag}(\lambda_1, \lambda_2, \ldots)$ ,$U(x) = \{u_1(x), u_2(x), \ldots\}$, and $V(x) = \{v_1(x), v_2(x), \ldots\}$

$$\begin{aligned} A(x) &= f(x)^{-1/2} U(x) D_\lambda, \\ B(y) &= g(y)^{-1/2} V(y) D_\lambda, \end{aligned} \tag{12}$$

$$\begin{aligned} \mathrm{corr}\{a_k(X), a_\ell(X)\} &= \mathrm{corr}\{b_k(Y), b_\ell(Y)\} = 0, \qquad k \neq \ell \geq 1, \\ \mathrm{corr}\{a_k(X), b_k(Y)\} &= \lambda_k, \qquad k \geq 1. \end{aligned}$$

Using the integral operation $\circ$, the relation between $A(x)$ and $B(y)$ is as follows.

**Proposition 1.**

$$\begin{aligned} A(x) &= f(x)^{-1}\{h(x,y) - f(x)g(y)\} \circ B(y) D_\lambda^{-1}, \\ B(y) &= g(y)^{-1}\{h(x,y) - f(x)g(y)\} \circ A(x) D_\lambda^{-1}. \end{aligned} \tag{13}$$

**Proof.** From (11) and (12),

$$\begin{aligned} A(x) &= f(x)^{-1/2}\{\lambda_1 u_1(x), \lambda_2 u_2(x), \ldots\} \\ &= f(x)^{-1/2} \sum_{k \geq 1} \lambda_k u_k(x) v_k(y) \circ \{v_1(y), v_2(y), \ldots\} \\ &= f(x)^{-1/2} f(x)^{-1/2}\{h(x,y) - f(x)g(y)\} g(y)^{-1/2} \circ g(y)^{1/2} B(y) D_\lambda^{-1}. \end{aligned}$$

Introducing the conditional densities

$$f(y/x) = h(x,y)/f(x), \quad x \in [a,b], \qquad f(x/y) = h(x,y)/g(y), \quad y \in [c,d],$$

(13) can be expressed in a simpler form.

**Proposition 2.**

$$\begin{aligned} A(x) &= f(y/x) \circ B(y) D_\lambda^{-1}, \\ B(y) &= f(x/y) \circ A(x) D_\lambda^{-1}. \end{aligned} \tag{14}$$

**Proof.**

$$\begin{aligned} g(y) \circ B(y) &= \int_c^d g(y) B(y) dy \\ &= \int_c^d g(y) g(y)^{-1}\{h(x,y) - f(x)g(y)\} dy \circ A(x) D_\lambda^{-1} \\ &= 0, \end{aligned}$$

and similarly for $x$.

The standard coordinates, see (5), are

$$\begin{aligned} A_0(x) &= f(x)^{-1/2}U(x), \\ B_0(y) &= g(y)^{-1/2}V(y). \end{aligned} \tag{15}$$

Then $A_0(x) = \{a_1^0(x), a_2^0(x), \ldots\}$ and $B_0(y) = \{b_1^0(y), b_2^0(y), \ldots\}$ are related by

$$\begin{aligned} a_k^0(x) &= \lambda_k^{-1} \int_c^d f(y/x) b_k^0(y) dy, \\ b_k^0(y) &= \lambda_k^{-1} \int_a^b f(x/y) a_k^0(x) dx. \end{aligned}$$

In other words

$$\begin{aligned} A_0(x) D_\lambda &= E(B_0(Y) \mid x), \\ B_0(y) D_\lambda &= E(A_0(X) \mid y). \end{aligned}$$

The above relations are the basis of the ACE (Alternating Conditional Expectations) algorithm. ACE seeks the optimal transformations of $X$ and $Y$ for prediction. See Breiman and Freiman [3], Buja [4] and Hastie and Tibshirani [16].

Noting that

$$A_0(x) D_\lambda B_0'(y) = \{h(x,y) - f(x)g(y)\}/\{f(x)g(y)\},$$

and using standard coordinates, we have

$$\begin{aligned} h(x,y) &= f(x)g(y)\{1 + \sum_{k \geq 1} \lambda_k a_k^0(x) b_k^0(y)\} \\ &= f(x)g(y)\{1 + A_0(x) D_\lambda B_0'(y)\}. \end{aligned}$$

Hence $A_0(x) D_\lambda B_0'(y)$ contains the full stochastic dependence between $X$ and $Y$.

A global measure of dependence is the Pearson contingency coefficient $\phi^2$ defined by

$$\begin{aligned} \phi^2 &= \int_a^b \int_c^d \frac{\{h(x,y) - f(x)g(y)\}^2}{f(x)g(y)} dx dy \\ &= \int_a^b \int_c^d \left\{ \frac{h(x,y)}{f(x)g(y)} - 1 \right\}^2 f(x)g(y) dx dy. \end{aligned} \tag{16}$$

**Proposition 3.**

$$\phi^2 = \sum_{k \geq 1} \lambda_k^2. \tag{17}$$

**Proof.**

$$\begin{aligned}\phi^2 &= \int_a^b \int_c^d f(x)g(y)A_0(x)D_\lambda B_0'(y)A_0(x)D_\lambda B_0'(y)dxdy \\ &= \int_a^b \int_c^d \Big\{ \sum_{k,\ell\geq 1} \lambda_k\lambda_\ell u_k(x)u_\ell(x)v_k(y)v_\ell(y)\Big\} dx\,dy \\ &= \sum_{k\geq 1} \lambda_k^2.\end{aligned}$$

**Proposition 4.** $X$ and $Y$ are independent iff $\phi^2 = 0$, i.e., $\lambda_k = 0$, $k \geq 1$.

## 7.5 Continuous Weighted Metric Scaling

In this section, we generalise and interpret the chi-square distance to random variables, and give a continuous extension of weighted metric scaling. As in Section 3, we consider the likelihood ratio which appears in the alternative form of the contingency coefficient $\phi^2$, see (16):

$$r(x,y) = \frac{h(x,y)}{f(x)g(y)}$$

and the symmetric function

$$K(x,x') = E_Y[r(x,Y)(r(x',Y)].$$

Recall that $h$ is absolutely continuous with respect to $fg$. Thus $r$ is the Radon-Nikodym derivative of $h$ with respect to $fg$.

**Proposition 5.** $E_X K(X,x') = 1$, where $x'$ is fixed.

**Proof.**

$$\begin{aligned}E_X K(X,x') &= \int_a^b \int_c^d \{h(x,y)/f(x)g(y)\}\{h(x',y)/f(x')g(y)\}f(x)g(y)dxdy \\ &= \int_a^b \int_c^d \{h(x,y)/g(y)\}\{h(x',y)/f(x')\}dxdy \\ &= \int_a^b f(x/y)dx \int_c^d f(y/x')dy \\ &= 1.\end{aligned}$$

$K(x,x')$ is the continuous version of $K = QD_c^{-1}Q$. In CA and in metric scaling, to find the principal coordinates, $K$ must be double-centered. Its continuous version is the double-centering of $K(x,x')$

$$\begin{aligned}G(x,x') &= K(x,x') - E_X K(X,x') - E_{X'} K(x,X') + E_X E_{X'} K(X,X') \\ &= K(x,x') - 1 - 1 + 1.\end{aligned}$$

This gives

$$G(x,x') = E_Y[\{r(x,Y)-1\}\{r(x',Y)-1\}],$$

the continuous version of $(I - 1r')QD_c^{-1}Q'(I - r1')$, see (9). $G(x, x')$ is symmetric, positive semidefinite and satisfies

$$\begin{aligned} E_X G(X,x') &= E_{X'}G(x,X') = 0, \qquad \forall x,x' \in [a,b], \\ E_X G(X,X) &= \phi^2. \end{aligned}$$

From (11), we have

$$\begin{aligned} & f(x)^{-1/2}\{h(x,y) - f(x)g(y)\}g(y)^{-1}\{h(x',y) - f(x')g(y)\}g(y)^{-1/2} \\ & = \sum_{k\geq 1} \lambda_k u_k(x) v_k(y) \sum_{\ell \geq 1} \lambda_\ell u_\ell(x') v_\ell(y), \qquad (18) \end{aligned}$$

which may be compared with (8). We can also write (11) as

$$f(x)^{1/2}(r(x,y)-1)g(y)^{1/2} = \sum_{k\geq 1} \lambda_k u_k(x) v_k(y).$$

Thus (18) can be written as

$$f(x)^{1/2}\{r(x,y)-1\}g(y)\{r(x',y)-1\}f(x')^{1/2} = \sum_{k\geq 1} \lambda_k u_k(x) v_k(y) \sum_{\ell\geq 1} \lambda_\ell u_\ell(x') v_\ell(y),$$

and integration with respect to $y$ gives

$$\begin{aligned} f(x)^{1/2}G(x,x')f(x')^{1/2} &= \sum_{k\geq 1} \lambda_k^2 u_k(x) u_k(x') \\ &= U(x) D_\lambda^2 U'(x), \qquad (19) \end{aligned}$$

the continuous counterpart of (9).

Let us now extend (10). The profiles of $x, x' \in [a, b]$ are $f(y/x), f(y/x')$. The chi-square distance between profiles is

$$\begin{aligned} \delta^2(x,x') &= \int_c^d \frac{\{f(y/x) - f(y/x')\}^2}{g(y)} dy \\ &= E_Y[\{r(x,Y) - r(x',Y)\}^2]. \qquad (20) \end{aligned}$$

**Proposition 6.** $\delta^2(x,x') = G(x,x) + G(x',x') - 2G(x,x')$.

**Proof.**

$$\delta^2(x,x') = E_Y\left([\{r(x,Y)-1\} - \{r(x',Y)-1\}]^2\right).$$

Now the role of the principal coordinates $A(x)$ is clear:

**Proposition 7.** The ordinary Euclidean distance between $A(x)$ and $A(x')$ is the chi-square distance, i.e.,

$$\delta^2(x,x') = \sum_{k\geq 1} \{a_k(x) - a_k(x')\}^2.$$

**Proof.**

$$\begin{aligned}\sum_{k\geq 1} a_k(x)a_k(x') &= \sum_{k\geq 1} \lambda_k^2 f(x)^{-1/2}u_k(x)f(x')^{-1/2}u_k(x') \\ &= G(x,x').\end{aligned}$$

## 7.6 Geometric Variability, Proximity Function and Isometries

The *geometric variability* of the random variable $X$ with respect to a distance $\delta$ is defined as

$$V_\delta(X) = \frac{1}{2}\int_a^b\int_a^b \delta^2(x,x')f(x)f(x')dxdx'.$$

The *proximity function* of an observation $x_0$ of $X$ to the population defined by $X$ is defined by

$$D_\delta^2(x_0) = \int_a^b \delta^2(x_0,x)f(x)dx - V_\delta(X).$$

When $\delta$ is the chi-square distance, we have

**Proposition 8.** $V_\delta(X) = \phi^2$.

**Proof.**

$$\begin{aligned}\phi^2+1 &= \int_a^b\int_c^d r^2(x,y)f(x)g(y)dxdy, \\ 2V_\delta(X) &= \int_a^b\int_a^b\int_c^d \{r(x,y)-r(x',y)\}^2 f(x)f(x')g(y)dxdx'dy \\ &= \phi^2+1+\phi^2+1-2\int_a^b\int_a^b\int_c^d r(x,y)r(x',y)f(x)f(x')g(y)dxdx'dy \\ &= 2\phi^2+1+1-2.\end{aligned}$$

**Proposition 9.** $D_\delta^2(x_0) = G(x_0,x_0)$.

**Proof.**

$$\begin{aligned}G(x_0,x_0) &= \int_c^d (r(x_0,y)-1)^2 g(y)dy \\ &= \int_c^d r^2(x_0,y)g(y)dy + 1 - 2\int_c^d r(x_0,y)g(y)dy, \\ D^2(x_0,X) &= \int_a^b\int_c^d \{r(x_0,y)-r(x,y)\}^2 f(x)g(y)dxdy - (\phi^2-1) \\ &= \int_c^d r^2(x_0,y)g(y)dy - 2\int_a^b\int_c^d r(x_0,y)r(x,y)f(x)g(y)dxdy + 1 \\ &= \int_c^d r^2(x_0,y)g(y)dy - 2\int_c^d r(x_0,y)g(y)dy\int_a^b f(x/y)dx + 1.\end{aligned}$$

The geometric variability and the proximity function are concepts used in several areas: continuous metric scaling (Cuadras and Fortiana [7]), discriminant analysis (Cuadras, Fortiana and Oliva [10]) and constructing probability densities from distances (Cuadras, Atkinson and Fortiana [11]).

Clearly, we can define another chi-square distance $\delta'$ between the profiles $f(x/y), f(x/y')$ of $y, y' \in [c, d]$ and obtain similar results. For instance, the geometric variability of $\delta'$ is also $\phi^2$. Using the notation $\|A(x)\|^2 = \sum_{k\geq 1} a_k(x)^2$, etc., we can summarise these results as follows.

*Given two random variables $(X, Y)$ and considering the chi-square distances $\delta$, $\delta'$, there exists, for every $(x, y)$, a representation $(x, y) \to \{A(x), B(y)\}$ and a linear transformation $T : A(x) \to B(y)$ such that:*

$$\delta^2(x, x') = \|A(x) - A(x')\|^2, \quad \delta'^2(y, y') = \|B(y) - B(y')\|^2,$$

*and*

$$\begin{aligned} \delta^2(x,x') &= E_Y[\{r(x,Y) - r(x',Y)\}^2], & V_\delta(X) &= \phi^2, \\ D_\delta^2(x) &= \|A(x)\|^2, & A(x) &= E(B(Y) \mid x) D_\lambda{}^{-1}, \\ \delta'^2(y,y') &= E_X[\{r(X,y) - r(X,y')\}^2], & V_{\delta'}(Y) &= \phi^2, \\ D_{\delta'}^2(y) &= \|B(y)\|^2, & B(y) &= E(A(X) \mid y) D_\lambda{}^{-1}. \end{aligned}$$

The representations $A(x)$, $B(y)$ are *isometries*, expressing $\delta$, $\delta'$ as Euclidean distances, and are used in CA to represent data (usually categorical data as in Greenacre [14], and Lebart, Morineau and Warwick [21]), while $T$ relates optimal transformations of $X$ and $Y$ for correlation and regression, see Breiman and Freiman [3]. The representation using the first $d$ coordinates of $A(x)$ implicitly supposes the approximation of the density $h$ by another density $h_d$. This is justified in the next sections.

## 7.7 The FGM Family of Distributions and Correspondence Analysis

Let us write $f$ for $f(x)$, $g$ for $g(x)$, etc. If the variables $X, Y$ have cdfs $F, G$, the Farlie–Gumbel–Morgenstern (FGM) family of distributions is a one-parameter bivariate distribution $H$ with marginals $F, G$ defined by

$$H = FG\{1 + \theta(1 - F)(1 - G)\}, \quad -1 \leq \theta \leq 1,$$

(see Hutchinson and Lai [17]). The density is then

$$h = fg\{1 + \theta(1 - 2F)(1 - 2G)\}, \quad -1 \leq \theta \leq 1.$$

Thus

$$f^{-1/2}(h - fg)g^{-1/2} = \frac{\theta}{3}\sqrt{3}f^{1/2}(1 - 2F)\sqrt{3}g^{1/2}(1 - 2G)$$

is the singular value decomposition, where there is one "dimension", with singular value $\lambda = \theta/3$ and singular vectors

$$\begin{aligned} u(x) &= \sqrt{3}f^{1/2}(x)\{1-2F(x)\}, \\ v(y) &= \sqrt{3}g^{1/2}(y)\{1-2G(y)\}. \end{aligned}$$

Hence the maximum absolute value for the correlation is 1/3. We can also show, see (15), that

$$\begin{aligned} a^0(x) &= \sqrt{3}\{1-2F(x)\}, \\ b^0(y) &= \sqrt{3}\{1-2G(y)\}. \end{aligned}$$

The best approximation of $b^0(y)$ given $X = x$ is $(\theta/3)a^0(x)$. Further, $F(X)$, $G(Y)$ are the functions of $X, Y$ providing maximal correlation.

## 7.8 A Generalised FGM Family

Increasing the dimension, a generalised FGM family, GFGM say, is

$$h_G = fg\Big\{1 + \sum_{k\geq 1}\gamma_k L_k(F)L_k(G)\Big\}, \tag{21}$$

where $L_1(x) = \sqrt{3}(2x-1)$, $L_2(x) = \sqrt{5}(6x^2-6x+1)$, $L_3(x) = \sqrt{7}(20x^3-30x^2+12x-1)$, etc., are the shifted Legendre polynomials on $(0,1)$ (Hutchinson and Lai [17, p. 221[]). The orthonormal functions with respect to the inner product $\int f_1(x)f_2(x)dx$, see (11), are

$$\begin{aligned} u_k(x) &= f^{1/2}(x)L_k\{F(x)\}, \\ v_k(y) &= g^{1/2}(y)L_k\{G(y)\}. \end{aligned} \tag{22}$$

Assuming $h_G \geq 0$, note that

$$\int_a^b f(x)L_k\{F(x)\}dx = \int_0^1 L_k(u)du = 0, \tag{23}$$

and similarly for $Y$. Thus (21) is a probability density. Using cdfs, the GFGM can be expressed as

$$H_G = FG\Big\{1 + \sum_{k\geq 1}\gamma_k L_k^*(F)L_k^*(G)\Big\},$$

where

$$\begin{aligned} L_k^*(u) &= \int_0^u L_k(t)dt \\ &= \{L_{k+1}(u)/\sqrt{2k+3} - L_{k-1}(u)/\sqrt{2k-1}\}/\sqrt{2k+1}. \end{aligned}$$

In general, expansion (21) is not the singular value decomposition for $h$. It is then of interest to fit $h$ to $h_d$, the $d$-dimensional GFGM distribution obtained by taking only $d$ terms in (21). Considering the expansion of $f^{-1/2}(h-fg)g^{-1/2}$ in terms of

$$\sum_{k=1}^{d} \gamma_k f^{1/2} L_k(F) g^{1/2} L_k(G),$$

we have $\gamma_k = \int_a^b \int_c^d (h - fg) L_k(F) L_k(G) dx dy$, and, from (23),

$$\gamma_k = \int_a^b \int_c^d h L_k(F) L_k(G) dx dy.$$

The probability integral transformation $u = F$, $v = G$, finally gives

$$\gamma_k = \int_0^1 \int_0^1 r\{F^{-1}(u), G^{-1}(v)\} L_k(u) L_k(v) du dv,$$

where $r(x, y) = h(x, y)/f(x)g(y)$.

However, it turns out that the orthonormal set $\{u_k(x)v_k(x), k \geq 1\}$ is not complete. The complete orthonormal basis is $\{u_k(x)v_\ell(y), k, \ell \geq 1\}$. Then a yet more general family is

$$h_\Gamma = fg\Big\{1 + \sum_{k,\ell \geq 1} \gamma_{k\ell} L_k(F) L_\ell(G)\Big\}.$$

With the notation $L(\cdot) = \{L_1(\cdot), L_2(\cdot), \ldots\}$, $\Gamma = (\gamma_{k\ell})$, we can write

$$h_\Gamma = fg\{1 + L(F)\Gamma L(G)'\}.$$

Then if $h$ is approximated by $h_{\widehat{\Gamma}}$, where $\widehat{\Gamma}$ is now an $m \times n$ finite matrix, the coefficients are given by

$$\gamma_{k\ell} = \int_0^1 \int_0^1 r\{F^{-1}(u), G^{-1}(v)\} L_k(u) L_\ell(v) du dv,$$

and can be interpreted as association coefficients (see Cuadras [5] and Joe [18] for other methods of construction).

It is easily proved (see Proposition 3) that the contingency coefficient for $h_{\widehat{\Gamma}}$ is

$$\widehat{\phi}^2 = \sum_{k=1}^{m} \sum_{\ell=1}^{n} \gamma_{k\ell}^2.$$

If we consider the full expansion of $f^{-1/2}(h - fg)g^{-1/2}$ providing $h = h_\Gamma$, then from Parseval's equality, the contingency coefficient for $h$ is given by

$$\phi^2 = \sum_{k=1}^{\infty} \sum_{\ell=1}^{\infty} \gamma_{k\ell}^2.$$

As this coefficient is also the geometric variability, the quality (from a continuous CA perspective) of the representation of $h$ by $h_{\widehat{\Gamma}}$, can be measured by the ratio

$$0 < \frac{\widehat{\phi}^2}{\phi^2} \leq 1.$$

The use of Legendre polynomials can be generalised to other functions. If the marginal distributions are normal, the appropriate functions are Hermitian polynomials, but then the distribution obtained is not a generalisation of the FGM family. The construction with these polynomials can be useful to relate dimensions (e.g., Guttman effects) in CA, when the latent distribution of the frequency matrix is bivariate normal. (see Baccini, Caussinus and de Falguerolles [1]).

## References

[1] Baccini, A., H. Caussinus and A. de Falguerolles, (1993), Analysing Dependence in Large Contingency Tables: Dimensionality and Patterns in Scatter Plots, pp. 245–263 in C.M. Cuadras and C.R. Rao (eds.), *Multivariate Analysis Future Directions 2,* Elsevier Science Publishers (North-Holland), Amsterdam.

[2] Benzécri, J.P., (1973), *L'Analyse des Données II, L'Analyse des Correspondances,* Dunod, Paris.

[3] Breiman, L., and J.H. Friedman, (1985), Estimating Optimal Transformations for Multiple Regression and Correlation, *Journal or the American Statistical Association,* 80, 61–619.

[4] Buja, A., (1990), Remarks on Functional Canonical Variates, Alternating Least-Squares Methods and ACE, *Annals of Statistics,* 18, 1032–1069.

[5] Cuadras, C.M., (1992), Probability Distributions with Given Multivariate Marginals and Given Dependence Structure, *Journal of Multivariate Analysis,* 42, 51–66.

[6] Cuadras, C.M., and J. Fortiana, (1993), Continuous Metric Scaling and Prediction. pp. 47–66 in C.M. Cuadras and C.R. Rao (eds.), *Multivariate Analysis Future Directions 2,* Elsevier Science Publishers (North-Holland), Amsterdam.

[7] Cuadras, C.M., and J. Fortiana, (1995), A Continuous Metric Scaling Solution for a Random Variable, *Journal of Multivariate Analysis,* 52, 1–14.

[8] Cuadras, C.M., and J. Fortiana, (1996), Weighted Continuous Metric Scaling, pp. 27–40 in A. K. Gupta and V. L. Girko (eds.), *Multidimensional Statistical Analysis and Theory of Random Matrices,* VSP, Zeist, The Netherlands.

[9] Cuadras, C.M., and J. Fortiana, (1997), Continuous Scaling on a Bivariate Copula, pp. 137–142 in Viktor Benes and Josef Stepan (eds.), *Distributions with Given Marginals and Moment Problems,* Kluwer Academic Publishers, Dordrecht.

[10] Cuadras, C.M., J. Fortiana and F. Oliva, (1997), The Proximity of an Individual to a Population with Applications in Discriminant Analysis, *Journal of Classification,* 14, 117–136.

[11] Cuadras, C.M., R.A. Atkinson and J. Fortiana, (1997), Probability Densities from Distances and Ddiscriminant Analysis, *Statistics and Probability Letters,* 33, 405–411.

[12] Domenges, D., and M. Volle, (1979), Analyse Factorielle Sphérique: une Exploration, *Annales de l'INSEE,* 35, 3–83.

[13] Goodman, L.A., (1986), Some Useful Extensions of the Usual Correspondence Analysis Approach and the Usual Log-linear Models Approach in the Analysis of Contingency Tables, *International Statistical Review,* 54, 243–309.

[14] Greenacre, M., (1984), *Theory and Applications of Correspondence Analysis,* Academic Press, London.

[15] Greenacre, M., and T.J. Hastie, (1987), The Geometric Interpretation of Correspondence Analysis, *Journal of the American Statistical Association,* 82, 437–447.

[16] Hastie, T.J., and R.J. Tibshirani, (1990), *Generalised Additive Models,* Chapman and Hall, London.

[17] Hutchinson, T.P., and C.D. Lai, (1991), *The Engineering Statistician's Guide to Continuous Bivariate Distributions,* Rumsby Scientific Publishers, Adelaide.

[18] Joe, H., (1997), *Multivariate Models and Dependence Concepts,* Chapman and Hall, London.

[19] Koudou, A.E., (1996), Probabilités de Lancaster. *Expositiones Mathematicae,* 14, 247–275.

[20] Lancaster, H.O., (1969), *The Chi-Squared Distribution,* John Wiley, New York.

[21] Lebart, L., A. Morineau, and K.M. Warwick, (1984), *Multivariate Descriptive Statistical Analysis,* John Wiley, New York.

[22] Meyer, R., (1991), Canonical Correlation Analysis as a Starting Point for Extensions of Correspondence Analysis, *Statistique et Analyse des Données,* 16, 55–77.

[23] Naouri, J.C., (1970), Analyse Factorielle des Correspondances Continues, *Publications de l'Institut de Statistique de la Université de Paris*, XIX(1), 1–100.

[24] Nishisato, S., (1980), *Analysis of Categorical Data: Dual Scaling and its Applications*, University of Toronto Press, Toronto.

[25] Rao, C.R., (1995), A Review of Canonical Coordinates and an Alternative to Correspondence Analysis using Hellinger Distance, *Qüestiio*, 19, 23–63.

CHAPTER 8

# Utility Maximisation and Mode of Payment

Ruud H. KONING and Geert RIDDER[1]

## 8.1 Introduction

Most microeconometric demand and supply studies are based on the assumption of utility maximisation over a budget set. It is well-known that the hypothesis of utility maximisation subject to a linear budget constraint implies certain restrictions on the parameters of the demand (or supply) function. These conditions are the so-called "integrability conditions"(see *inter alia* Deaton and Muellbauer [5]). If we let $g_i(x,p), i = 1,\ldots,I$ denote a system of (Marshallian) demand functions, where $x$ denotes total expenditure and $p$ an $I$-vector of prices, then these conditions can be stated as

$$\sum_i \frac{p_i g_i(x,p)}{x} = 1, \tag{I-1}$$

$$g_i(x,p) = g_i(\lambda x, \lambda p), \quad \lambda > 0, \tag{I-2}$$

$$C = C', \tag{I-3}$$

$$C \leq 0, \tag{I-4}$$

where the matrix of compensated price responses $C$ is defined as

$$C_{ij} = \frac{\partial g_i}{\partial x} g_j(x,p) + \frac{\partial g_i}{\partial p_j}.$$

These integrability conditions have an analogue in discrete choice models. In these models, the demand for discrete alternatives is modelled instead of the

[1]Research of the first author was made possible by a grant of the Royal Dutch Academy of Arts and Sciences. Research that resulted in this paper started when both authors were affiliated with the Department of Economics of the Vrije Universiteit, Amsterdam. Without implicating them in remaining errors, we would like to thank participants at the European Economic Association meeting in Prague, the Winter Meeting of the Econometric Society in Copenhagen, and at a seminar at New York University for their comments.

demand for some continuously varying good. Let an economic agent have to choose between $I$ alternatives, with a specific alternative being denoted by $i$ ($i = 1, \ldots, I$). Alternative $i$ has utility $u_i$ and alternative $i$ is chosen if and only if $u_i > u_j, j \neq i = 1, \ldots, I$. The econometrician, however, does not observe $u_i$. Instead, he expresses it as

$$u_i = -v_i + \varepsilon_i, \qquad i = 1, \ldots, I. \tag{1}$$

Here, $-v_i$ is the average (over all agents) utility attached to alternative $i$ and $\varepsilon_i$ is a mean 0 stochastic disturbance term that measures utility variation among the agents. $\varepsilon_i$ is assumed to be known to the agent, but not to the econometrician. The choice is therefore deterministic from the agent's point of view. An alternative interpretation is that $\varepsilon_i$ is unknown to the agent and that before making a choice the agent obtains $\varepsilon_i$. This introduces randomness in the choice. Without repeated observations of choices made by the same agent, we can not distinguish between these interpretations. If the distribution function $F(\varepsilon)$ and the mean utilities $v$ are known, one can assign probabilities to the event that alternative $i$ is chosen:

$$P(i \text{ is chosen}) \equiv P_i(v) = P(-v_i + \varepsilon_i > -v_j + \varepsilon_j, j \neq i = 1, \ldots, I). \tag{2}$$

Usual specifications for choice probabilities are the (multinomial) logit and (multinomial) probit model. These choice probabilities are derived under the assumption of utility maximisation, that is, the choice probabilities $P_i(v)$ are derived by assuming some particular distribution for $\varepsilon$ and by proceeding from the right-hand side of (2) to the left-hand side. However, it is also possible to fit a flexible functional form to the data and then ask the question whether the observed choice probabilities are compatible with utility maximisation of preferences of the form in (1), that is, whether we can proceed from the left-hand side to the right-hand side of (2). This question has been discussed in a number of papers: Daly and Zachary [4], Williams [18], McFadden [14] and more recently in Börsch-Supan [1], Koning and Ridder [9] and Koning and Ridder [10]. Local conditions for utility maximisation on an interval have been applied by Kling and Herriges [7] who also give a Bayesian interpretation. They use a nested multinomial logit model as a specification for the choice probabilities.

In this chapter we discuss different tests of the hypothesis that the estimated choice probabilities $P_i(v)$ are compatible with maximisation of preferences of the form in (1). The average utility levels $v_i$ are usually related to socioeconomic characteristics of the agent and to characteristics of the alternatives. The specification of the choice probabilities depends on the specification of this relationship and on the specification of $P(\cdot)$ that relates the choice probabilities to the average utility levels. If we choose a parametric representation for $P(\cdot)$, then this representation may be inconsistent with utility maximisation for a set of parameter values. For instance, the multinomial logit and probit specification are consistent with utility maximisation for all values of their parameters, but the nested multinomial logit model is inconsistent with utility maximisation if one if its parameters is in a certain range. Note that a test based on $P(\cdot)$ ignores potential misspecification of the average utility levels $v_i$. If we are prepared to make assumptions on the shape of $v_i$ as a function of the socioeconomic and alternative-specific characteristics, then we could use the procedure of Matzkin [12], [13] to obtain nonparametric estimates of $v_i$. Without shape assumptions we must assume that our (parametric) specification of $v_i$ is correct. A check of the sensitivity of the test result to changes in the specification of $v_i$ is then useful.

Our specific application is to the choice of mode of payment for an over-the-counter purchase. When someone pays for an over-the-counter purchase, he has the choice between different modes of payment, for example, payment in cash, by means of a cheque, a credit card, etc. Insight in the choice of mode of payment is important, because of its implications for the demand of token money, the development of new monetary products, etc. A structural interpretation of the choice probabilities as provided by the random utility model (1) is a necessary condition for the use of the model to predict the use of new modes of payments and welfare analysis. Even though a complete analysis of the choice of mode of payment is interesting in its own right and because of the reasons given above, we will focus here on the different tests for rationality. A detailed analysis of the choice of mode of payment based on the dataset that we use has been published by Mot, Cramer, and van der Gulik [15].

The remainder of this chapter is organised as follows. Conditions for compatibility of choice probabilities with stochastic utility maximisation are recapitulated in section 8.2. In section 8.3 we estimate two models for the choice of payment mode. We examine whether the estimated choice probabilities are compatible with stochastic utility maximisation in section 8.4. In section 8.5 we estimate the choice probability semiparametrically and we will test for rationality in that model as well. We summarise and conclude in section 8.6.

## 8.2 Compatibility of Choice Probabilities with Stochastic Utility Maximisation

In this section we briefly recapitulate the relations between the hypothesis of stochastic utility maximisation and choice probabilities. We do not give proofs of statements, nor do we provide an extensive discussion of the theory but we focus on the testable implications of the hypothesis of stochastic utility maximisation. The interested reader is referred to McFadden [14], Koning and Ridder [10] and Koning [8] for detailed discussions and proofs.

We consider a population of agents who each choose one alternative out of $I$ possible alternatives. Let $u_{it}$ be the utility of alternative $i$ to agent $t$. The utilities vary in the population. To be specific we assume that the $I$-vector of utilities is a draw from an $I$-variate distribution with mean $-v_t$. Hence, without loss of generality we can write

$$u_{it} = -v_{it} + \varepsilon_{it}, \qquad i = 1, \ldots, I. \tag{3}$$

This representation is known as the *additive random utility model.* In (3), $-v_{it}$

is the mean of utility attached to alternative $i$ and $\varepsilon_{it}$ is a mean zero random variable. The distribution of $\varepsilon_t$ (the $I$-vector with typical element $\varepsilon_{it}$) is non-defective, independent of $v_t$ (the $I$-vector with typical element $v_{it}$) and assumed to be continuous with respect to Lebesgue measure. The continuity ensures that the probability of a tie is 0. In the sequel we omit the subscript $t$.

Suppose for the moment that $v$ is known (or estimated). Since the econometrician does not observe $u$ (the $I$-vector with typical element $u_i$) he is unable to predict the exact choice made by the household. He can assign choice probabilities instead:

$$P(i \text{ is chosen}) \equiv P_i(v) = P(-v_i + \varepsilon_i \geq -v_j + \varepsilon_j, j \neq i = 1, \dots, I). \qquad (4)$$

The choice probabilities in (4) are derived under the assumption of utility maximisation. Compatibility with stochastic utility maximisation is now defined as follows. A set of choice probabilities $P_i(v), i = 1, \dots, I$ is compatible with stochastic utility maximisation on a set $\mathcal{V}$ if for all $v \in \mathcal{V}$ we can write[2]

$$P_i(v) = P(-v_i + \varepsilon_i \geq -v_j + \varepsilon_j, j \neq i = 1, \dots, I), \qquad i = 1, \dots, I, \qquad (5)$$

with $\varepsilon$ a stochastic $I$-vector that has a non-defective, absolutely continuous distribution that does not depend on $v$. If $\mathcal{V} = \mathcal{R}^I$ we speak of global compatibility and if $\mathcal{V}$ is a strict subset of $\mathcal{R}^I$ we speak of local (on $\mathcal{V}$) compatibility.

First we discuss the case of global compatibility ($\mathcal{V} = \mathcal{R}^I$). It is well-known (see Daly and Zachary [4], Williams [18] and McFadden [14]) that the hypothesis of stochastic utility maximisation implies the following restrictions on the choice probabilities:

$$P_i(v) \geq 0, \quad \sum_{i=1}^{I} P_i(v) = 1, \qquad \text{(C–1)}$$

$$P_i(v) = P_i(v + c\iota_I) \quad \text{for all} \quad c \in \mathcal{R} \quad \text{(translation invariance)}, \qquad \text{(C–2)}$$

$$\lim_{v_i \to -\infty} P_i(v) = 1, \qquad \lim_{v_j \to -\infty} P_i(v) = 0, \qquad \text{(C–3)}$$

$$P_i(v) \text{ is differentiable with respect to } v^i \text{ and } \frac{\partial^{(I-1)} P_i}{\partial v^i}(v) \geq 0$$

$$\text{(non-negativity)}, \qquad \text{(C–4)}$$

$$\frac{\partial P_i}{\partial v_j}(v) = \frac{\partial P_j}{\partial v_i}(v) \quad \text{(symmetry)}. \qquad \text{(C–5)}$$

In (C–4) $v^i$ is the $I-1$-vector obtained by omitting component $v_i$. On the other hand, if conditions (C–1) to (C–5) hold for some arbitrary set of functions $P_i(v)$ for all $v \in \mathcal{R}^I$, there exists a random variable $\varepsilon$ such that these choice probabilities can be written as in (4) and hence, the choice probabilities are

[2]Some of the conditions about the distribution of $\varepsilon$ may be relaxed slightly if $\mathcal{V}$ is taken to be a finite set or an $I$-dimensional interval.

globally compatible with maximisation of the utility function (3). In other words, conditions (C–1) to (C–5) are necessary and sufficient for compatibility with stochastic utility maximisation.

In practice, it might be overly restrictive to require that the conditions (C–1) to (C–5) hold for *all* $v \in \mathcal{R}^I$. If one of the conditions is violated for some points 'far away' from the observed $v_i$, rejection of the hypothesis of stochastic utility maximisation may be due to extrapolation of the model far beyond the observations. Following Börsch-Supan [1], Koning and Ridder [10] define the concept of local compatibility with stochastic utility maximisation on a set $\mathcal{V}$. They distinguish between two cases for $\mathcal{V}$: Koning and Ridder [10] take $\mathcal{V}$ to be a closed and finite $I$-dimensional interval ($\mathcal{V}_2 = [a, b]$), and in Koning and Ridder [9] $\mathcal{V}$ is taken to be a (finite) set of points ($\mathcal{V}_3 = \{v_1, \ldots, v_T\}$). It can be shown that conditions (C–1), (C–2), (C–4) and (C–5) are necessary and sufficient conditions for compatibility of choice probabilities with stochastic utility maximisation if $v \in \mathcal{V}_2$.

Koning and Ridder [9] also give necessary and sufficient conditions for compatibility on a finite set of points. These conditions are hard to check (see Koning [8]). A necessary implication of stochastic utility maximisation is simple to check, though. If the choice probabilities $P_i(v)$ are compatible with stochastic utility maximisation on $\mathcal{V}_3$, then for each pair of observations $(t, t')$ it must hold

$$v_t - v_{it}\iota_I \leq v_{t'} - v_{it'}\iota_I \to P_i(v_t) \leq P_i(v_{t'}), \qquad i = 1, \ldots, I. \tag{6}$$

In other words, if for some observation $t'$ alternative $i$ is more attractive when compared to all other alternatives than for observation $t$ (remember that the systematic part of person's $t$'s utility is $-v_t$ so $v_t - v_{it}\iota_i$ measures for each alternative to what extent alternative $i$ is better), then the probability that $t'$ chooses $i$ is not lower than the probability that $t$ chooses $i$. This condition is generally not sufficient for compatibility with stochastic utility maximisation on a finite set, though, a counter example is given in Koning and Ridder [9]. However, for the case of only two alternatives, condition (6) is both necessary and sufficient for compatibility. Since we can set the utility of one alternative (say, the first) to 0 without loss of generality, condition (6) requires that the choice probability of the first alternative is not increasing in $v_2$.

## 8.3 Choice of Mode of Payment

We will now illustrate the issue of compatibility using a discrete choice model for the mode of payment. In this section we discuss a very simple model of choice of mode of payment and we discuss the data and some estimation results.

Consider an agent who makes an over-the-counter purchase (we assume that the purchase decision has been made already). The agent has indirect utility function $\nu(Y, p; \theta)$ with $Y$ denoting income, $p$ relative prices, and $\theta$ the parameters that captures the preference structure. He can choose between three modes of payment: cash payment, payment by giro cheque, and payment by bank cheque. Let transactions costs $T_i$ are associated with payment by mode $i$.

Indirect utility associated with each choice is now $\nu(Y - T_i, p; \theta)$. From now on we will suppress the dependence of the indirect utility function on the vector with relative prices $p$ because we are dealing with a cross-section and relative prices do not vary in our dataset. Transaction costs $T_i$ may be both of a monetary and non-monetary nature. As an example of the first we mention the costs associated with keeping a checking account, and the costs of using more than a certain number of cheques each quarter. Examples of non-monetary aspects of transaction costs are time required to complete the transaction (paying cash is usually quicker, especially for small amounts), the inconvenience of using cheques, and the risk of losing cash or cheques. It is clear that these transaction costs cannot be constant, since only one alternative (the one with the lowest transaction costs) would be chosen in that case. The transaction costs may vary by mode of payment and will certainly vary with the amount of the transaction so we can write $T_i = a_i(AMT)$ with $AMT$ the amount of the transaction. The mode of payment that maximises utility is now given by the alternative that maximises

$$\nu\{Y - a_i(AMT); \theta\}. \tag{7}$$

The data that we use are taken from the Intomart Bestedingen Index,[3] a panel of slightly more than 1000 households. In that panel various kinds of expenditures are recorded, together with information on the household. The alternatives (payment by a giro cheque, payment by a bank cheque[4] and cash payment) are labelled 1, 2, and 3. Hardly any transaction with modern forms of payment like a credit card, a PIN-card, or a chip-card were recorded, so these observations are deleted from the analysis.

We use the 1987 wave of the Intomarkt Expenditure Index. The households in this wave have on average slightly more than 2 members older than 12 years. Each household records its expenditures daily, in total 2161 over-the-counter purchases are recorded. We restrict ourselves to households that both have a bank and a giro account, and to one transaction per household. This leaves us with 225 observations. We could have used all observations but that would complicate the statistical model because the number of observed purchases per household varies between 1 and 48, and the independence between observations which we make is hard to maintain if multiple transactions per household are recorded. We have selected randomly one transaction if more than one transaction is observed for a particular household. For each transaction the mode of payment is recorded, as well as the amount paid ($AMT$), size of the household ($SIZE$), age of the head of the household ($AGE$), gender of the person making the transaction ($SEX$) (this dummy is 1 for females and 0 for males) and household income ($INC$). All these variables are generic: they do not vary with any of the alternatives. The dataset does not contain any alternative-specific variables.

[3]Which will be translated as the Intomart Expenditure Index.

[4]This category consists of two subcategories, viz. Eurocheques and green bank cheques. All types of cheques are guaranteed.

Table 1: Distribution of modes of payment.

| | | | |
|---|---|---|---|
| Payment by cheque | | | 39 |
| | giro cheques | 20 | |
| | bank cheques | 19 | |
| Payment in cash | | | 186 |

The distribution of the observations over the mode of payment is given in table 1.

It is expected that cheques are used to pay larger amounts, which is confirmed by the data. The average amount per transaction paid in cash is Dfl. 33.30 and the average amount paid with giro cheques is Dfl. 131.20 and with bank cheques Dfl. 84.58.

We model the choice of mode of payment using two different functional forms for the choice probabilities. The first model we estimate is the multinomial logit model (MNL) (see Maddala [11]), where the choice probabilities are given by

$$P_i(v_t) = \frac{\exp(-v_{ti})}{\exp(-v_{t1}) + \exp(-v_{t2}) + \exp(-v_{t3})}, \quad i = 1,2,3. \tag{8}$$

It is easily established that the MNL choice probabilities satisfy the Daly–Zachary–Williams conditions (C–1) to (C–5) for all $v \in \mathcal{R}^3$, and hence, that they are globally compatible with stochastic utility maximisation, if only because of the *a priori* chosen functional form.

The distribution of $(\,\varepsilon_1 \quad \varepsilon_2 \quad \varepsilon_3\,)'$ which generates the choice probabilities (8) is

$$F(\varepsilon) = \exp\{-\exp(-\varepsilon_1) - \exp(-\varepsilon_2) - \exp(-\varepsilon_3)\}.$$

Note that $\varepsilon_1$, $\varepsilon_2$ and $\varepsilon_3$ are independently distributed.

We choose alternative 1 as the reference alternative, *i.e.* we set $v_{t1} = 0$ for all individuals, and our approximation to the indirect utility function in (7) is

$$v_{t2} = \beta_2' x_t, \tag{9}$$
$$v_{t3} = \beta_3' x_t. \tag{10}$$

The indirect utility of a particular alternative depends on the log of the amount of the transaction ($\ln AMT$), the size of the household ($SIZE$), the age of the head of the household ($AGE/10$), the log of household income ($\ln INC/10000$) and sex of the person making the transaction ($SEX$).

The estimation results for this specification are given in tables 2. Only the log amount in the cash-equation differs significantly from 0. Since all variables are generic, it is difficult to interpret the coefficients. The usual interpretation as a partial derivative is not valid since it is not possible to increase, say, household size in the equation for the utility of bank cheques while keeping it constant

Table 2: Estimation results, multinomial logit model (standard errors in parentheses).

| Parameter | Giro cheques | | Bank cheques | | Cash | |
|---|---|---|---|---|---|---|
| $\beta_0$ | – | – | −0.76 | (2.43) | −5.91 | (1.98) |
| ln *AMT* | – | – | 0.38 | (0.34) | 1.72 | (0.34) |
| *SIZE* | – | – | −0.14 | (0.26) | 0.014 | (0.21) |
| *AGE*/10 | – | – | 0.097 | (0.32) | −0.22 | (0.23) |
| *SEX* | – | – | 0.56 | (0.68) | −0.37 | (0.56) |
| ln *INC*/10000 | – | – | −0.95 | (0.66) | −0.73 | (0.55) |

Table 3: Quasi elasticities, multinomial logit model.

| Parameter | Giro cheques | | Bank cheques | | Cash | |
|---|---|---|---|---|---|---|
| *AMT* | 0.22 | 0.056 | 0.14 | 0.051 | −0.36 | −0.11 |
| *SIZE* | −0.012 | 0.00078 | 0.077 | 0.019 | −0.065 | −0.020 |
| *AGE*/10 | −0.089 | −0.031 | −0.15 | −0.056 | 0.24 | 0.087 |
| *SEX* | −0.037 | −0.018 | −0.19 | −0.059 | 0.22 | 0.077 |
| *INC* | −0.12 | −0.024 | 0.056 | 0.0098 | 0.062 | 0.015 |

for the other alternatives. Therefore, more information on the effect of the regressors is obtained from quasi-elasticities (the percentage-point change of a choice probability due to a 1-percent increase in a regressor). Exact formulae for the quasi-elasticities for this model and the nested multinomial logit model are available on request. The first column of each alternative in table 3 gives the quasi-elasticities evaluated in using the mean values of all variables of the payments by cheque (either bank cheque of giro cheque) and the second column uses the mean values of cash payments. From the quasi-elasticities we see that an increase of the amount of the transactions leads to an increase of payment by either type of cheque. Furthermore, older people tend to use cash to pay for transactions and an increase of income leads to an increase of payment by cash or bank cheque and to a decrease of payment by giro cheque.

The multinomial logit specification (8) is rather restrictive: it suffers from the 'independence of irrelevant alternatives' (IIA) property: the odds of paying with a bank cheque over paying cash is independent of the presence and characteristics of the third alternative, viz. paying with a giro cheque. Another drawback of the MNL model is that rationality of the choice is imposed *a priori*. A specification of the choice probabilities which recognises the fact that the first two alternatives are rather similar is the nested multinomial logit (NMNL) specification. In the nested multinomial logit specification the IIA-assumption

Table 4: Estimation results, nested multinomial logit model (standard errors in parentheses).

| Parameter | Giro cheques | | Bank cheques | | Cash | |
|---|---|---|---|---|---|---|
| $\beta_0$ | – | – | −1.74 | (2.60) | −0.90 | (2.84) |
| $\ln AMT$ | – | – | 0.61 | (0.46) | 0.78 | (0.53) |
| $SIZE$ | – | – | −0.33 | (0.31) | 0.51 | (0.30) |
| $AGE/10$ | – | – | 0.35 | (0.38) | −0.63 | (0.31) |
| $SEX$ | – | – | 0.79 | (0.77) | −1.76 | (0.78) |
| $\ln INC/10000$ | – | – | −1.40 | (0.76) | 1.58 | (1.18) |
| $\theta$ | | −2.54 | (1.50) | | | |

Table 5: Quasi elasticities, nested multinomial logit model.

| Parameter | Giro cheques | | Bank cheques | | Cash | |
|---|---|---|---|---|---|---|
| $AMT$ | 0.042 | −0.0057 | −0.19 | −0.073 | 0.15 | 0.079 |
| $SIZE$ | 0.32 | 0.13 | 0.20 | 0.083 | −0.51 | −0.21 |
| $AGE/10$ | −0.45 | −0.24 | −0.29 | −0.16 | 0.74 | 0.39 |
| $SEX$ | −0.40 | −0.20 | −0.29 | −0.14 | 0.69 | 0.35 |
| $INC$ | 0.31 | 0.13 | 0.16 | 0.076 | −0.47 | −0.21 |

can be tested because it corresponds to one parameter being equal to 1. The functional form for the choice probabilities of the nested multinomial logit model is:

$$P_i(v_t) = \frac{\exp(-v_{ti}/\theta)\Big\{\exp(-v_{t1}/\theta) + \exp(-v_{t2}/\theta)\Big\}^{(\theta-1)}}{\Big\{\exp(-v_{t1}/\theta) + \exp(-v_{t2}/\theta)\Big\}^{\theta} + \exp(-v_{t3})}, \qquad i = 1, 2,$$

$$P_3(v_t) = \frac{\exp(-v_{t3})}{\Big\{\exp(-v_{t1}/\theta) + \exp(-v_{t2}/\theta)\Big\}^{\theta} + \exp(-v_{t3})}. \tag{11}$$

The parameter $\theta$ is known as the association parameter. It is approximately equal to the correlation coefficient between $\varepsilon_1$ and $\varepsilon_2$. Again, we choose the first alternative as the reference alternative.

The estimation results of the nested multinomial logit model are given in tables 4 (point estimates) and 5 (quasi-elasticities). None of the variables are significantly different from 0. Again we see that older people tend to pay using cash and in this specification an increase of income leads to an increase of the probability of paying with either type of cheque.

From the choice probabilities in (11) it is seen that the NMNL logit model reduces to the MNL logit model if $\theta = 1$. The null hypothesis $H_0 : \theta = 1$ against the two-sided alternative $H_1 : \theta \neq 1$ rejected.

The estimation results clearly indicate that the choice between either type of cheque is difficult to model. Other specifications did not lead to an improvement of fit or significance of variables. In fact, one can wonder whether both types of cheques should not be grouped together into a single alternative. This problem is examined in Cramer and Ridder [3]. In a different specification, they firmly reject the null hypothesis that both types of cheques are perceived as equivalent alternatives.

We now turn to the question whether the estimated NMNL choice probabilities are compatible with stochastic utility maximisation.

## 8.4 Compatibility with Utility Maximisation

In this section, we examine whether the estimated NMNL model of the previous section is compatible with stochastic utility maximisation for three choices of the set of non-random utility components $\mathcal{V}$.

First, we examine whether the model is globally compatible with utility maximisation. If the choice probabilities are compatible, the density function of the random utility components $(\,\eta_1 \quad \eta_2\,)' = (\,\varepsilon_2 - \varepsilon_1 \quad \varepsilon_3 - \varepsilon_1\,)'$ which generates the choice probabilities (11) is given by

$$\begin{aligned} h_1(\eta_1, \eta_2) &= \exp(-\eta_2)\exp(-\eta_1/\theta)\frac{\{1+\exp(-\eta_1)\}^{(\theta-2)}}{\exp(-\eta_2)+\{1+\exp(-\eta_1/\theta)\}^{\theta}} \\ &\quad \times \left[\frac{2\{1+\exp(-\eta_1/\theta)\}^{\theta}}{\exp(-\eta_2)+\{1+\exp(-\eta_1/\theta)\}^{\theta}} - \frac{\theta-1}{\theta}\right] \\ &\quad \times \frac{1}{\exp(-\eta_2)+\{1+\exp(-\eta_1/\theta)\}^{\theta}}. \end{aligned}$$

It is easily checked that the NMNL-specification of the choice probabilities satisfies all conditions (C–1), (C–2), (C–3) and (C–5) for all values of $(\eta_1, \eta_2)$. Condition (C–4) is satisfied for all $(\eta_1, \eta_2) \in \mathcal{R}^2$ if and only if this density function is nonnegative. This density is nonnegative for all values $\eta \in \mathcal{R}^2$ if and only if $0 < \theta \leq 1$. Our estimate for $\theta$ is $-2.54$, so we conclude that the NMNL model is not globally compatible with utility maximisation, that is, it is not compatible for all values $v \in \mathcal{R}^3$.

One possible reason for rejection of rationality of choice is that we require the Daly–Zachary–Williams conditions to hold for *all* values of $v$, even if they are far away from the observed values. A less ambitious question is whether the model is compatible with utility maximisation on a closed interval $\mathcal{V}_2 = [a, b]$. According to theorem 1 in Koning and Ridder [10], the same Daly–Zachary–Williams conditions applied on an interval[5] are necessary and sufficient conditions for

[5]Except for (C–3) which becomes vacuous in the case of local compatibility.

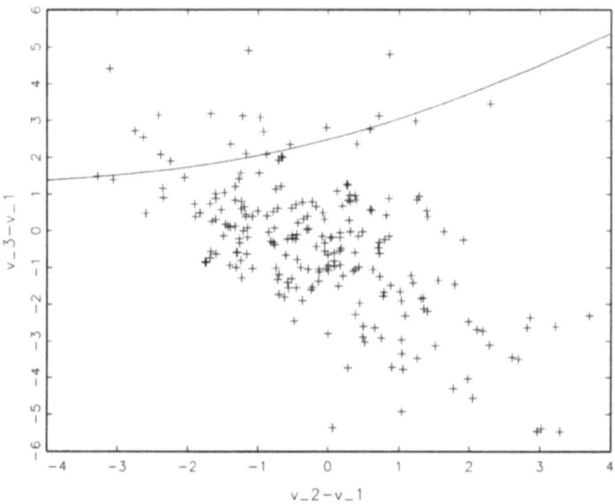


Figure 1: Estimated non-random utility components and the curve $h_1(\eta_1, \eta_2) = 0$.

compatibility with stochastic utility maximisation on that interval. If all observed non-random utility components lie in that interval, we can conclude that the model is compatible with utility maximisation on that interval.

We plot the curve $h_1(\eta_1, \eta_2) = 0$ and the estimated non-random utility components in figure 1. The curve is represented by the solid line and is given by

$$\eta_2 = -\ln\frac{\theta+1}{\theta-1} - \theta\ln\{1+\exp(-\eta_1/\theta)\}, \tag{12}$$

with $\theta < -1$ or $\theta > 1$.[6] For the line in figure 1 we have taken $\theta = -2.54$. $h_1(\eta_1, \eta_2)$ is negative beneath the curve.

A substantial number of observations are situated in the area where $h_1(\eta_1, \eta_2)$ is negative and hence the model is not compatible with utility maximisation on an interval.

Finally we check whether the estimated choice probabilities are compatible with utility maximisation on the finite set $\mathcal{V}_3 = \{v_1, \ldots, v_T\}$ of observed utility components.[7] Because of the computational complexity of examining the necessary and sufficient condition for compatibility on a finite set, we check a necessary condition for compatibility first. For all pairs of observations $(t, t')$ there must hold:

$$v_t - v_{it}\iota_3 \leq v_{t'} - v_{it'}\iota_3 \rightarrow P_i(v_t) \leq P_i(v_{t'}) \qquad i = 1, 2, 3. \tag{13}$$

[6]The function $h_1(\eta_1, \eta_2)$ is negative for all values $\eta \in \mathcal{R}^2$ if $-1 \leq \theta < 0$.

[7]In Koning and Ridder [9] an example is given where choice probabilities are not compatible on an interval but are on a finite set. Hence, this test is not superfluous.

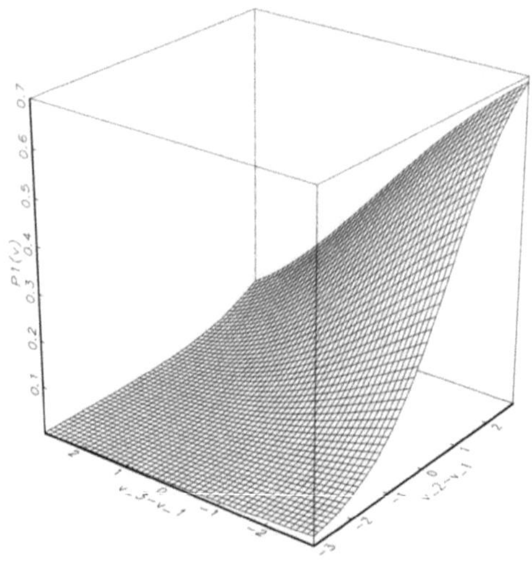


Figure 2: $P_1(v)$ with $\theta = -2.54$.

We have checked[8] condition (13) for the estimated NMNL model. The condition was violated in 6.02% of all comparisons. This means that the nested multinomial logit model is not locally compatible with stochastic utility maximisation on the set $\mathcal{V}_3$. There are observations where one alternative is more attractive than both other alternatives (as measured by $v$), but that alternative has a lower estimated choice probability.

This result raises the question whether one obtains a model compatible with utility maximisation if pairs of observations which violate the necessary condition (13) are excluded. We have re-estimated the nested multinomial logit model twice. First, we have deleted observations for which the necessary condition was violated more than 25 times, which leaves us with 201 observations. The estimate for $\theta$ in this case is $-1.46$. This estimate is increased further if we select only those observations which violate condition (13) 15 times or less. In the latter case, we find $\hat{\theta} = -0.83$. A natural step would be to select only those observations which do not violate condition (13) at all. This, however, leaves us with too few observations to estimate the parameters of the model with any reasonable degree of accuracy.

The choice probability of the first alternative is drawn in figures 2–4. We see that the surface that fits the data best (figure 2) decrease in $v_3$. Fixing the utility of the second alternative, one sees that the probability of choosing the first alternative decreases even though it becomes more attractive when compared to the third alternative. This, of course, reflects the rejection of compatibility.

In figure 3 the same surface is shown for $\theta = -0.83$. $P_1(v)$ decreases even more clearly in $v_3 - v_1$. For comparison, $P_1(v)$ is drawn in figure 4 for $\theta = 0.50$.

[8]Details are given in Appendix 8.6.

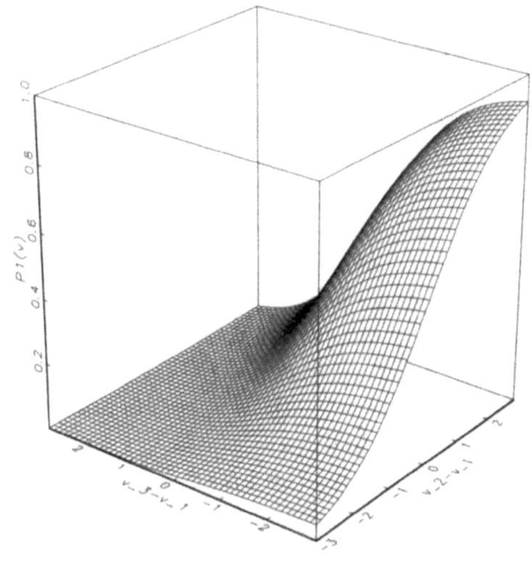


Figure 3: $P_1(v)$ with $\theta = -0.83$.

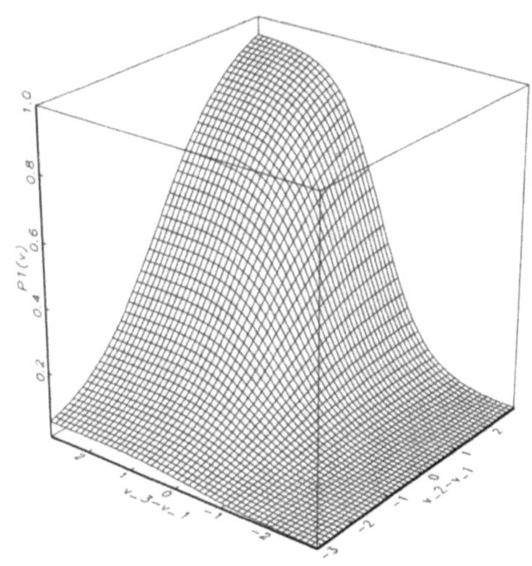


Figure 4: $P_1(v)$ with $\theta = 0.50$.

The NMNL model is compatible with stochastic utility maximisation for this value of $\theta$.

## 8.5 Semiparametric Estimation of the Choice Model

So far, we have examined compatibility of the choice of mode of payment with stochastic utility maximisation using a flexible parametric form for the choice probabilities. An alternative to this approach is to estimate the choice probabilities semiparametrically, we will pursue this approach in this section. The semiparametric method we use does not impose any restrictions on the choice probability $P(\cdot)$ but it maintains linearity (in explanatory variables) of the utility index. An advantage of a non-parametric approach to the estimation of $P(\cdot)$ is that this allows us to test whether $P(\cdot)$ is nondecreasing in its argument more directly. Rejection of stochastic utility maximisation cannot be due to misspecification of the functional form of the choice probabilities. Note that most semi-parametric methods for the binary choice model impose compatibility with utility maximisation by restricting the distribution function of $\varepsilon_2 - \varepsilon_1$ to be nondecreasing in its argument (Cosslett [2], Klein and Spady [6]).

We will use a particular attractive semiparametric technique—average derivative estimation—to estimate the functional form of the choice probabilities. Because of the relatively small sample size and the complexity of a choice model with three alternatives, we will estimate a choice model for the decision to pay by cheque or not.

Let the utility of paying with a cheque be $u_1 = -v_1 + \varepsilon_1$ and the utility of paying cash $u_2 = -v_2 + \varepsilon_2$ and let $D$ be a dummy with value 0 if transaction is paid with a cheque and 1 if it is paid cash. The probability of paying cash is now:

$$P(D = 1) = P(\varepsilon_1 - \varepsilon_2 \leq v_1 - v_2) = F(v_1 - v_2). \tag{14}$$

In (14), $F$ is the distribution function of $\varepsilon_1 - \varepsilon_2$. The expectation of $D$ is, of course, equal to $P(D = 1) = F(v_1 - v_2)$. If we let the first alternative be the reference alternative ($v_1 = 0$), and parametrise $v_2$ as $\beta' x$, we obtain a so-called single-index model:

$$ED = P(D = 1) = G(\beta' x). \tag{15}$$

$E$ denotes the expectation operator, the term $\beta' x$ is known as the index and $G(\cdot)$ is the link function between the expectation of $D$ and the index. If the hypothesis of stochastic utility maximisation is correct, this function is equal to the distribution function $F$ of $\varepsilon_1 - \varepsilon_2$ and hence, it is nondecreasing in its argument.

We have estimated model (15) using the average derivative approach, as discussed by Stoker [16] and [17]. The average derivative in model (15) is defined as the expectation of

$$m(x) = \frac{\partial E(D \mid x)}{\partial x} = G'(\beta' x)\beta \tag{16}$$

Table 6: Estimation results, single index model and logit model (standard errors in parentheses)

| Parameter | Single index | | Logit | |
|---|---|---|---|---|
| $\beta_0$ | – | – | 2.06 | (0.25) |
| $\ln AMT$ | –0.19 | (0.031) | –1.33 | (0.24) |
| $AGE$ | 0.097 | (0.050) | 0.41 | (0.26) |
| $\ln INC$ | 0.068 | (0.052) | 0.064 | (0.22) |

with respect to the distribution of $x$:

$$\delta = E_x m(x) = E_x G'(b'x)\beta. \tag{17}$$

In the sequel we will redefine $G$ such that $EG' = 1$ so that the average impact of a one unit change in the index $\beta'x$ results in a one unit change in the mean of $D$. This method is consistent for both the parameters and the function $G$ if the true model is of the form (15), $m(\cdot)$ is twice differentiable, the regressors $x$ are continuous, and if some boundedness and regularity conditions hold. Estimation proceeds in two steps. In the first step a scaled version of the vector of coefficients $\beta$ is estimated (with its standard errors). Using the estimated value of $\beta$ we regress nonparametrically $D$ on the estimated index $\hat{\beta}'x$—see equation (15). We estimated $\delta$ by the indirect slope estimator (see Stoker [17]). The point estimates for $\delta$ are given in table 6. For comparison we give the estimates of a logit model as well. Because the scale of $\delta$ is not identified, the estimates of the single-index model are not directly comparable to the estimates of the logit model, only their relative values are. The estimate[9] of $G$ is drawn in figure 5. The bandwidth used in estimating $\delta$ is 0.065 and we used that same bandwidth in the estimation of $G$. To examine sensitivity to the choice of bandwidth, $G$ was estimated with a smaller bandwidth ($h = 0.032$) and a much larger bandwidth ($h = 0.129$).

In table 6 we see that the qualitative conclusions drawn form the single index model estimates and the logit model coincide: increasing the amount of the transaction decreases the probability of paying it in cash and older and people with higher incomes tend to pay in cash. From figure 5 it is clear that the hypothesis of stochastic utility maximisation is rejected: $\hat{G}$ is decreasing on intervals. In order not to be distracted by two spikes for very small values of the estimated index and some erratic behaviour for very large values of the index we deleted the 10 smallest and 10 largest values of the index. For the remaining observations we see that the estimated choice function is not monotonic in its argument, even if we oversmooth ($h = 0.129$). The choice of mode of payment is not compatible with stochastic utility maximisation even with this very flexible functional form for the choice probability.

[9] $G$ was estimated using the Nadaraya–Watson kernel regression estimator with a biweight kernel.

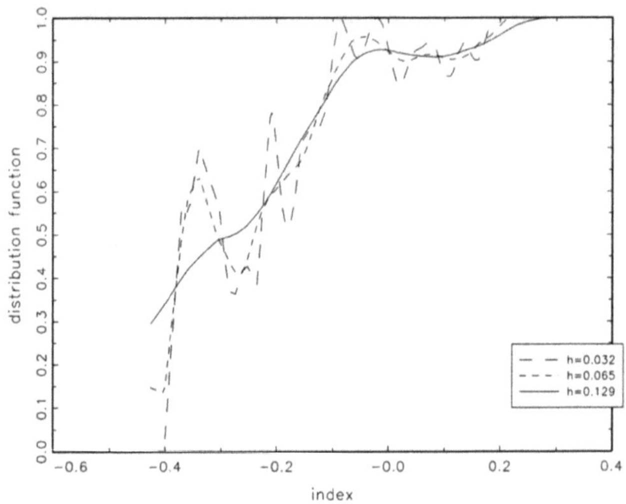


Figure 5: Nonparametric regression of choice (cash payment=1) on index.

## 8.6 Conclusion

In this chapter we have used the theory developed in Koning and Ridder [9] and Koning and Ridder[10] to examine whether the choice of mode of payment is rational. Rationality can be imposed by specifying a specific functional form for the choice probabilities that satisfies the Daly–Zachary–Williams conditions, as the MNL model. This model, however, is rejected against a more flexible functional form for the choice probabilities, viz. the NMNL model. The NMNL model is not guaranteed to be compatible with stochastic utility maximisation; indeed we reject rationality for three choices of sets of non-random utility components.

Of course, the nested multinomial logit model might be rejected against an even more flexible functional form for the choice probabilities. Methods for semiparametric estimation of discrete choice models with more than two alternatives that impose no restrictions on the functional form of the choice probabilities are rather complex. For the simpler case of only two alternatives (paying with a cheque or paying cash) we estimated the model using average derivative estimation. The estimated choice probabilities were clearly incompatible with stochastic utility maximisation.

## Appendix A: Descriptive Statistics

Table 7: Descriptive statistics.

| Variable | mean | standard dev. | min | max |
|---|---|---|---|---|
| all cases ($n = 225$) | | | | |
| *AMT* | 46.33 | 88.70 | 2.00 | 849.00 |
| *SIZE* | 3.08 | 1.35 | 1.00 | 7.00 |
| *AGE* | 43.03 | 14.17 | 20.00 | 83.00 |
| *SEX* | 0.61 | 0.49 | 0.00 | 1.00 |
| *INC* | 58.27 | 35.39 | 12.00 | 126.00 |
| giro cheque ($n = 20$) | | | | |
| *AMT* | 131.20 | 184.08 | 18.00 | 765.00 |
| *SIZE* | 3.15 | 1.27 | 1.00 | 5.00 |
| *AGE* | 38.15 | 12.10 | 25.00 | 66.00 |
| *SEX* | 0.55 | 0.5 | 0.00 | 1.00 |
| *INC* | 50.85 | 33.52 | 25.50 | 126.00 |
| bank cheque ($n = 19$) | | | | |
| *AMT* | 84.58 | 84.44 | 19.00 | 300.00 |
| *SIZE* | 3.47 | 1.22 | 1.00 | 5.00 |
| *AGE* | 34.74 | 8.63 | 20.00 | 56.00 |
| *SEX* | 0.42 | 0.51 | 0.00 | 1.00 |
| *INC* | 67.66 | 41.23 | 25.50 | 126.00 |
| cash ($n = 186$) | | | | |
| *AMT* | 33.30 | 65.21 | 2.00 | 849.00 |
| *SIZE* | 3.03 | 1.38 | 1.00 | 7.00 |
| *AGE* | 44.40 | 14.46 | 20.00 | 83.00 |
| *SEX* | 0.64 | 0.48 | 0.00 | 1.00 |
| *INC* | 58.11 | 34.94 | 12.00 | 126.00 |

## Appendix B: Check of the Necessary Condition

We have written a computer program[10] which checks condition (13) for all different pairs of points. In the case with three alternatives, every two observations imply exactly two restrictions of the form (13) on the choice probabilities. This is illustrated in figure 6. Consider for example one observation $t$. The other observation $t'$ lies in either one of the areas $A_1, A_2, \ldots, A_6$ and each case implies two restrictions on the choice probabilities, see table 8.

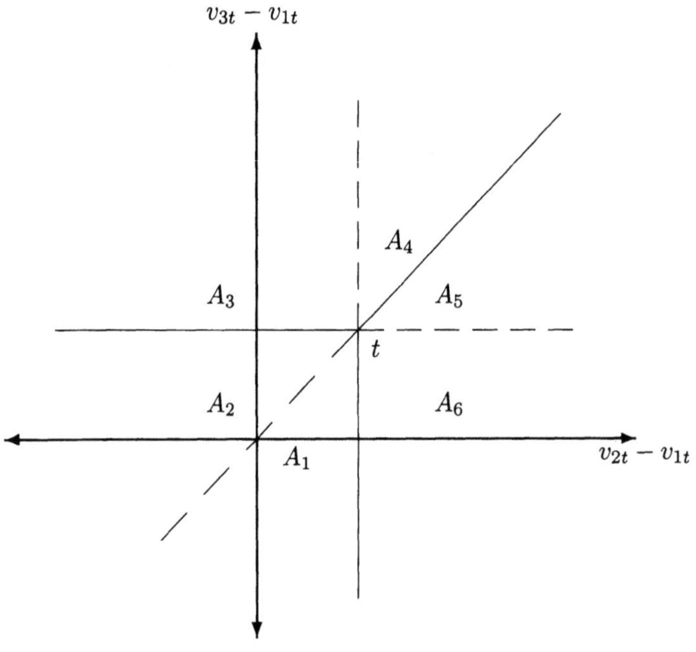


Figure 6: Two restrictions per observation

Table 8: Restrictions on choice probabilities.

| | Restriction on probabilities | |
|---|---|---|
| $t' \in A_1$ | $P_1(v_t) \geq P_1(v_{t'})$ | $P_3(v_t) \leq P_3(v_{t'})$ |
| $t' \in A_2$ | $P_1(v_t) \geq P_1(v_{t'})$ | $P_2(v_t) \leq P_2(v_{t'})$ |
| $t' \in A_3$ | $P_3(v_t) \geq P_3(v_{t'})$ | $P_2(v_t) \leq P_2(v_{t'})$ |
| $t' \in A_4$ | $P_3(v_t) \geq P_3(v_{t'})$ | $P_1(v_t) \leq P_1(v_{t'})$ |
| $t' \in A_5$ | $P_2(v_t) \geq P_2(v_{t'})$ | $P_1(v_t) \leq P_1(v_{t'})$ |
| $t' \in A_6$ | $P_2(v_t) \geq P_2(v_{t'})$ | $P_3(v_t) \leq P_3(v_{t'})$ |

[10]This program is written in the matrix programming language GAUSS and is available on request.

## References

[1] Börsch-Supan, A., (1990), On the Compatibility of Nested Logit Models with Utility Maximisation, *Journal of Econometrics,* 43, 371–387.

[2] Cosslett, S.R., (1983), Distribution-free Maximum Likelihood Estimator of the Binary Choice Model, *Econometrica,* 51, 765–782.

[3] Cramer, J.S., and G. Ridder, (1991), Testing for Equivalent Alternatives in Discrete Choice Models, Unpublished manuscript, Groningen University.

[4] Daly, A.J., and S. Zachary, (1979), Improved Multiple Choice Models, pp. 337–357 in D.A. Hensher and Q. Dalvi (eds.), *Determinants of Travel Choice,* London, Teakfield.

[5] Deaton, A., and J. Muellbauer, (1980), *Economics and Consumer Behaviour,* Cambridge, Cambridge University Press.

[6] Klein, R.W., and R.H. Spady, (1993), An Efficient Semiparametric Estimator for Binary Response Models, *Econometrica,* 61, 387–421.

[7] Kling, C.L., and J.A. Herriges, (1995), An Empirical Investigation of the Consistency of Nested Logit Models with Utility Maximisation, *American Journal of Agricultural Economics,* 77, 875–884.

[8] Koning, R.H., (1995), *Essays on Applied Microeconometrics,* Capelle aan de IJssel: Labyrint Publication.

[9] Koning, R.H., and G. Ridder, (1991), Discrete Choice and Stochastic Utility Maximisation, Research Memorandum 414, Department of Economics, Groningen University.

[10] Koning, R.H., and G. Ridder, (1994), On the Compatibility of Nested Logit Models with Utility Maximisation, *Journal of Econometrics,* 63, 389–396.

[11] Maddala, G.S., (1983), *Limited-dependent and Qualitative Variables in Econometrics,* Cambridge, Cambridge University Press.

[12] Matzkin, R.L., (1993), Nonparametric Identification and Estimation of Polychotomous Choice Models, *Journal of Econometrics,* 58, 137–168.

[13] Matzkin, R.L., (1994), Restrictions on Economic Theory in Nonparametric models, in R.F. Engle and D.L. McFadden (eds.), *Handbook of Econometrics, Volume IV*, Amsterdam, pp. 2523–2558, North-Holland Publishing Co..

[14] McFadden, D., (1981), Econometric Models of Probabilistic Choice, in C.F. Manski and D. McFadden (eds.), *Structural Analysis of Discrete Data with Econometric Applications,* pp. 198–272. Cambridge, MIT Press.

[15] Mot, E.S., J.S. Cramer, and E.M. van der Gulik, (1989), Choice of Mode of Payment, Rapport 228, Stichting voor Economisch Onderzoek, In Dutch.

[16] Stoker, T.M., (1991), Equivalence of Direct, Indirect and Slope Estimators of Average Derivatives, in W.A. Barnett, J.L. Powell, and G. Tauchen (Eds.), *Nonparametric and Semiparametric Methods in Econometrics and Statistics,* Cambridge, Cambridge University Press.

[17] Stoker, T.M., (1992), *Lectures on Semiparametric Econometrics,* Louvain-la-Neuve, CORE Foundation.

[18] Williams, H., (1977), On the Formation of Travel Demand Models and Econometric Evaluation Measures, *Environment and Planning A,* 9, 285–344.

CHAPTER 9

# Gibbs Sampling in B-VAR Models with Latent Variables

Wolfgang POLASEK

## 9.1 Introduction

This chapter analyses different types of AR (autoregressive) and VAR (vector autoregressive) models with a latent variables or errors-in-variables (eiv) structure from a Bayesian perspective. A hierarchical prior distribution which imposes constraints in the form of tightness (or smoothness) on the lag distribution is assumed. To find the posterior distributions of the hyperparameters, a Gibbs sampling approach is proposed. All the full conditional distributions (f.c.d.s) which are necessary for the efficient numerical calculation of the posterior distribution are derived. First, the single and multiple tightness models are analysed and then the extension to a VAR model is given. The simulated posterior distribution also allows for the simulation of the predictive distribution.

Because the forecasting properties have been found to outperform those of the corresponding classical models, Bayesian VAR models have been used increasingly for macroeconomic forecasting in the last decade. Since most economic time series are observed with errors, there is increasing demand for time series methods (classical or Bayesian), like AR or VAR models, with an errors-in-variables or latent variable structure.

The assumption of a tightness or smoothness structure for the lag coefficients implies a hierarchy of prior parameters, also called hyperparameters. In the approach used by Litterman (1986) the hyperparameters for the tightness model are fixed by the modeller. Using a full Bayesian specification, we can find the posterior distribution of all parameters by the Gibbs sampler as in Gelfand and Smith (1990), a simulation which utilises all full conditional distributions. In a conjugate framework, the full conditional distributions can be derived in closed form which also allows an efficient generation of random numbers.

We will show that the Gibbs sampler can be used in a similar way as in the tightness models to find the posterior distribution of all parameters. The Gibbs sampler has been applied to time series models by Chib (1993), Marriot et al.

(1992), and McCulloch and Tsay (1991). Errors in variables models have been analysed from a Bayesian perspective by Lindley and ElSayyad (1968), Leamer (1978, 1987), and Klepper and Leamer (1984). A Gibbs sampling perspective was given in Polasek and Krause (1992), and in Polasek (1993).

Classical eiv treatments for time series are given in Maravall and Aigner (1977). Observation rules for latent variables can be quite different, but it has been demonstrated that the Gibbs sampler can treat such models quite easily—see Albert and Chib (1993). Assuming an informative prior distribution for the eiv component in the VAR model, one can avoid the classical identification problem. The effect of a possible identification problem can be seen indirectly. If the data are non-informative for certain parameters, then the posterior distribution does not differ too much from the prior. More discussion on the Bayesian identification problem can be found in Leamer (1978).

The plan for the paper is as follows: In the second section, we develop the "single", i.e. simple univariate tightness model, including a partial tightness model, and the multiple tightness model with more than one time series as regressor. The third section discusses the latent variable or errors-in-variables AR tightness with an exogenous variables (BEV-ARX) model. In section 4, we give the multivariate extension to the BEV-VAR model with smoothness prior. The calculus we propose is general enough to cover further extensions of the eiv model, like, for example, the Chib (1993) regression model with AR errors. A final section discusses some related topics.

## 9.2 The AR($p$) Model with Latent Variables

The Bayesian univariate latent variable (or Bayesian eiv AR model) BEV-AR($p$) model is a one-equation model with one dependent time series on the left and the past of the same time series on the right. If there is more than one time series with measurement errors on the right hand side of the regression, then we call this approach multiple latent variable or errors-in-variables or simply the "BEV-ARX model". These models will be analysed in section 3.

### 9.2.1 The Univariate Latent Variables AR($p$) Model with Tightness Prior

The univariate latent variables tightness or univariate Bayesian eiv model (BEV-AR($p$) model) can be formulated as an extension of the Bayesian AR model—see Polasek (1993). Let $X = (1_T, \xi_{-1}, \ldots, \xi_{-p})$, where each $\xi_{-i}$ represents a lagged latent time series with lag $i$, where the last $i$ elements are eliminated at the end and, instead, $i$ unknown starting values (parameters) are inserted at the beginning of the $T \times 1$ vector. This leads to the basic regression model $\xi = Xb + \epsilon$; and the total model has the following structure of a BEV-AR($p$) model:

$$x \sim N(\xi, \phi I_T), \qquad \xi \sim N(Xb, \sigma^2 I_T), \tag{2.1}$$

$$b \sim N(b*, \lambda H*), \qquad \sigma^{-2} \sim G(s_*^2, n_*),$$
$$\lambda^{-1} \sim G(\lambda^*, l^*), \qquad x_- \sim N(x_-^*, \Psi^*), \qquad \phi^{-1} \sim G(\phi^*, \nu^*).$$

The idea of the BEV-AR model (2.1) is simple: The underlying AR process is driven by the latent variables $\{\xi_t\}$, but we observe only the process $\{x_t = \xi_t + v_t\}$, i.e. the latent variable $\xi_t$ plus an error term $v_t$. The latent variables $\xi_t$ are assumed to be recursively generated from the starting values $x_-$. Note that $1_T$ in model (2.1) is a vector of ones of length $T$ and $G$ stands for the gamma-2 distribution as in Leamer (1978).

The variance $\phi$ of the error term is assumed to be constant for all time points. For the time series model, we assume a tightness structure which was analysed for AR($p$) models in Polasek and Jin (1994). The starting values of an AR($p$) process are denoted by $x_-$ and can be incorporated in the Gibbs sampling simulation. Note that the last line in model (2.1) formulates the prior information for the errors-in-variables parameters. All parameters having a star (*) are known hyperparameters of the prior distribution and have to be provided by the user.

Here are some hints for finding good, partially data-based starting values. The Gibbs sampler can be invoked with the starting values based on simple OLS estimates: $b_*^{(0)} = b_{OLS} = (X'X)^{-1}X'x$, where we impose the following conditions on the $p$ starting values, $\sigma_*^{2\,(0)} = \sigma_{OLS}^2$ and $\lambda^{(0)} = (b_{OLS} - b_*)'H_*^{-1}(b_{OLS} - b_*)$. The $(p+1) \times (p+1)$ prior covariance matrix in this tightness model is specified as $H_* = \text{diag}(d_{*0}, D_*)$, where $D_*$ is the known diagonal $p \times p$ matrix $D_* = \text{diag}(1, 1/2, \ldots, 1/p)$ and $d_{*0}$ is any prior (positive) variance for the intercept. We also use the tightness matrix $\Psi_* = D_*$ for the covariance matrix of the starting parameters. (The diagonal structure is justified if we assume a white noise prior for $b$.) The tightness structure on the starting parameters can be viewed as a compromise between a diffuse prior and an AR model where the starting parameters are 0.

The joint distribution of the time series $x$ and the parameters

$$\theta = (b, \sigma^{-2}, \lambda, \phi, \xi, x_-)$$

of the BEV-AR model is

$$\begin{aligned} p(x, \theta) &= N(x \mid \xi, \phi I_T) N(\xi \mid Xb, \sigma^2 I_n) N(b \mid b_*, \lambda H_*) \\ &\quad \times G(\sigma^{-2} \mid \sigma_*^2, n_*) G(\phi^{-1} \mid \phi_*, \nu_*) \\ &\quad \times G(\lambda^{-1} \mid \lambda_*, l_*) N(x_- \mid x_-^*, \Psi^*). \end{aligned} \tag{2.2}$$

The full conditional distribution (f.c.d.)

(a) for the regression coefficients is

$$p(b \mid x, \theta^c) = N(b \mid b_{**}, H_{**}), \tag{2.3}$$

(where $\theta^c$ denotes all parameters in $\theta$ without $b$), and it follows a normal distribution with the parameters

$$H_{**}^{-1} = \lambda^{-1} H_*^{-1} + \sigma^{-2} X'X,$$

$$b_{**} = H_{**}(\lambda^{-1}H_*^{-1}b_* + \sigma^{-2}X'\xi).$$

(b) The f.c.d. for the residual precision is

$$\begin{aligned} p(\sigma^{-2} \mid x, \theta^c) &\propto (\sigma^{-2})^{T/2} \exp\{-\frac{1}{2\sigma^2}(\xi - Xb)'(\xi - Xb)\} \\ &\quad \times (\sigma^{-2})^{n_*/2-1} \exp\{-\frac{1}{2\sigma^2}n_* s_*^2\} \\ &\propto G(\sigma^{-2} \mid \sigma_{**}^2, n_{**}), \end{aligned} \tag{2.4}$$

a gamma distribution with the parameters $n_{**} = n_* + T$ and

$$n_{**}\sigma_{**}^2 = n_*\sigma_*^2 + (\xi - Xb)'(\xi - Xb).$$

(c) For the latent variable $\xi$ of the time series

$$\begin{aligned} p(\xi \mid x, \theta^c) &\propto \exp\{-\frac{1}{2\sigma^2}(\xi - Xb)'(\xi - Xb)\} \\ &\quad \times \exp\{-\frac{1}{2}\phi(x - \xi)'(x - \xi)\} \\ &\propto N(\xi \mid \xi_{**}, G_{**}), \end{aligned} \tag{2.5}$$

we derive a normal distribution with the parameters

$$G_{**}^{-1} = \phi^{-1}I_T + \sigma^{-2}P_1'P_1, \quad \xi_{**} = G_{**}(\phi^{-1}x + \sigma^{-2}P_1'a),$$

where the location vector $a$ is given by $a = \alpha 1_T + A_0 x_-$ and $P_1 = I_T - A_1$. The $T \times (T + p)$ matrix $A$ is partitioned into $A = (A_0, A_1)$, i.e. a $T \times p$ block $A_0$ and a $T \times T$ block $A_1$:

$$A = \begin{pmatrix} b_p & b_{p-1} & \dots & b_1 & 0 & \dots & 0 \\ 0 & b_p & \dots & 0 & b_1 & \dots & 0 \\ \vdots & \vdots & \ddots & \vdots & \vdots & \ddots & \vdots \\ 0 & 0 & \dots & b_p & b_{p-1} & \dots & b_1 \end{pmatrix}, \tag{2.6}$$

where the $p$ diagonal bands are built up by the regression coefficient vector $b$.

**Proof.** The basic eiv model with the starting values collected in the $p \times 1$ vector $x_- = (\xi_{-p}, \xi_{-p-1}, \dots, \xi_{-1})'$ (note the time ordering) has the form

$$\xi = Xb + \epsilon = \alpha 1_T + \begin{pmatrix} \xi_{-1} & \xi_{-2} & \dots & \xi_{1-p} & \xi_{-p} \\ \xi_1 & \xi_{-1} & \dots & \xi_{2-p} & \xi_{1-p} \\ \vdots & \vdots & \ddots & \vdots & \vdots \\ \xi_{p-1} & \xi_{p-2} & \dots & \xi_1 & \xi_{-1} \\ \vdots & \vdots & \ddots & \vdots & \vdots \end{pmatrix} \begin{pmatrix} b_1 \\ b_2 \\ \vdots \\ b_{p-1} \\ b_p \end{pmatrix} + \epsilon.$$

To obtain the f.c.d., we have used the following decomposition of the eiv regression model:

$$\xi = \alpha 1_T + A\xi_0 + \epsilon,$$

where $\xi_0' = (x_-', \xi')$ and the $T \times (T+p)$ coefficient matrix $A$ is given as in (2.6). We need a further decomposition of $A = (A_0, A_1)$ into the $(T \times p)$ block $A_0$ and the $(T \times T)$ matrix $A_1$:

$$\begin{aligned} \xi &= \alpha 1_T + A_0 x_- + A_1 \xi + \epsilon, \qquad \text{or} \\ (I_T - A_1)\xi &= a + \epsilon. \end{aligned} \tag{2.7}$$

Now we can apply the usual combination of quadratic forms to obtain the full conditional distribution.

(d) The f.c.d. for the variance of the latent variables $\phi$ is

$$\begin{aligned} p(\phi^{-1} \mid x, \theta^c) &\propto |\phi^{-1}|^{T/2} \exp\{-\frac{1}{2\phi}\sum_{t=1}^{T}(x_t - \xi_t)^2\} \\ &\quad \times (\phi^{-1})^{\nu_*/2-1} \exp\{-\frac{1}{2}\phi^{-1}\phi_*\nu_*\} \\ &\propto G(\phi \mid \phi_{**}, \nu_{**}), \end{aligned} \tag{2.8}$$

a gamma distribution with the parameters $\nu_{**} = \nu_{**} + T$ and

$$\nu_{**}\phi_{**} = \nu_*\phi_* + \sum_{t=1}^{T}(x_t - \xi_t)^2.$$

(e) The f.c.d. for the tightness parameter $\lambda$ is

$$p(\lambda^{-1} \mid x, \theta^c) = G(\lambda \mid \lambda_{**}, l_{**}),$$

a gamma distribution with the parameters $l_{**} = l_* + K$ and

$$l_{**}\lambda_{**} = l_*\lambda_* + (b - b_*)'D^{-1}(b - b_*).$$

(f) The f.c.d. for the distribution of the starting parameters $x_-$ is given by

$$p(x_- \mid x, \theta^c) = N(x_-^{**}, \Psi_{**}), \tag{2.9}$$

a normal distribution with the hyperparameters

$$\begin{aligned} \Psi_{**}^{-1} &= \Psi_*^{-1} + \sigma^{-2}A_0'A_0, \\ y_-^{**} &= \Psi_{**}\{\Psi_*^{-1}x_-^* + \sigma^{-2}A_0'(\xi_0 - a_0)\}. \end{aligned}$$

The first $p$ latent variables of the time series are collected in $\xi_0$. Furthermore, $a_0 = X_0 b + a 1_p$. and the other matrices are given by

$$X_0 = \begin{pmatrix} 0 & 0 & \dots & 0 & 0 \\ \xi_1 & 0 & \dots & 0 & 0 \\ \xi_2 & \xi_1 & \dots & 0 & 0 \\ \vdots & \vdots & \ddots & \vdots & \vdots \\ \xi_{p-1} & \xi_{p-2} & \dots & \xi_1 & 0 \end{pmatrix} \quad \text{and} \tag{2.10}$$

$$A_0 = \begin{pmatrix} b_p & b_{p-1} & \dots & b_2 & b_1 \\ 0 & b_p & \dots & b_3 & b_2 \\ \vdots & \vdots & \ddots & \vdots & \vdots \\ 0 & 0 & \dots & b_p & b_{p-1} \\ 0 & 0 & \dots & 0 & b_p \end{pmatrix}.$$

**Proof.** We write the eiv regression model in (2.7) as $\xi = a_0 + A_0 x_- + \epsilon$, with $a_0 = \alpha 1_T + A_1 \xi$. Then the f.c.d. can be written as

$$\begin{aligned} p(x_- \mid x, \theta^c) \quad &\propto \quad \exp\{-\frac{1}{2\sigma^2}(\xi_0 - a_0 - A_0 x_-)'(\xi_0 - a_0 - A_0 x_-)\} \\ &\quad \times \exp\{-\frac{1}{2}(x_- - x_-^*)'\Psi_*^{-1}(x_- - x_-^*)\}. \end{aligned}$$

Since the first quadratic form contains many zeros, one can work only with the non-zero part. Thus, the above transformation can be written as

$$(\xi_0 - a_0 - A_0 x_-)'(\xi_0 - a_0 - A_0 x_-) = (\xi_0 - X_0 b)'(\xi_0 - X_0 b),$$

where $a_0, A_0$ and $X_0$ are given as above.

In the next section, we will generalise this approach to $K$ time series.

## 9.3 The Multiple ARX($p$) Model with Latent Variables

The multiple latent variables tightness model with $K$ latent time series $\xi_1, \dots, \xi_K$ and with normal errors can be formulated as

$$\xi_1 = X_1 b_1 + \dots + X_K b_K + \epsilon, \tag{3.1}$$

where the first block $X_1 = (1_T, \xi_{1,-1}, \dots, \xi_{1,-p})$ contains in the first column the constant and the other blocks, $X_k = (\xi_k, \xi_{k,-1}, \dots, \xi_{k,-p})$ for $k = 2, \dots, K$, contain the current and the lagged values of the latent time series $\xi_1$ to $\xi_K$. Equation (3.1) can be written in compact form as $\xi_1 = Zb + \epsilon$ or in distributional form as

$$\xi_1 \sim N(Zb, \sigma^2 I_T) \quad \text{with} \quad Z = (X_1, X_2, \dots, X_K). \tag{3.2}$$

Let $b = \text{vec}B = (b_1, \dots, b_K)$ be the stacked coefficient vector for $K$ time series responses, where $B$ is a $(p+1) \times K$ coefficient matrix. The priors for $b$ and $\sigma^{-2}$ are a normal and a gamma distribution, respectively:

$$b \sim N(b_*, D_\lambda \otimes H_*), \quad \sigma^{-2} \sim G(\sigma_*^2, n_*).$$

Let the $(p+1) \times (p+1)$ prior covariance matrix in this tightness model be the block-diagonal matrix $H_* = \text{diag}(d_{*0}, D_*)$, where $D_*$ is a known $p \times p$ matrix $D_* = \text{diag}(1, 1/2, \dots, 1/p)$. By specifying the precision matrix $H_{k*}^{-1} = \{\text{diag}(d_{*k}, D_*)\})^{-1} = \text{diag}(d_{*k}^{-1}, 1, 2, \dots, p)$ and by setting $d_{*k}^{-1} = 0$ for $k = 1, 2, \dots, K$, one could also use a non-informative prior for the intercept in the first block and the level responses in the remaining blocks. Assume a Wishart prior distribution for the $K \times K$ matrix of tightness coefficients $D_\lambda^{-1}$ (which might be a diagonal matrix):

$$D_\lambda^{-1} \sim W_p(D_\lambda^*, \nu_*).$$

For the $pK$ starting values, we assume the matrix normal distribution to be

$$X_- \sim N_{p\times K}(X_-^*, D_{*\lambda} \otimes \Psi_*),$$

where $X_-^*$ is the known mean and $D_{*\lambda} \otimes \Psi_*$ the known covariance structure of the starting parameters. Setting $X_-^* = 0$ and $D_{*\lambda} = 0$ implies the widely used customs of setting the starting values to zero.

The observation rule for the latent time series matrix $\Xi = (\xi_1, \xi_2, \dots, \xi_K)$ is, in our case, simply

$$X \sim N_{T\times K}(\Xi, \phi \otimes I_T), \quad \phi^{-1} \sim W_k(\phi_*, \nu_*).$$

Compactly formulated, the multiple tightness BEV-ARX($p$) model has the following hierarchical linear model structure:

$$\begin{array}{ll} X \sim N_{T\times K}(\Xi, \phi \otimes I_T), & \xi_1 \sim N(Zb, \sigma^2 I_T), \\ b \sim N(b_*, H_*), & \sigma^{-2} \sim G(s_*^2, n_*), \\ D_\lambda^{-1} \sim W_p(D_\lambda^*, l_*), & X_- \sim N_{p\times K}(X_-^*, D_{*\gamma} \otimes \Psi_*), \\ \xi_k \sim N(\xi_{*k} 1_T, \tau_{*k} I_T), k = 2, \dots, K; & \phi^{-1} \sim W_K(\phi_*, \nu_*). \end{array} \tag{3.3}$$

To facilitate the notation and the subsequent analysis, we assume that each latent variable comes from a (independent) normal distribution with constant mean and variance. Therefore we can write the prior distribution for the latent variables compactly as

$$\Xi \sim N_{T\times K}(1_T \xi_*', \tau_* \otimes I_T) \quad \text{with} \quad \tau_*^{-1} = \text{diag}(0, \tau_*^2, \dots, \tau_*^{\text{K}}),$$

where the $\tau_*^k$ are known prior precisions (inverse variances). Note that a prior distribution of the form

$$\text{vec}\Xi \sim N_{T\times K}\{\text{vec}(Zb : 1_T \xi_*'), \tau_* \otimes I_T\}$$

is a more complicated alternative, since it assumes a known variance matrix $\tau_*$ or a further prior distribution for $\tau_*$ (see Polasek (1994) for this possibility).

The BEV-ARX model can be viewed as a straightforward multivariate extension of the univariate model (2.1) in the previous section. The joint distribution for the data in the observed $T \times k$ matrix $X = (x_1, x_2, \dots, x_K)$ and the parameter in $\theta = (b, \sigma^2, \lambda, X_-, \Xi, \phi^{-1})$ is

$$\begin{aligned} p(\theta, X) &= N_{T\times K}(X \mid \Xi, \phi \otimes I_T) N_{T\times K}(\xi_1 \mid Zb, \sigma^2 I_T) N(b \mid b_*, D_\lambda \otimes H_*) \\ &\quad \times W(D_\lambda^{-1} \mid D_\lambda^*, l_*) G(\sigma^{-2} \mid \sigma_*^2, n_*) N_{p\times K}(X_- \mid X_-^*, D_{*\gamma} \otimes \Psi_*) \\ &\quad \times W(\phi^{-1} \mid \phi_*, \nu_*) N_{T\times K}(\Xi \mid 1_T \xi_*', \tau_* \otimes I_T). \end{aligned} \tag{3.4}$$

Note that this distribution is structurally identical to the joint distribution of the single tightness model, except for the fact that we have a larger $X$ matrix and more regression coefficients. The full conditional distributions are:

(a) For the regression coefficients $b$,

$$\begin{aligned} p(b \mid X, \theta^c) &\propto (\sigma^{-2})^{T/2} \exp\{-\frac{1}{2\sigma^2}(\xi_1 - Zb)'(\xi_1 - Zb)\} \\ &\quad \times |H_*|^{-1/2} \exp\{-\frac{1}{2}(b - b_*)' H_*^{-1}(b - b_*)\} \\ &\propto N(b \mid b_{**}, H_{**}), \end{aligned} \tag{3.5}$$

a normal distribution with the parameters

$$H_{**}^{-1} = H_*^{-1} + \sigma^{-2} Z'Z, \quad b_{**} = H_{**}(H_*^{-1} b_* + \sigma^{-2} Z' \xi_1).$$

(b) The f.c.d. for the residual precision $\sigma^{-2}$ is a gamma distribution

$$p(\sigma^{-2} \mid X, \theta^c) \propto G(\sigma^{-2} \mid \sigma_{**}^2, n_{**}) \tag{3.6}$$

with the parameters

$$n_{**} = n_* + T \quad \text{and} \quad n_{**}\sigma_{**}^2 = n_* \sigma_*^2 + (\xi_1 - Zb)'(\xi_1 - Zb).$$

(c) The f.c.d. for the tightness parameter matrix $D_\lambda$ is

$$\begin{aligned} p(D_\lambda^{-1} \mid X, \theta^c) &\propto |D_\lambda \otimes H_*|^{-1/2} \exp\{-\frac{1}{2}\mathrm{vec}'(B - B_*)(D_\lambda \otimes H_*)^{-1} \\ &\quad \times \mathrm{vec}(B - B_*)\} |D_\lambda^{-1}|^{(l_* - K - 1)/2} \exp\{-\frac{1}{2}\mathrm{tr} D_\lambda^{-1} D_\lambda^*\} \\ &\propto W(D_\lambda^{-1} \mid D_\lambda^{**}, \nu_{**}), \end{aligned} \tag{3.7}$$

a Wishart distribution with the parameters

$$\nu_{**} = \nu_* + p + 1 \quad \text{and} \quad D_\lambda^{**} = D_\lambda^* + (B - B_*) H_*^{-1} (B - B_*)'.$$

As before, the starting values for the Gibbs sampler can be found using the OLS estimates.

(d) The f.c.d. for the distribution of the starting parameters $X_-$:

The first elements of an AR($p$) process suffer from the "starting value problem". In Bayesian terms, starting values can be simply viewed as unknown parameters which can be estimated from the data. The convenient feature of the Gibbs sampler is that the full conditional distribution of the starting values can be also expressed in closed form as a normal distribution from which we can draw samples. So the "starting value problem" adds just another round of iteration in a Gibbs sampling step.

The f.c.d. for the $p \times K$ matrix $X_- = (x_{1-}, x_{2-}, \ldots, x_{K-})$ can be computed as

$$p(X_- \mid X, \theta^c) = N(X_-^{**}, \Psi_{**}), \tag{3.8}$$

a normal distribution with the parameters

$$\Psi_{**}^{-1} = D_\gamma^* \otimes \Psi_*^{-1} + \sigma^{-2} D_0' D_0 = \text{diag}(\gamma_1^* \Psi_*^{-1} + A_{01}, \ldots, \gamma_M^* \Psi_*^{-1} + A_{0M}),$$

$$\text{vec} X_-^{**} = \Psi_{**}\{(D_\gamma^* \otimes \Psi_*^{-1}) \text{vec} X_-^* + \sigma^{-2} D_0'(\xi_{01} - Z_0 b)\}.$$

Again, we can only use the first $p$ observations of the ARX model, which is decomposed as

$$\xi_1 = \alpha 1_T + A_{01} x_{1-} + \cdots + A_{0K} x_{K-} + A_{11}\xi_{1-} + \cdots + A_{1K}\xi_{K-} + \epsilon_1.$$

By collecting the first $p$ values in $\xi_{01}$, the reduced time series model is given by $\xi_{01} = Z_0 b + D_0 \text{vec} X_- + \epsilon_1$, where the two components separate the starting values from the other parameters:

$$Z_0 b = (X_{01}, \ldots, X_{0K}) b, \quad D_0 = \text{diag}(A_{01}, \ldots, A_{0K})$$

and the other matrices are given by

$$X_{0k} = \begin{pmatrix} 0 & 0 & \ldots & 0 & 0 \\ \xi_{1k} & 0 & \ddots & 0 & 0 \\ \vdots & \vdots & \ddots & \vdots & \vdots \\ \xi_{p-2,k} & \xi_{p-3,k} & \ldots & 0 & 0 \\ \xi_{p-1,k} & \xi_{p-2,k} & \ldots & \xi_{1k} & 0 \end{pmatrix} \quad \text{and}$$

$$A_{0k} = \begin{pmatrix} b_{p,k} & b_{p-1,k} & \ldots & b_{2,k} & b_{1,k} \\ 0 & b_{p,k} & \ldots & b_{3,k} & b_{2,k} \\ \vdots & \vdots & \ddots & \vdots & \vdots \\ 0 & 0 & \ldots & 0 & b_{p,k} \end{pmatrix}.$$

Note that the posterior mean of the starting parameters for each time series can be written, with $\Psi_k^{**} = (\gamma_k^* \Psi_*^{-1} + A_{0k})^{-1}$, as

$$x_{k-}^{**} = \Psi_k^{**}\{(\gamma_k^* \Psi_*^{-1}) x_{k-}^* + \sigma^{-2} A_{0k}'(\xi_{01} - Z_0 b)\}, \quad k = 1, \ldots, K.$$

**Proof.** We now write the eiv regression model as $\xi = \alpha 1_T + A_0 x_- + A_1 \xi + \epsilon = a_0 + A_0 x_- + \epsilon$, with $a_0 = \alpha 1_T + A_1 \xi$. Then the f.c.d. can be written as

$$\begin{aligned} p(x_- \mid x, \theta^c) \quad &\propto \quad \exp\{-\frac{1}{2\sigma^2}(\xi_{01} - a_0 - A_0 x_-)'(\xi_{01} - a_0 - A_0 x_-)\} \\ &\quad \times \exp\{-\frac{1}{2}(x_- - x_-^*)' \Psi_*^{-1}(x_- - x_-^*)\}. \end{aligned}$$

Since the first quadratic form contains many zeros, we can concentrate our analysis on the non-zero part. Thus, the above transformation can be written as

$$\begin{aligned} &(\xi_{01} - a_0 - A_0 x_-)'(\xi_{01} - a_0 - A_0 x_-) \\ &\qquad = \quad (\xi_{01} - a_{00} - A_{00} x_-)'(\xi_{01} - a_{00} - A_{00} x_-) \\ &\qquad = \quad (\xi_{01} - X_0 b)'(\xi_{01} - X_0 b), \end{aligned}$$

where $a_{00}, A_0$ and $X_0$ are given as above.

(e) The f.c.d. for the latent variables vec$\Xi$:
As can be seen from the joint density, the f.c.d. consists of 3 parts, i.e.

$$\begin{aligned} p(\text{vec}\Xi \mid Y, \theta^c) \quad \propto \quad & \exp\{-\frac{1}{2\sigma^2}(\xi_1 - Zb)'(\xi_1 - Zb)\} \\ & \times \exp\{-\frac{1}{2}\text{vec}'(X-\Xi)(\phi^{-1} \otimes I_T)\text{vec}(X-\Xi)\} \\ & \times \exp\{-\frac{1}{2}\text{vec}'(\Xi - 1_T\xi_*')(\tau_*^{-1} \otimes I_T)\text{vec}(\Xi - 1_T\xi_*')\}, \end{aligned} \tag{3.9}$$

with $\tau_*^{-1} = \text{diag}(0, \tau_*^2, \ldots, \tau_*^K)$. Since $Z$ is built up by the latent variables in $\Xi$, we have to reformulate the regression equation in the following way:

$$\xi_1 = \alpha_1 1_T + A_{01}x_{1-} + \cdots + A_{0K}x_{K-} + A_{11}\xi_{1-} + \cdots + A_{1K}\xi_{K-} + \epsilon_1,$$

where each $A_j$ matrix is further decomposed as $A_j = (A_{0j}, A_{1j})$ into the $T \times p$ block $A_{0j}$ and the $T{\times}T$ matrix $A_{1j}$. The coefficient matrix $A_j$ is band structured as in (2.6) consisting of the $j$-th variable regression coefficient $b_j, j = 1, \ldots, K$. The above regression equation can be compactly written as

$$D_1 \text{vec}\Xi = a_1 + \epsilon_1, \tag{3.10}$$

with $D_1 = [(I_T - A_{11}), A_{12}, \cdots, A_{1K})]$ and $a_1 = \alpha 1_T + A_{01}x_{1-} + A_{02}x_{2-} + \cdots + A_{0K}x_{K-}$. Thus, the f.c.d. for the latent variables matrix $\Xi$ is given by

$$p(\text{vec}\Xi \mid X, \theta^c) = N(\Xi \mid \Xi_{**}, G_{**}), \tag{3.11}$$

a normal distribution with the parameters

$$\begin{aligned} G_{**}^{-1} &= (\phi^{-1} + \tau_*^{-1}) \otimes I_T + \sigma^{-2}D_1'D_1, \\ \text{vec}\Xi_{**} &= G_{**}\{(\phi^{-1} \otimes I_T)\text{vec}X + (\tau_*^{-1} \otimes I_T)\text{vec}1_T\xi_*' + \sigma^{-2}D_1'a_1\}, \\ &= G_{**}\{\text{vec}(X\phi^{-1} + 1_T\xi_*'\tau_*^{-1}) + \sigma^{-2}D_1'a_1\}. \end{aligned}$$

The f.c.d. for an individual $\xi_k, k = 1, \ldots, K$ is given in section 3.1.

(f) The f.c.d. for the variance matrix $\phi$ of the latent variables is

$$\begin{aligned} p(\phi^{-1} \mid X, \theta^c) \quad \propto \quad & |\phi^{-1}|^{T/2} \exp\{-\frac{1}{2}\text{vec}'(\Xi - X)(\phi^{-1} \otimes I_T)\text{vec}(\Xi - X)\} \\ & \times |\phi^{-1}|^{f_*/2-1} \exp\{-\frac{1}{2}\text{tr}\phi^{-1}\phi_*\} \\ \propto \quad & W(\phi \mid \phi_{**}, f_{**}), \end{aligned} \tag{3.12}$$

a Wishart distribution with the parameters $f_{**} = f_* + T$ and $\phi_{**} = \phi_* + (\Xi - X)'(\Xi - X)$. The Gibbs sampler can be invoked with the starting values which are based on simple OLS estimates:

$$\begin{aligned} b_*^{(0)} &= b_{OLS} = (Z'Z)^{-1}Z'x_1, \\ \sigma_*^{2(0)} &= \sigma_{OLS}^2 \quad \text{and} \\ \lambda^{(0)} &= (b_{OLS} - b_*)'H_*^{-1}(b_{OLS} - b_*). \end{aligned}$$

As initial parameters for the starting values, we suggest the first observation or zero for first differences with a very informative (small) covariance matrix.

### 9.3.1 The f.c.d.s for the Individual Latent Factors

Simulating from the f.c.d. for the latent variables jointly as in (3.11) might be too big in size. Therefore we show in this section how the f.c.d.s of individual latent factors can be derived. First, we derive the f.c.d. for $\xi_1$ then for the remaining $\xi_1$s. The conditional distribution $p(\xi_1 \mid \xi_1^c, \phi, Y)$, where $\xi_1^c$ stands for all other latent variables except $\xi_1$, is given by

$$p(\xi_1 \mid \xi_1^c, \phi, Y) = N(\xi_1 \mid \xi_{1.2}, \phi_{1.2} \otimes I_T), \tag{3.13}$$

with the parameters

$$\begin{aligned} \phi_{1.2} &= \phi_{11} - \phi_{12}\phi_{22}^{-1}\phi_{21}, \\ \xi_{1.2} &= x_1 + (\phi_{12}\phi_{22}^{-1} \otimes I_T)(\Xi_1^c - X_1^c) \\ &= x_1 + \text{vec}(\Xi_1^c - X_1^c)\phi_{22}^{-1}\phi_{12}' \\ &= x_1 + (\Xi_1^c - X_1^c)\phi_{22}^{-1}\phi_{21}, \end{aligned}$$

where $\phi$ is partitioned as

$$\phi = \begin{pmatrix} \phi_{11} & \phi_{12} \\ \phi_{21} & \phi_{22} \end{pmatrix}$$

and $\Xi_1^c$ or $X_1^c$ are the $\Xi$ or $X$ matrices without the first column, respectively.

We have to distinguish between two cases: (1) the first latent variable—of the left hand side—and (2) the other latent variables in (3.1).

**Case 1.** We write the regression model as

$$\xi_1 = \alpha 1_T + A_{01}x_{1-} + A_{11}\xi_1 + \mu_1 + \epsilon_1,$$

where $x_{1-}$ are the starting values of the first latent variable and $\mu_1 = X_2b_2 + \cdots + X_Kb_K$ summarises the effects of all other variables in the system. Then the equation can be written briefly as

$$(I_T - A_{11})\xi_1 = a_1 + \epsilon_1,$$

with $a_1 = \alpha 1_T + A_{01}x_{1-} + \mu_1$ and $P_1 = I_T - A_{11}$. Thus, we find in analogy to the AR($p$) model as f.c.d.

$$p(\xi_1 \mid Y, \theta^c) = N(\xi_1 \mid \xi_1^{**}, G_1^{**}), \tag{3.14}$$

a normal distribution with the parameters

$$\begin{aligned} G_1^{**-1} &= \phi_{1.2}^{-1} I_T + \sigma^{-2} P_1' P_1, \\ \xi_1^{**} &= G_1^{**}(\phi_{1.2}^{-1}\xi_{1.2} + \sigma^{-2} P_1' a_1). \end{aligned}$$

The matrix $A_1 = (A_{11}, A_{12})$ is a band structured matrix as in (2.6) consisting of the first variable regression coefficient $b_1$.

**Case 2.** To find the f.c.d. of the other latent variables, we show how to get the distribution for $\xi_2$. The f.c.d. for the remaining $\xi_k, k = 3, \ldots, K$ can be obtained in the same way (just exchange places with the second variable). We write the regression model in (3.1) as

$$\xi_1 = A_{02} x_{2-} + A_{12}\xi_2 + \mu_2 + \epsilon_1,$$

where $x_{2-}$ are the starting values of the second latent variable and $\mu_2 = Zb - X_2 b_2$ summarises the effects of all other variables in the system. Then the equation can be written briefly as

$$z_1 = A_{12}\xi_2 + \epsilon_1,$$

with

$$z_1 = \xi_1 - A_{02} x_{2-} - \mu_2.$$

Thus, the f.c.d. for $\xi_2$ is given by

$$p(\xi_2 \mid Y, \theta^c) = N(\xi_2 \mid \xi_2^{**}, G_2^{**}), \tag{3.15}$$

a normal distribution with the parameters

$$\begin{aligned} G_2^{**-1} &= \phi_{2.1}^{-1} I_T + \sigma^{-2} A_{12}' A_{12}, \\ \xi_2^{**} &= G_2^{**}(\phi_{2.1}^{-1}\xi_{2.1} + \sigma^{-2} A_{12}' Z_1), \end{aligned}$$

where $\phi_{2.1}$ and $\xi_{2.1}$ are the parameters of the conditional distribution of the second variable in (3.13) given all the other variables.

The full VAR model can be written compactly in matrix notation in the following way. Let $Y = [y_1, \ldots, y_M]$ be the $T \times M$ matrix of the multivariate time series $y_t = (y_{t1}, \ldots, y_{tM})$, and let $Z$ be the $MT \times MKp'$ ($p' = p + 1$) block-diagonal matrix consisting of the $M$ blocks of lagged time series $Z = \text{diag}(X_1, \ldots, X_M)$. Each time series in the B-VAR system can be written as multiple tightness model

$$\begin{aligned} y_m &= X_{1m} b_{1m} + X_{2m} b_{2m} + \cdots + X_{Km} b_{Km} + \epsilon_m \\ &= X_m \text{vec} B_m + \epsilon_m, \end{aligned} \tag{3.16}$$

where $X_m = [X_{1m}, X_{2m}, \ldots, X_{Km}]$. The diagonal blocks of the regressor matrix $X$, i.e. $X_{mm} = [1_T, L^p y_m]$, consist of the constant and the lagged values of the left hand variable $y_m$. As before, $L^p$ is the matrix lag operator $L^p y_m =$

$[y_{m,-1},\ldots,y_{m,-p}]$. Each block $X_k = [y_k, L^p y_k]$ contains the current and the lagged values of the time series $y_k = \{y_{kt}\}$ where $t = 1,\ldots,T$. By convention, the constant (of the intercept) is the first column of the diagonal block $X_{mm}$. Therefore the $X$ matrix has the following form:

$$Z = \mathrm{diag}[X_1, X_2, \ldots, X_M] = \begin{pmatrix} [X_{11}\ldots X_{K1}] & \cdots & 0 \\ \vdots & \ddots & \vdots \\ 0 & \cdots & [X_{1M}\ldots X_{KM}] \end{pmatrix}. \tag{3.17}$$

The tightness B-VAR model for the time series system with latent variables is defined as the following BEV-VAR($p$) model:

$$\begin{aligned}
&\mathrm{vec}\Xi \sim N_{n\times KM}(Z\mathrm{vec}B, \Omega\otimes I_T), \\
&\mathrm{vec}Y \sim N_{T\times M}(\mathrm{vec}\Xi, \phi\otimes I_T), && \Xi \sim N(1_T\xi'_{*}, \tau_{*}\otimes I_T), \\
&B \sim N_{Kp'\times M}(B_*, \Sigma\otimes\Lambda\otimes H_*), && \Omega^{-1} \sim W_M(\Omega_*, n_*), \\
&\Lambda^{-1} \sim W_M(\Lambda_*, l_*), && \Sigma^{-1} \sim W_K(\Sigma_*, \nu_*), \\
&Y_- \sim N(Y_-^*, D_\gamma^*\otimes\Psi_*), && \phi^{-1} \sim W_M(\phi_*, f_*).
\end{aligned} \tag{3.18}$$

Note that the coefficient matrix can be written as $B = (\mathrm{vec}B_1, \ldots, \mathrm{vec}B_M)$, where each $B_m$, $m = 1,\ldots,M$, is the coefficient matrix of the multiple problem of the previous section. This simplification might also be useful for the prior specification for $B_m^*$. The prior covariance is again $H_* = \mathrm{diag}(d_{*0}, D_*)$ with diffuse $d_{*0}$, i.e. $d_{*0}^{-1} = 0$.

The joint distribution for observed data, i.e. the time series matrix $Y$ and the parameter $\theta = \{B, \Omega^{-1}, \Sigma^{-1}, \Lambda^{-1}, Y_-, \Psi, \Xi, \phi^{-1}\}$, is just the product of the above distributions:

$$\begin{aligned}
p(Y,\theta) &= N_{T\times M}(Y \mid \Xi, \phi\otimes I_T) N_{T\times KM}(\mathrm{vec}\Xi \mid X\mathrm{vec}B, \Omega\otimes I_T) \\
&\times N_{p'K\times M}(B \mid B_*, \Sigma\otimes\Lambda\otimes H_*) W_M(\Lambda^{-1} \mid \Lambda_*, l_*) W_M(\Omega^{-1} \mid \Omega_*, n_*) \\
&\times W_K(\Sigma^{-1} \mid \Sigma_*, \nu_*) W_M(\phi^{-1} \mid \phi_*, f_*) N_{p\times K}(Y_- \mid Y_-^*, D_\gamma^*\otimes\Psi_*).
\end{aligned} \tag{3.19}$$

The full conditional distributions are:

(a) For the regression coefficients $b = \mathrm{vec}B$:

$$\begin{aligned}
p(B \mid Y, \theta^c) &\propto \exp\{-\frac{1}{2}\mathrm{vec}'(\Xi - Zb)(\Omega^{-1}\otimes I_T)\mathrm{vec}(\Xi - Zb)\} \\
&\quad \times \exp\{-\frac{1}{2}\mathrm{vec}'(B - B_*)(\Sigma\otimes\Lambda\otimes H_*)^{-1}\mathrm{vec}(B - B_*)\} \\
&= Np'_{K\times M}(B \mid B_*, H_{**}),
\end{aligned} \tag{3.20}$$

a normal distribution with the parameters

$$\begin{aligned}
H_{**}^{-1} &= (\Sigma\otimes\Lambda\otimes H_*)^{-1} + Z'(\Omega^{-1}\otimes I_T)Z, \\
\mathrm{vec}B_{**} &= H_{**}\{(\Sigma\otimes\Lambda\otimes H_*)^{-1}\mathrm{vec}B_* + Z'(\Omega^{-1}\otimes I_n)\mathrm{vec}\Xi\} \\
&= H_{**}\{\mathrm{vec}(\Lambda\otimes H_*)^{-1}B_*\Sigma^{-1} + Z'\mathrm{vec}\Xi\Omega^{-1}\} \\
&= H_{**}\{\mathrm{vec}([\mathrm{vec}H_*^{-1}B_{*1}\Lambda^{-1}, \ldots, \mathrm{vec}H_*^{-1}B_{*M}\Lambda^{-1}]\Sigma^{-1}) + Z'\mathrm{vec}\Xi\Omega^{-1}\}.
\end{aligned}$$

Since we may have to draw samples from a very large normal distribution, we present in the appendix a conditional procedure which allows to cut down the simulation process to the $Kp'$ regression coefficients in each of the $M$ equations.

(b) For the across equations covariance matrix $\Omega$:
As in the bivariate model, the posterior distribution is given by

$$p(\Omega^{-1} \mid Y, \theta^c) = W(\Omega^{-1} \mid \Omega_{**}, n_{**}), \tag{3.21}$$

i.e. a Wishart distribution with the parameters

$$\nu_{**} = \nu_* + T \quad \text{and} \quad \Omega_{**} = \Omega_* + E'E,$$

with the residual matrix $E = (e_1, \ldots, e_M)$ and $e_m = \xi_m - M_m b_m$, with $m = 1, \ldots, M$.

(c) For the cross coefficient covariance matrix $\Sigma^{-1}$:
As in the bivariate model, we obtain as f.c.d.

$$p(\Sigma^{-1} \mid Y, \theta^c) = W(\Sigma^{-1} \mid \Sigma_{**}, \nu_{**}), \tag{3.22}$$

a Wishart distribution with the parameters $n_{**} = n_* + Kp'$ and

$$\Sigma_{**} = \Sigma_* + (B - B_*)'(\Lambda \otimes H_*)^{-1}(B - B_*).$$

(d) For the tightness matrix $\Lambda^{-1}$:
As in the bivariate model, we obtain as full conditional distribution

$$\begin{aligned} p(\Lambda^{-1} \mid Y, \theta^c) \;&\propto\; |\Lambda \otimes H_*|^{-1/2} \exp\{ -\frac{1}{2}\text{vec}'(B - B_*)(\Lambda \otimes H_*)^{-1} \\ &\quad \times \text{vec}(B - B_*)\} |\Lambda^{-1}|^{(l_* - K - 1)/2} \exp(-\frac{-1}{2}\text{tr}\Lambda^{-1}\Lambda_*) \\ &\propto\; |\Lambda^{-1}|^{(l_* - K - 1 + Mp')/2} \exp\Big[ -\frac{1}{2}\{\text{tr}\Lambda^{-1}\Lambda_* \\ &\quad + \text{tr}\Lambda_*^{-1}(B - B_*)'\Lambda^{-1}(B - B_*)\}\Big] \\ &\propto\; W(\Lambda^{-1} \mid \Lambda_{**}, \nu_{**}), \end{aligned} \tag{3.23}$$

a Wishart distribution with the parameters $\nu_{**} = \nu_* + Mp'$ and

$$\begin{aligned} \Lambda_{**} &= \Lambda_* + \sum_{i=1}^{M} S_i'(B - B_*)'\Sigma^{-1}(B - B_*)S_i \\ &= \Lambda_* + \sum_{i=1}^{M} (B_i - B_{*i})'\Sigma^{-1}(B_i - B_{*i}), \end{aligned}$$

where $S_i = (e_i \otimes I_T)K_{Kp'}, i = 1, \ldots, M$ is a selection matrix which selects the block diagonal elements of a partitioned matrix, $e_i$ is the $i$-th unit vector of order $M$ and $K_{Kp'}$ is the commutation matrix of order $Kp'$.

(e) For the VAR starting parameters $Y_-$:

We derive the f.c.d. as for the multiple AR($p$) process. Since we have $p$ starting values for $K$ time series, we collect them in the $p \times 1$ vectors $y_{k-} = (y_{k,-p}, \ldots, y_{k,-1})', k = 1, \ldots, K$ and the $p \times K$ matrix $Y_- = [y_{1,-}, \ldots, y_{K,-}]$.

We have to apply the starting value transformation to each equation $m = 1, \ldots, M$. Note that the starting value matrix $Y_-$ and the initial triangular data matrices $A_{0j}$ stay the same for the estimation in $m$ equations. The information across the $M$ equations have to be appropriately put together for the conditional distribution,

$$\begin{aligned}
y_{0m} &= \alpha_m 1_T + X_{1m} b_{1m} + \cdots + X_{Km} b_{Km} + \epsilon_{0m} \\
&= \alpha_m 1_T + A_{01} b_{1m} + A^\circ_{1m} y_{1,-} + \cdots + A_{0K} b_{Km} + A^\circ_{Km} y_{K,-} + \epsilon_{0m} \\
&= \alpha_m 1_T + [A^\circ_{1m}, \ldots, A^\circ_{pm}] \text{vec}[y_{1,-}, \ldots, y_{K,-}] + \epsilon_{0m} \\
&= a_m + A^\circ_m \text{vec} Y_- + \epsilon_{0m},
\end{aligned} \tag{3.24}$$

with $a_m = A_{01} b_{1m} + \cdots + A_{0K} b_{Km} + \alpha_m 1_T$ and the $p \times pK$ matrix $A^\circ_m = [A^\circ_{1m}, \ldots, A^\circ_{pm}]$. Each $A_{kjm}$ is a upper triangular matrix with regression coefficients of the corresponding $b_{kjm}, k = 1, \ldots, K$.

Therefore the posterior f.c.d. for $Y_-$ is given by

$$\begin{aligned}
p(Y_- \mid Y, \theta^c) \;\propto\; & \exp\{-\frac{1}{2} \text{vec}'(Y_0 - X_0 b)(\Omega^{-1} \otimes I_p) \text{vec}(Y_0 - X_0 b)\} \\
& \times \exp\{-\frac{1}{2} \text{vec}'(Y_- - Y^*_-)(D_{*\gamma} \otimes \Psi_*)^{-1} \text{vec}(Y_- - Y^*_-)\}.
\end{aligned}$$

Note that $b = \text{vec} B = \text{vec}[B_1, B_2, \ldots, B_K]$, that $Y_0 = [y_{01}, \ldots, y_{0K}]$ is a $p \times M$ matrix and that $X_0$ is a $pM \times p'KM$ matrix. Now the first quadratic form can be written by the "starting value" transformation as

$$\begin{aligned}
& \text{vec}'(Y_0 - X_0 b)(\Omega^{-1} \otimes I_p) \text{vec}(Y_0 - X_0 b) \\
& \quad = \{\text{vec}(Y_0 - A) - A^\circ \text{vec} Y_-\}'(\Omega^{-1} \otimes I_p)\{\text{vec}(Y_0 - A) - A^\circ \text{vec} Y_-\},
\end{aligned}$$

where $A = [a_1, \ldots, a_M] : p \times M$ and the $Mp \times M$ matrix

$$A^\circ = \begin{pmatrix} A^\circ_{11} & \ldots & A^\circ_{1M} \\ \vdots & & \vdots \\ A^\circ_{M1} & \ldots & A^\circ_{MM} \end{pmatrix}. \tag{3.25}$$

Therefore the full conditional distribution is

$$p(Y_- \mid Y, \theta^c) = N_{p \times K}(Y^{**}_-, \Psi_{**}), \tag{3.26}$$

a normal distribution with the parameters

$$\begin{aligned}
\Psi^{-1}_{**} &= (D_{*\gamma} \otimes \Psi_*)^{-1} + A^{\circ\prime}(\Omega^{-1} \otimes I_p) A^\circ, \\
\text{vec} Y^*_- &= \Psi_{**}\{(D_{*\gamma} \otimes \Psi_*)^{-1} \text{vec} Y^*_- + A^{\circ\prime}(\Omega^{-1} \otimes I_p) \text{vec}(Y_0 - A)\} \\
&= \Psi_{**}[\text{vec} \Psi^{-1}_* Y^*_- D^{-1}_{*\gamma} + A^{\circ\prime} \text{vec}\{(Y_0 - A)\Omega^{-1}\}].
\end{aligned}$$

(f) For the latent variables $\Xi$:
Unlike the ARX model, the f.c.d. is given by:

$$\begin{aligned} p(\text{vec}\Xi \mid Y, \theta^c) &\propto \exp\{-\frac{1}{2\sigma^2}\text{vec}'(\Xi - Zb)\text{vec}(\Xi - Zb)\} \\ &\quad \times \exp\{-\frac{1}{2}\text{vec}'(X - \Xi)(\phi^{-1} \otimes I_T)\text{vec}(X - \Xi)\}. \end{aligned}$$

The first quadratic form also contains the latent variables $\Xi$ in the regressor matrix $Z$. Therefore we rewrite all $M$ equations in the same form as in (3.10) for the BEV-ARX model:

$$\begin{aligned} D_1\text{vec}\Xi &= a_1 + \epsilon_1, \\ &\vdots \\ D_M\text{vec}\Xi &= a_M + \epsilon_M, \end{aligned}$$

or compactly as

$$D\text{vec}\Xi = \text{vec}A + \text{vec}E. \tag{3.27}$$

The following matrices are used: $A = (a_1, a_2, \dots, a_M)$, which consists of the first time series elements; $D = \text{diag}(D_1, \dots, D_K)$ where the $D_k$ are stacked triangular matrices with elements of $b_k$ and $E = (\epsilon_1, \epsilon_2, \dots, \epsilon_M)$ is the matrix of residuals. Note that $D = I_{MK} - A_-^1$ , where $A_-^1$ is the matrix $A^0$ as in (3.25), though the off diagonal block elements have negative signs.

Thus, combining the quadratic forms, we get a normal distribution

$$p(\text{vec}\Xi \mid Y, \theta^c) = N(\Xi \mid \Xi_{**}, G_{**}) \tag{3.28}$$

with the parameters

$$\begin{aligned} G_{**}^{-1} &= \phi^{-1} \otimes I_T + \sigma^{-2}D'D, \\ \text{vec}\Xi_{**} &= G_{**}\{(\phi^{-1} \otimes I_T)\text{vec}X + (\tau_*^{-1} \otimes I_T)\text{vec}1_T\xi_*' + \sigma^{-2}D'\text{vec}A\}, \\ &= G_{**}\{\text{vec}(X\phi^{-1}) + \sigma^{-2}D'\text{vec}A\}. \end{aligned}$$

(g) For the variance matrix of the latent variables $\phi$:

$$\begin{aligned} p(\phi^{-1} \mid Y, \theta^c) &\propto |\phi^{-1}|^{T/2}\exp\{-\frac{1}{2}\text{vec}'(\Xi - X)(\phi^{-1} \otimes I_T)\text{vec}(\Xi - X)\} \\ &\quad \times |\phi^{-1}|^{f_*/2-1}\exp\{-\frac{1}{2}\phi^{-1}\phi_*\} \\ &\propto W(\phi \mid \phi_{**}, f_{**}), \end{aligned} \tag{3.29}$$

a Wishart distribution with the parameters $f_{**} = f_* + T$ and

$$\phi_{**} = \phi_* + (\Xi - X)'(\Xi - X).$$

In the next section, we turn to a related B-VAR concept, the smoothness prior model.

## 9.4 Further topics

### 9.4.1 The BEV-VAR Model with Smoothness Prior

If we use a smoothness covariance matrix instead of a tightness covariance matrix $\Psi_*$, we can use the same model set-up to estimate a BEV-VAR model with smoothness prior. The $\lambda$-tightness constants are then interpreted as smoothness constants. The prior values for the smoothness constants have to be elicited in the same way as for the tightness constants: What size differences between the coefficients are considered to be "small"?

The Bayesian smoothness model (see Shiller (1973) for the original model) can be formulated as

$$\begin{aligned} y &\sim N(Xb, \sigma^2 I_n), \\ Rb &\sim N(0, \lambda I_p), \quad \sigma^{-2} \sim G(s_*^2, n_*), \\ \lambda^{-1} &\sim G(\lambda_*, l_*), \end{aligned}$$

where $R$ is a $(p-1) \times p$ differencing matrix of appropriate order: Define

$$R = R_1 = \begin{pmatrix} 1 & -1 & 0 & \dots & 0 \\ 0 & 1 & -1 & \dots & 0 \\ \vdots & \vdots & \vdots & \ddots & \vdots \\ 0 & 0 & 0 & \dots & -1 \end{pmatrix}.$$

and a $d$-th order differencing matrix $R_d$ can be produced by $d$-fold multiplication: $R_d = R_1^d$.

### 9.4.2 Prediction and Order Selection

For forecasting purposes, we will use the predictive distribution

$$p(x_{+1} \mid x) = \int p(x_{+1} \mid x, \theta) p(\theta \mid x) d\theta.$$

It can be found easily by calculating the average over conditional predictive distributions, induced by the posterior sample of the coefficients. Let $M$ be the number of sample points in the Gibbs iteration: the predictive density for the next observation $x_{+1}$ is then simply an average of normals:

$$p(x_{+1} \mid x) = \frac{1}{M} \sum_{i=1}^{M} N(Xb^{(i)}, \sigma^{2(i)}).$$

The subscript $(i)$ denotes the $i$-th sample of the Gibbs iteration.

Using the predictive distribution for each time point, we can calculate the conditional predictive ordinates (CPOs). They can be used for the determination of the order of the AR($p$) processes as in Polasek (1994) or Polasek and Jin (1994).

## 9.5 Conclusions

Classical and Bayesian eiv models are rather rare in applications. With this Gibbs version of latent variables models, we hope to encourage more applications. The paper has demonstrated that, when using an ordinary parametrisation, the f.c.d.s are given in closed form. This allows for an easy implementation of a Gibbs sampling routine since random variables can be drawn easily from the f.c.d.s. The model covers the estimation of the starting values and the hyperparameters which govern the tightness priors. The concept of the tightness prior is a more general concept than the one originally proposed by Litterman (1986) and is no longer data-based.

In a similar way, we have indicated how to formulate the multivariate smoothness model with errors-in-variables. In a further paper, we will give the eiv extension of the Chib (1993) regression model with AR errors.

## Appendix: A Conditional Sampling Scheme for Regression Matrices

To calculate the full conditional posterior distribution for large systems, i.e. for large $M$ or $K$, one needs large storage space, therefore a smaller dimension of the f.c.d.s would be preferable. From the following partitioned normal distribution

$$\begin{pmatrix} y_1 \\ y_2 \end{pmatrix} \sim N\left[\begin{pmatrix} X_1 b_1 \\ X_2 b_2 \end{pmatrix}, \Sigma = \begin{pmatrix} \Sigma_{11} & \Sigma_{12} \\ \Sigma_{21} & \Sigma_{22} \end{pmatrix}\right]$$

and the rules for partitioned matrices, we can get the conditional distribution $y_1 \mid y_2 \sim N(\mu_{1.2}, \Sigma_{1.2})$ with the parameters

$$\Sigma_{1.2} = \Sigma_{11} - \Sigma_{12}\Sigma_{22}^{-1}\Sigma_{21} = \sigma_{1.2} I_T,$$

where the conditional variance, which is $\sigma_{1.2} = \sigma_{22} - \sigma_{21}^2/\sigma_{11}$, follows from the simplified covariance structure $\Sigma \otimes I_T$. The conditional mean is

$$\mu_{1.2} = X_1 b_1 + \Sigma_{12}\Sigma_{22}^{-1}(y_2 - X_2 b_2).$$

Now the quadratic form can be transformed as

$$(y_1 - \mu_{1.2})'\Sigma_{1.2}^{-1}(y_1 - \mu_{1.2}) \propto (y_{1*} - X_1 b_1)'\Sigma_{1.2}^{-1}(y_{1*} - X_1 b_1),$$

with $y_{1*} = y_1 + \Sigma_{12}\Sigma_{22}^{-1}X_2 b_2$. The full conditional distribution can be written as

$$\begin{aligned} p(b_1 \mid Y, \theta^c) &\propto \exp\{-\frac{1}{2}(y_{1*} - X_1 b_1)'\Sigma_{1.2}^{-1}(y_{1*} - X_1 b_1)\} \\ &\quad \times \exp\{-\frac{1}{2}(b_1 - b_1^*)'\lambda_1^{-1}D_*^{-1}(b_1 - b_1^*) \\ &\propto \exp\{-\frac{1}{2}(b_1 - b_1^{**})'H_1^{**}(b_1 - b_1^{**})\} \\ &= N(b_1^{**}, H_1^{**}), \end{aligned}$$

a normal distribution with the parameters

$$\begin{aligned} b_1^{**} &= H_{**}(\lambda_1^{-1} D_*^{-1} b_1^* + X_1' \Sigma_{2.1}^{-1} y_{1*}) \\ &= H_{**}\{\lambda_1^{-1} D_*^{-1} b_1^* + X_1'(\Sigma_{11} y_1 - \Sigma_{12} X_2 b_2)\} \end{aligned}$$

and

$$\begin{aligned} H_1^{**-1} &= \lambda_1^{-1} D_*^{-1} + X_1' \Sigma_{2.1}^{-1} X_1 \\ &= \lambda_1^{-1} D_*^{-1} + X_1' \Sigma^{11} X_1. \end{aligned}$$

The $\Sigma^{ij}$ are the inverse elements of the partitioned inverse of $\Sigma$. The proof of the second line follows by virtue of the partitioned inverse rule—see, for exaxmple, Magnus and Neudecker (1988, p. 11):

$$\Sigma^{-1} = \begin{pmatrix} \Sigma_{11} & \Sigma_{12} \\ \Sigma_{21} & \Sigma_{22} \end{pmatrix} = \begin{pmatrix} \Sigma_{1.2}^{-1} & -\Sigma_{1.2}^{-1}\Sigma_{12}\Sigma_{22}^{-1} \\ -\Sigma_{22}^{-1}\Sigma_{21}\Sigma_{1.2}^{-1} & \Omega_{22} \end{pmatrix}$$

with $\Omega_{22} = \Sigma_{22}^{-1} - \Sigma_{22}^{-1}\Sigma_{21}\Sigma_{1.2}^{-1}\Sigma_{12}\Sigma_{22}^{-1}$. It is easy to see that, by repeating this conditional sampling procedure for all $M$ time series in the VAR system, one can obtain the sample for the whole coefficient matrix B. For $m = 2, \ldots, M$, the above formula has to be applied repeatedly. The ordering of the variable changes while the structure of the conditional normal distribution stays the same.

## References


[1] Albert, J., and S. Chib, (1993), Bayesian Inference of Autoregressive Time Series with Mean and Variance Subject to Markov Jumps, *Journal of Business and Economic Statistics,* 11, 1–15.

[2] Chib, S., (1993), Bayes Regression with Autoregressive Errors: A Gibbs Sampling Approach, *Journal of Econometrics,* 58, 275–294.

[3] Gelfand, A.E., and D.K. Dey, (1992), Bayesian Model Choice: Asymptotics and Exact Calculation, University of Connecticut, to appear in *Journal of the Royal Statistical Society, Series B.*

[4] Gelfand, A.E., and A.F.M. Smith, (1990), Sampling Based Approaches to Calculating Marginal Densities, *Journal of the American Statistical Association,* 85, 398–409.

[5] Klepper, S., and E.E. Leamer, (1984), Consistent Sets of Estimates for Regressions with all Variables Measured with Error, *Econometrica,* 52,163–183.

[6] Leamer, E.E., (1978), *Specification Searches,* John Wiley, New York.

[7] Leamer, E.E., (1987), Errors-in-Variables in Linear Systems, *Econometrica,* 55, 893–909.

[8] Leeuw, J.v.d., (1994), The Covariance Matrix of ARMA Errors in Closed Form, *Journal of Econometrics,* 63, 397–405.

[9] Lindley, D., and A.F.M. Smith, (1972), Bayes Estimates for the Linear Model, *Journal of the Royal Statistical Society, Series B,* 34, 1–41.

[10] Lindley, D., and G.M. ElSayyad, (1968), The Bayesian Estimation of a Linear Functional Relationship, *Journal of the Royal Statistical Society, Series B,* 30, 190–202.

[11] Littermann, R.B., (1986), A Statistical Approach to Economic Forecasting, *Journal of Business and Economic Statistics,* 4, 1–24.

[12] Magnus, J.R., and H. Neudecker, (1988), *Matrix Differential Calculus with Applications in Statistics and Econometrics,* John Wiley, New York.

[13] Maravall, A., and D.J. Aigner, (1977), Identification of the Dynamic Shock-Error Model, Chapter 18 in D.J. Aigner and A.S. Goldberger (eds.), *Latent Variables Models in Socioeconomic Models,* North-Holland, Amsterdam.

[14] Marriott, J., N. Ravishanker, A.E. Gelfand and J. Pai, (1995), Bayesian Analysis of ARMA Processes: Complete Sampling Based Inference Under Full Likelihood, pp. 241–254 in D. Berry et al. (eds.) *Bayesian Statistics and Econometrics: Essays in Honour of Arnold Zellner,* John Wiley, New York,

[15] McCulloch, R.E., and R.S. Tsay, (1994), Bayesian Analysis of Autoregressive Time Series via the Gibbs Sampler, *Journal of Time Series Analysis,* 15, 235–250.

[16] Press, W.H., B.P. Flannery, S.A. Teukolsky and W.T. Vetterling, (1986), *Numerical Recipes, The Art of Scientific Computing,* Cambridge University Press, New York.

[17] Polasek, W., (1995), Bayesian Generalised Errors-in-Variables (GEIV) Models for Censored Regressions, in Mammitzsch and Schneeweiss (eds.), *Symposia Gaussiana,* Conf. B, Walter de Gruyter, Berlin, New York, 261–279.

[18] Polasek, W., and S. Jin, (1996), Gibbs Sampling in AR Models with Random Walk Priors, pp. 86–93 in Gaul W. and D. Pfeifer (eds.), *From Data to Knowledge,* Springer Verlag, Heidelberg.

[19] Polasek, W., and S. Jin, (1994), *Gibbs Sampling in AR Models with Random Walk Prior,* mimeo, University of Basel.

[20] Reinsel, G., (1983), Some Results on Multivariate Index Models, *Biometrika,* 70, 145–56.

[21] Shiller, R.J., (1973), A Distributed Lag Estimator Derived from Smoothness Priors, *Econometrica,* 41, 775–788.

CHAPTER 10

# Least-Squares Autoregression with Near-unit Root

Jan R. MAGNUS and Thomas J. ROTHENBERG[1]

## 10.1 Introduction

In autoregressive models, assumptions about the initial conditions seem to matter a lot when one of the roots of the characteristic polynomial is near the unit circle. In particular, the fixed start-up model (where the first few observations are treated asymmetrically) behaves quite differently in small samples than the stationary model which assumes that all the observations have the same distribution.

In this chapter, we explore some implications for the sampling distributions of least-squares estimates, test statistics, and forecasts in the leading case of the first-order autoregressive model. This is accomplished by studying the limiting behaviour of the statistics, for fixed sample size, as the autoregressive parameter approaches unity. When regressions are fitted *without* intercept, the stationary and fixed start-up models produce strikingly different sampling distributions for the least-squares estimator as the autoregressive parameter tends to one in absolute value. When regressions are fitted *with* intercept, the two models produce similar sampling properties when the autoregressive parameter tends to plus one, but very different properties when the parameter tends to minus one.

Previous Monte Carlo and numerical integration calculations have demonstrated that least-squares estimates, test statistics and forecasts in autoregressive time-series models exhibit unusual features when one of the roots of the characteristic polynomial lies near the unit circle. In particular, the traditional approximate distribution theory based on normality is often unsatisfactory. Furthermore, small differences in the initial conditions sometimes yield noticeable differences in the sampling distributions even when the number of observations is rather large. See, e.g., Phillips (1977, 1987), Evans and Savin (1981, 1984)

[1]The first version of this paper was completed in July 1988 when both authors were at the London School of Economics.

and Magnus and Pesaran (1989, 1991). In this chapter, we provide further evidence and offer some simple explanations for these phenomena in the leading case of the AR(1) model

$$y_t = \alpha + \beta y_{t-1} + \sigma u_t, \quad t = 1, 2, \ldots, \tag{1}$$

where $\{u_t\}$ is a sequence of independent standard normal random variables.

We shall examine some statistics obtained from a least-squares regression of $y_t$ on its lagged value, using observations $y_0, \ldots, y_T$. Section 2 considers the case where $\alpha$ is known to be zero and the regression is performed without an intercept. Section 3 considers the case where $\alpha$ may be nonzero and the regression contains an intercept. The behaviour of the sampling distributions turns out to be rather different in the two cases when $\beta$ is near one. The sampling distributions are also sensitive to the specification of the initial observation $y_0$. We shall assume that $y_0$ has the representation

$$y_0 = \mu + \delta u_0, \tag{2}$$

where $u_0$ is standard normal and independent of $\{u_1, u_2, \ldots\}$

Specification (2) covers a number of possibilities that have been discussed in the literature. If, for example, $\alpha = (1-\beta)\mu$ and $\sigma^2 = (1-\beta^2)\delta^2$, we have the stationary model; that is, the $\{y_t\}$ are a stationary stochastic process with expectation $\mu$ and variance $\delta^2$. Alternatively, if $\delta = 0$ and $\mu = c$, the first observation $y_0$ is the constant $c$; Evans and Savin call this the "fixed start-up" model. If $\delta = \sigma, \mu = \alpha + \beta c$, and (1) also holds for $t = 0$, then the value of $y$ preceding $y_0$ is the constant $c$; this is another version of the fixed start-up assumption, used by Fuller (1976) and by Magnus and Pesaran (1989, 1991). (Note that Fuller and Magnus and Pesaran assume that the sample consists of observations $y_1, \ldots, y_n$. Hence our $T$ corresponds to their $n-1$.)

Our approach will be to investigate the limiting distributions of the least-squares estimates, test statistics and forecasts, for fixed sample size, as $\beta$ tends to one. Although some large-sample approximations will be employed to simplify the formulae, our main results are valid for all sample sizes. Indeed, one of our main purposes is to discover the small-sample properties of least-squares statistics when $\beta$ is close to one. Asymptotic analysis of dynamic models when the autocorrelation coefficient tends to unity has been employed previously by Chipman (1979) and Jensen (1986). Our results can be viewed as a generalisation of those found by Jensen.

The model represented by (1) and (2) is characterised by the five parameters $\alpha, \beta, \sigma^2, \mu$ and $\delta$. When examining sequences of models as $\beta$ approaches one, we must specify the behaviour of the other parameters. For example, in the stationary case, $\alpha$ and $\sigma^2$ could be held fixed while $\mu = \alpha/(1-\beta)$ and $\delta = \sigma/\sqrt{1-\beta^2}$ both tend to infinity. Alternatively, $\mu$ and $\delta$ could be held fixed while $\alpha$ and $\sigma$ tend to zero. In the nonstationary case, all sorts of dependency of $\alpha, \sigma^2, \mu$ and $\delta$ on $\beta$ could be examined. Luckily, not all of the possibilities lead to different results. In fact, only a few basic forms of limiting behaviour for the least-squares statistics turn out to be possible.

## 10.2 Regression without Intercept

When $\alpha$ is zero and the regression has no intercept, the least-squares estimates of $\beta$ and $\sigma^2$ based on $y_0, y_1, \ldots, y_T$ can be written as

$$\hat{\beta} = \frac{\sum_t y_{t-1}y_t}{\sum_t y_{t-1}^2} = \beta + \sigma\frac{\sum_t y_{t-1}u_t}{\sum_t y_{t-1}^2}, \tag{3}$$

$$\hat{\sigma^2} = \frac{1}{T-1}\sum_t (y_t - \hat{\beta}y_{t-1})^2 = \frac{\sigma^2}{T-1}\sum_t u_t^2 - \frac{(\hat{\beta}-\beta)^2}{T-1}\sum_t y_{t-1}^2, \tag{4}$$

where $\sum_t$ represents summation from 1 to $T$. Also of interest are the $t$-ratio for $\beta$,

$$t_\beta = (\hat{\beta}-\beta)\left[\sum_t y_{t-1}^2/\hat{\sigma^2}\right]^{1/2} = \frac{\sigma\sum_t y_{t-1}u_t}{\hat{\sigma}\left[\sum_t y_{t-1}^2\right]^{1/2}}, \tag{5}$$

and the error in the $s$-period ahead forecast,

$$e_{T+s} = \hat{\beta}^s y_T - y_{T+s} = (\hat{\beta}^s - \beta^s)y_T - \sigma\sum_{i=0}^{s-1}\beta^i u_{T+s-i}. \tag{6}$$

When $y_0$ is specified as in (2), the difference equation (1) may be solved as

$$y_t = \beta^t(\mu + \delta u_0) + \sigma\sum_{i=0}^{t-1}\beta^i u_{t-i} \qquad (t \geq 0), \tag{7}$$

where the summation term is taken to be zero when $t = 0$. For any $\beta$, the behaviour of $y_t$ (and hence that of $\hat{\beta}, \hat{\sigma^2}, t_\beta$ and $e_{T+s}$ depends on the relative magnitudes of the three parameters $\mu, \delta$ and $\sigma$. In fact, when considering the limit as $\beta$ tends to one, there are three possible cases:

*Case 1*: As $\beta$ approaches one, both $\mu/\delta$ and $\sigma/\delta$ approach zero. (This occurs, for example, when $\mu = 0$ and $\sigma^2 = \delta^2(1-\beta^2)$ so that the $y_t$ are stationary.) Then, as $\beta \to 1$,

$$y_t/\delta \to u_0 \quad \text{for all} \quad t \geq 0.$$

*Case 2*: As $\beta$ approaches one, $\sigma/\mu$ approaches zero and $\delta/\mu$ approaches the constant $\theta$. (This occurs, for example, when $\mu$ and $\delta$ are nonzero constants and $\sigma^2 = \delta^2(1-\beta^2)$; the $y_t$ are variance stationary but not mean stationary.) Then, as $\beta \to 1$,

$$y_t/\mu \to 1 + \theta u_0 \quad \text{for all} \quad t \geq 0.$$

*Case 3*: As $\beta$ approaches one, $\mu/\sigma$ approaches the constant $\bar{\mu}$ and $\delta/\sigma$ approaches the constant $\bar{\delta}$. (This occurs, for example, in either form of the fixed start-up model.) Then, as $\beta \to 1$,

$$y_t/\sigma \to \bar{\mu} + \bar{\delta}u_0 + u_1 + \cdots + u_t.$$

Clearly, the limiting behaviour in case 3 (where $y_t$ behaves like a random walk) is quite different from that in cases 1 and 2 (where $y_t$ does not depend on $t$). This carries over to the distributions of the least-squares statistics. Taking limits in equations (3)–(6), we find the following extensions of a result given in Jensen (1986, p. 139).

**Proposition 1.** In case 1, as $\beta$ tends to one, $\hat{\beta}$ converges in probability to one. Furthermore, for fixed $T$,

$$\text{(a)}\ T^{1/2}(\hat{\beta}-\beta)\frac{\delta}{\sigma} \to \frac{T^{1/2}\bar{u}}{u_0} \sim \text{ Cauchy},$$

$$\text{(b)}\ (T-1)\frac{\hat{\sigma}^2}{\sigma^2} \to \sum_t (u_t-\bar{u})^2 \sim \chi^2(T-1),$$

$$\text{(c)}\ t_\beta \to \sqrt{T(T-1)}\frac{\bar{u}}{\sqrt{\sum_t (u_t-\bar{u})^2}} \sim \text{ Student } (T-1),$$

$$\text{(d)}\ \frac{e_{T+s}}{\sigma} \to -(u_{T+1}+\cdots+u_{T+s})+s\bar{u} \sim N\left(0, s+\frac{s^2}{T}\right),$$

where $\bar{u} = T^{-1}\sum_t u_t$.

**Proposition 2.** In case 2, as $\beta$ tends to one, $\hat{\beta}$ converges in probability to one. For fixed $T$,

$$T^{1/2}(\hat{\beta}-\beta)\frac{\mu}{\sigma} \to \frac{T^{1/2}\bar{u}}{1+\theta u_0}.$$

The statistics $(T-1)\hat{\sigma}^2/\sigma^2, t_\beta$, and $e_{T+s}/\sigma$ approach the same random variables and therefore have the same limiting distributions as in Proposition 1.

The random walk case is considerably more complicated. Since $y_t/\sigma$ remains, in the limit, a linear function of all the errors, the least-squares estimator converges to a ratio of two quadratic functions of the errors. The exact distributions of the least-squares statistics have no simple expression, but large sample representations in terms of a continuous time process are available; see Phillips (1987). We have the following result comparable to those above.

**Proposition 3.** In case 3, as $\beta$ tends to one, $\hat{\beta}$ converges in probability to a nondegenerate random variable. The limiting distributions (as $\beta \to 1$) of the least-squares statistics have the large sample approximations:

$$\text{(a)}\ T(\hat{\beta}-\beta) \to \frac{\int BdB}{\int B^2} + \mathcal{O}(T^{-1/2}),$$

$$\text{(b)}\ (T-1)\frac{\hat{\sigma}^2}{\sigma^2} \to \sum_t (u_t-\bar{u})^2 + \mathcal{O}(T^0),$$

(c) $t_\beta \to \dfrac{\int B dB}{\left[\int B^2\right]^{1/2}} + \mathcal{O}(T^{-1/2})$,

(d) $\dfrac{e_{T+s}}{\sigma} \to -(u_{T+1} + \cdots + u_{T+s}) + \dfrac{s}{\sqrt{T}} \dfrac{B(1) \int B dB}{\int B^2} + \mathcal{O}(T^{-1})$,

where $B(r)$ is a Wiener process on the unit interval given by limit (as $T \to \infty$) of $(u_1 + \cdots + u_{[rT]})/\sqrt{T}$ and $[rT]$ denotes the integer part of $rT$.

The assumption about the initial value $y_0$ clearly matters a lot in determining the limiting behaviour, as $\beta \to 1$, of the least-squares estimator $\hat{\beta}$. In the stationary case, the standardised statistic $\sqrt{T}(\hat{\beta} - \beta)/\sqrt{1 - \beta^2}$ tends to a Cauchy variate; in the fixed start-up case, this statistic's dispersion grows without bound as $\beta$ tends to one. Or, put another way, under stationarity $\hat{\beta}$ converges rapidly to one for any sample size, whereas, under fixed start-up, $\hat{\beta}$ converges to a random variable with considerable dispersion unless $T$ is large.

At first glance, this finding seems to be at variance with previous Monte Carlo results. Phillips (1977), for example, reports that $\hat{\beta}$ has a highly skewed distribution in the stationary case when $\beta$ is large. Yet, as $\beta \to 1$, the limiting distribution is symmetric about 1. This suggests that the rate of convergence of $\sqrt{T}(\hat{\beta} - \beta)\delta/\sigma$ in Proposition 1 is very slow. To investigate this possibility, we calculated the percentiles of $\sqrt{T}(\hat{\beta} - \beta)/\sqrt{1 - \beta^2}$ for various values of $\beta$ under the stationary assumption that $\mu = 0$ and $\sigma^2 = \delta^2(1 - \beta^2)$. The value for $T$ was set at 14 and the Imhof routine was employed. The results are presented in the top panel of Table 1. The middle panel gives the percentiles of the variable $\hat{\beta} - \beta$. The lower panel gives the percentiles of the distribution of $\hat{\beta} - \beta$ under the fixed start-up assumption that $\mu = 0$ and $\delta = \sigma$. The last two columns give the interquartile range (IQR) and a skewness measure

$$\text{SK} = \frac{x_{.75} - x_{.50}}{x_{.50} - x_{.25}},$$

where $x_\alpha$ is the $\alpha$ percentile point. SK will be one for a symmetric distribution.

It is clear from the top panel that the rate of convergence to Cauchy is extremely slow. Although the standardisation factor $\delta/\sigma$ seems to stabilise the dispersion, the median and skewness are always far from the limiting values. Indeed, the largest departure for the median occurs at $\beta = 0.95$ and the largest departure for the skewness occurs at $\beta = 0.99$. The limiting distribution gives little indication of the shape of the true sampling distribution even for $\beta$ values as large as 0.999. On the other hand, a comparison of the middle and bottom panels suggests that the distribution of $\hat{\beta}$ in the fixed start-up model is a good approximation to the distribution of $\hat{\beta}$ in the stationary model for $\beta = 0.5$ but is not very good for $\beta$ greater than 0.9. This conclusion agrees well and, to some extent, explains the exact results on the mean-square forecast error obtained by Hoque, Magnus and Pesaran (1988).

Table 1: Distributions for regression without intercept in the model $y_t = \beta y_{t-1} + \sigma u_t (t = 1, \ldots, T), y_0 = \delta u_0$ for $T = 14$.

| $\beta$ | 2.5% | 10% | 25% | 50% | 75% | 90% | 97.5% | IQR | SK |
|---|---|---|---|---|---|---|---|---|---|
| | | | Percentage points of $\sqrt{T}(\hat{\beta} - \beta)\delta/\sigma,\ \delta = \sigma/\sqrt{1-\beta^2}$ | | | | | | |
| 0.50 | −2.499 | −1.613 | −0.868 | −0.130 | 0.493 | 0.956 | 1.381 | 1.362 | 0.844 |
| 0.80 | −3.545 | −2.175 | −1.150 | −0.272 | 0.355 | 0.761 | 1.169 | 1.505 | 0.713 |
| 0.90 | −4.568 | −2.690 | −1.385 | −0.367 | 0.287 | 0.728 | 1.227 | 1.672 | 0.643 |
| 0.95 | −5.847 | −3.279 | −1.630 | −0.438 | 0.280 | 0.786 | 1.400 | 1.910 | 0.603 |
| 0.99 | −9.942 | −4.713 | −1.912 | −0.397 | 0.404 | 1.092 | 2.157 | 2.317 | 0.528 |
| 0.999 | −17.945 | −5.395 | −1.524 | −0.206 | 0.642 | 1.729 | 4.072 | 2.166 | 0.644 |
| 1.00 | −12.706 | −3.078 | −1.000 | 0.000 | 1.000 | 3.078 | 12.706 | 2.000 | 1.000 |
| | | | Percentage points of $\hat{\beta} - \beta,\ \delta = \sigma/\sqrt{1-\beta^2}$ | | | | | | |
| 0.50 | −0.578 | −0.373 | −0.201 | −0.030 | 0.114 | 0.221 | 0.320 | 0.315 | 0.844 |
| 0.80 | −0.568 | −0.349 | −0.184 | −0.044 | 0.057 | 0.122 | 0.187 | 0.241 | 0.713 |
| 0.90 | −0.532 | −0.313 | −0.161 | −0.043 | 0.033 | 0.085 | 0.143 | 0.195 | 0.643 |
| 0.95 | −0.488 | −0.274 | −0.136 | −0.037 | 0.023 | 0.066 | 0.117 | 0.159 | 0.603 |
| 0.99 | −0.375 | −0.178 | −0.072 | −0.015 | 0.015 | 0.041 | 0.081 | 0.087 | 0.528 |
| 0.999 | −0.214 | −0.064 | −0.018 | −0.002 | 0.008 | 0.021 | 0.049 | 0.026 | 0.644 |
| 1.00 | 0.000 | 0.000 | 0.000 | 0.000 | 0.000 | 0.000 | 0.000 | 0.000 | 0.000 |
| | | | Percentage points of $\hat{\beta} - \beta,\ \delta = \sigma$ | | | | | | |
| 0.50 | −0.585 | −0.378 | −0.204 | −0.031 | 0.115 | 0.223 | 0.322 | 0.319 | 0.844 |
| 0.80 | −0.600 | −0.372 | −0.198 | −0.048 | 0.059 | 0.128 | 0.198 | 0.258 | 0.711 |
| 0.90 | −0.595 | −0.361 | −0.190 | −0.052 | 0.037 | 0.096 | 0.161 | 0.227 | 0.648 |
| 0.95 | −0.590 | −0.353 | −0.184 | −0.054 | 0.029 | 0.083 | 0.145 | 0.212 | 0.630 |
| 0.99 | −0.585 | −0.344 | −0.177 | −0.053 | 0.024 | 0.074 | 0.132 | 0.202 | 0.624 |
| 1.00 | −0.584 | −0.342 | −0.175 | −0.053 | 0.023 | 0.072 | 0.129 | 0.199 | 0.619 |

## 10.3 Regression with Intercept

When the regression contains an intercept, the relevant least-squares statistics based on $y_0, y_1, \ldots, y_T$ can be expressed as

$$\tilde{\beta} = \beta + \sigma\frac{\sum_t (y_{t-1} - \bar{y})u_t}{\sum_t (y_{t-1} - \bar{y})^2}, \tag{8}$$

$$\tilde{\sigma}^2 = \frac{\sigma^2}{T-2}\sum_t (u_t - \bar{u})^2 - (\tilde{\beta} - \beta)^2 \frac{1}{T-2}\sum_t (y_{t-1} - \bar{y})^2, \tag{9}$$

$$\tilde{t}_\beta = \frac{\sigma \sum_t (y_{t-1} - \bar{y})u_t}{\tilde{\sigma}\left[\sum_t (y_{t-1} - \bar{y})^2\right]^{1/2}}, \tag{10}$$

$$\begin{aligned}\tilde{e}_{T+s} &= -\sigma\sum_{i=0}^{s-1}\beta^i u_{T+s-i} + \sigma\bar{u}\frac{1-\beta^s}{1-\beta} \\ &\quad + (\tilde{\beta}^s - \beta^s)(y_T - \bar{y}) + \left[\frac{1-\tilde{\beta}^s}{1-\tilde{\beta}} - \frac{1-\beta^s}{1-\beta}\right]\frac{y_T - y_0}{T},\end{aligned} \tag{11}$$

where $\bar{y} = T^{-1}\sum_t y_{t-1}$. (We shall not examine the least-squares estimator or $t$-statistic for $\alpha$, whose behaviour is somewhat different.) The analysis is similar to that conducted in the previous section with $y_t - \bar{y}$ replacing $y_t$. By successive substitution in (1), we find

$$y_t = \alpha\frac{1-\beta^t}{1-\beta} + \beta^t(\mu + \delta u_0) + \sigma\sum_{i=0}^{t-1}\beta^i u_{t-i} \quad (t \geq 0)$$

and

$$\bar{y} = \frac{\alpha}{1-\beta} + \frac{1}{T}\frac{1-\beta^T}{1-\beta}\left(\mu + \delta u_0 - \frac{\alpha}{1-\beta}\right) + \frac{\sigma}{T}\sum_{j=1}^{T-1}\frac{1-\beta^{T-j}}{1-\beta}u_j,$$

and hence, for $t \geq 0$,

$$\begin{aligned}y_t - \bar{y} &= \frac{\beta^t - T^{-1}(1-\beta^T)/(1-\beta)}{1-\beta}\left[\delta(1-\beta)u_0 - \{\alpha - \mu(1-\beta)\}\right] \\ &\quad + \sigma\sum_{j=1}^{t}\beta^{t-j}u_j - \frac{\sigma}{T}\sum_{j=1}^{T-1}\frac{1-\beta^{T-j}}{1-\beta}u_j.\end{aligned} \tag{12}$$

The behaviour of $y_t - \bar{y}$ is rather different from that of $y_t$ when $\beta$ is near one. Now everything depends on the relative magnitudes of $\alpha - \mu(1-\beta)$, $\delta(1-\beta)$, and $\sigma$. Only one case seems to be of any interest (and covers both the fixed start-up and stationary models).

*Case 4*: As $\beta$ approaches one, $\delta(1-\beta)/\sigma$ approaches zero and $[\alpha - \mu(1-\beta)]/\sigma$ approaches the constant $\lambda$. Then, as $\beta \to 1$,

$$\frac{y_t - \bar{y}}{\sigma} \to \lambda\Big(t - \frac{T-1}{2}\Big) + \sum_{j=1}^{t} u_j - \sum_{j=1}^{T-1}\Big(1 - \frac{j}{T}\Big)u_j. \tag{13}$$

Again, when $T$ is large, the weighted sum of errors can be approximated by a functional on a Wiener process. We have, as $\beta$ tends to one and $t = rT$,

$$\frac{y_t - \bar{y}}{\sigma\sqrt{T}} \to \lambda\Big(r - \frac{1}{2}\Big)\sqrt{T} + B(r) - \int B + \mathcal{O}(T^{-1/2}). \tag{14}$$

The sampling distributions of the least-squares statistics depend crucially on $\lambda$. If the $y_t$ process is mean stationary (i.e., $Ey_t$ does not depend on $t$), then $\lambda$ is necessarily zero. Likewise, if $\alpha = 0$ and $\mu$ does not depend on $\beta$, then $\lambda$ will be zero. In these cases, we have

**Proposition 4.** If, as $\beta \to 1, \delta(1-\beta)/\sigma$ and $[\alpha - \mu(1-\beta)]/\sigma$ both tend to zero, then $\tilde{\beta}$ converges in probability to a nondegenerate random variable. The limiting distributions (as $\beta \to 1$) of the least-squares statistics have the large sample approximations

(a) $T(\tilde{\beta} - \beta) \to \dfrac{\int BdB \;-\; B(1)\int B}{\int B^2 - (\int B)^2} + \mathcal{O}(T^{-1/2})$,

(b) $(T-2)\dfrac{\tilde{\sigma}^2}{\sigma^2} \to \chi^2(T-2) + \mathcal{O}(T^0)$,

(c) $\tilde{t}_\beta \to \dfrac{\int BdB - B(1)\int B}{\left[\int B^2 - (\int B)^2\right]^{1/2}} + \mathcal{O}(T^{-1/2})$,

(d) $$\frac{\tilde{e}_{T+s}}{\sigma} \to -(u_{T+1} + \cdots + u_{T+s}) + \frac{s}{\sqrt{T}}B(1) + \frac{s}{\sqrt{T}}\frac{\int BdB - B(1)\int B}{\int B^2 - (\int B)^2}\left\{B(1) - \int B\right\} + \mathcal{O}(T^{-1}).$$

If the $\{y_t\}$ are not mean stationary, then $\lambda$ will generally be nonzero. In this case $y_t$ behaves like a time trend (plus noise) when $\beta$ is near one. We have

**Proposition 5.** If, as $\beta \to 1$, $\delta(1-\beta)/\sigma$ tends to zero but $[\alpha - \mu(1-\beta)]/\sigma$ tends to the nonzero constant $\lambda$, then $\tilde{\beta}$ converges in probability to a nondegenerate random variable. The limiting distributions (as $\beta \to 1$) of the least-squares statistics have the large sample approximations

(a) $\dfrac{\lambda}{\sqrt{12}}T\sqrt{T}(\tilde{\beta} - \beta) \to N(0,1) + \mathcal{O}(T^{-1/2})$,

(b) $(T-2)\frac{\tilde{\sigma}^2}{\sigma^2} \to \chi^2(T-2) + \mathcal{O}(T^{-1/2})$,

(c) $\tilde{t}_\beta \to$ Student $(T-2) + \mathcal{O}(T^{-1/2})$,

(d) $\frac{\tilde{e}_{T+s}}{\sigma} \to N\Big(0, s + \frac{4s^2}{T}\Big) + \mathcal{O}(T^{-1})$.

These results indicate that, when $\beta$ is near one, regression with an intercept is very different from regression without an intercept. In the former case, the limiting behaviour of $\tilde{\beta}$ is quite sensitive to assumptions on $\mu$, but not sensitive to assumptions on $\delta$. In the latter case, the opposite holds. In regression with intercept but $\alpha = 0$ , the fixed start-up and the stationary models produce the same limiting distributions. In regression without intercept, the two models produce very different limiting distributions.

Some percentiles for the distribution of $\tilde{\beta}-\beta$ when $T$=14 are given in Table 2. The top panel is for the stationary model where $\alpha = \mu(1-\beta)$ and $\sigma^2 = \delta^2(1-\beta^2)$. The bottom panel is for the fixed start-up model where $\mu = \alpha = 0$ and $\delta = \sigma$. The two distributions are very similar, although not identical, when $\beta$ is close but not equal to one. For example, when $\beta = 0.90$, the dispersion and skewness of the two distributions are close, but the medians differ by about 8%. This agrees with the results obtained by Evans and Savin (1984, p. 1265) and Magnus and Pesaran (1989, 1991).

## 10.4 Negative Unit Root

Although perhaps of less practical importance, the situation where $\beta$ is near minus one also produces some unusual sampling behaviour. In regression without intercept, a symmetry argument employed by Fuller (1976) can be exploited. With $x_t = (-1)^t y_t, v_t = (-1)^t u_t$ and $\alpha = 0$, the model (1)–S(2) can be written as

$$x_t = \gamma x_{t-1} + \omega v_t, \tag{15}$$

$$x_0 = \mu + \delta v_0, \tag{16}$$

where $\gamma = -\beta$ and $\omega = \sigma$. Since the $v_t$ are independent standard normal variables, the results of Section 2 apply to this new model for $\gamma$ tending to one. Furthermore, a regression of $x_t$ on $x_{t-1}$ produces statistics $\hat{\gamma} = -\hat{\beta}, t_\gamma = -t_\beta, \hat{\omega}^2 = \hat{\sigma}^2$, and $\hat{x}_{T+s} - x_{T+s} = (-1)^{T+s}(\hat{y}_{T+s} - y_{T+s})$. Thus we have

**Proposition 6.** The limiting distributions of $-(\hat{\beta}-\beta), -t_\beta, \hat{\sigma}^2$, and $(-1)^{T+s}e_{T+s}/\sigma$ as $\beta \to -1$ are identical to those given in Propositions 1-3 (where the three cases now describe the behaviour of the parameters as $\beta \to -1$ and the $u$'s are replaced with the $v$'s.)

Regression with an intercept, however, does not demonstrate the same symmetries. From (12), the behaviour of $y_t - \bar{y}$ as $\beta$ tends to minus one depends

Table 2: Distributions for regression with intercept in the model $y_t = \alpha + \beta y_{t-1} + \sigma u_t (t = 1, \ldots, T), y_0 = \delta u_0$ for $T = 14$.

| $\beta$ | 2.5% | 10% | 25% | 50% | 75% | 90% | 97.5% | IQR | SK |
|---|---|---|---|---|---|---|---|---|---|
| | | | Percentage points of $\tilde{\beta} - \beta, \alpha = 0, \delta = \sigma/\sqrt{1-\beta^2}$ | | | | | | |
| 0.50 | −0.717 | −0.513 | −0.334 | −0.151 | 0.011 | 0.136 | 0.255 | 0.346 | 0.879 |
| 0.80 | −0.809 | −0.579 | −0.391 | −0.213 | −0.071 | 0.033 | 0.143 | 0.320 | 0.796 |
| 0.90 | −0.844 | −0.607 | −0.417 | −0.242 | −0.106 | −0.003 | 0.104 | 0.311 | 0.776 |
| 0.95 | −0.864 | −0.624 | −0.433 | −0.259 | −0.125 | −0.023 | 0.082 | 0.308 | 0.774 |
| 0.99 | −0.882 | −0.639 | −0.447 | −0.274 | −0.140 | −0.040 | 0.061 | 0.307 | 0.774 |
| 1.00 | −0.887 | −0.644 | −0.451 | −0.278 | −0.144 | −0.045 | 0.056 | 0.307 | 0.774 |
| | | | Percentage points of $\tilde{\beta} - \beta,\ \alpha = 0, \delta = \sigma$ | | | | | | |
| 0.50 | −0.723 | −0.519 | −0.339 | −0.154 | 0.010 | 0.135 | 0.256 | 0.349 | 0.880 |
| 0.80 | −0.831 | −0.601 | −0.410 | −0.228 | −0.082 | 0.024 | 0.135 | 0.329 | 0.801 |
| 0.90 | −0.867 | −0.631 | −0.441 | −0.263 | −0.124 | −0.020 | 0.090 | 0.316 | 0.779 |
| 0.95 | −0.881 | −0.642 | −0.451 | −0.277 | −0.142 | −0.038 | 0.070 | 0.310 | 0.774 |
| 0.99 | −0.887 | −0.645 | −0.453 | −0.280 | −0.146 | −0.045 | 0.058 | 0.307 | 0.775 |
| 1.00 | −0.887 | −0.644 | −0.451 | −0.278 | −0.144 | −0.045 | 0.056 | 0.307 | 0.774 |

on the relative magnitudes of $\alpha - \mu(1-\beta), \delta$, and $\sigma$. As noted by Fuller (1976), the results are quite different from the situation where $\beta$ tends to plus one. In fact, the results are close to those for regression without intercept. Again three cases can be distinguished.

**Proposition 7.** Suppose both $\sigma/\delta$ and $[\alpha - \mu(1-\beta)]/\delta$ tend to zero as $\beta$ approaches minus one; e.g., the $y_t$ are stationary. Then, as $\beta \to -1$,

$$\frac{y_t - \bar{y}}{\delta} \to \begin{cases} (-1)^t u_0 & (T \text{ even}) \\ \left\{(-1)^t - \frac{1}{T}\right\} u_0 & (T \text{ odd}) \end{cases}$$

and $\tilde{\beta}$ converges in probability to minus one. Furthermore, for fixed $T$,

(a) $\varphi_T^{1/2}(\tilde{\beta} - \beta)\frac{\delta}{\sigma} \to$ Cauchy,

(b) $(T-2)\frac{\tilde{\sigma}^2}{\sigma^2} \to \chi^2(T-2)$,

(c) $\tilde{t}_\beta \to$ Student $(T-2)$,

(d) $\frac{\tilde{e}_{T+s}}{\sigma} \to N(0, \omega_{T,s}^2)$,

where

$$\varphi_T = \begin{cases} T & (T \text{ even}) \\ T - \frac{1}{T} & (T \text{ odd}) \end{cases}$$

and

$$\omega_{T,s}^2 = \begin{cases} s + \dfrac{s^2}{T} & (T \text{ even}, s \text{ even}) \\ s + \dfrac{s^2+1}{T} & (T \text{ even}, s \text{ odd}) \\ s + \dfrac{s^2(T+s)^2}{T(T^2-1)} & (T \text{ odd}, s \text{ even}) \\ s + \dfrac{s^2(T+s)^2}{T(T^2-1)} + \dfrac{1}{T} & (T \text{ odd}, s \text{ odd}). \end{cases}$$

**Proposition 8.** Let $\mu_* = -\frac{1}{2}[\alpha - \mu(1-\beta)]$ and suppose $\sigma/\mu_*$ tends to zero and $\delta/\mu_*$ tends to the constant $\bar{\theta}$ as $\beta$ approaches minus one; e.g. the $y_t$ are variance stationary, but not mean stationary. Then, as $\beta \to -1$,

$$\frac{y_t - \bar{y}}{\mu_*} \to \begin{cases} (-1)^t(1+\bar{\theta}u_0) & (T \text{ even}) \\ \{(-1)^t - T^{-1}\}(1+\bar{\theta}u_0) & (T \text{ odd}), \end{cases}$$

and $\tilde{\beta}$ converges in probability to minus one. Furthermore, for fixed $T$,

$$\varphi_T^{1/2}(\tilde{\beta} - \beta)\frac{\mu_*}{\sigma} \to \frac{z}{1+\bar{\theta}u_0},$$

where $\varphi_T$ is defined in Proposition 7 and $z \sim N(0,1)$ is independent of $u_0$; the remaining statistics behave as in Proposition 7.

**Proposition 9.** Suppose $\delta/\sigma \to \bar{\delta}$ and $[\alpha - \mu(1-\beta)]/\sigma \to \lambda$ as $\beta$ approaches minus one; e.g., the starting value $y_0$ is fixed. Then, as $\beta \to -1$,

$$\frac{y_t - \bar{y}}{\sigma} \to (-1)^t\Big[\bar{\delta}u_0 - \frac{1}{2}\lambda + \sum_{j=1}^{t} v_j\Big] + \frac{1}{T}S_T,$$

where

$$S_T = \begin{cases} v_1 + v_3 + \cdots + v_{T-1} & (T \text{ even}) \\ -\left[\bar{\delta}v_0 - \frac{1}{2}\lambda + v_2 + v_4 + \cdots + v_{T-1}\right] & (T \text{ odd}). \end{cases}$$

If, for large $T$, $(v_1 + \cdots + v_t)/\sqrt{T}$ is approximated by the Wiener process $\bar{B}(r)$ where $r = t/T$, then

(a) $T(\tilde{\beta} - \beta) \to -\dfrac{\int \bar{B}d\bar{B}}{\int \bar{B}^2} + \mathcal{O}(T^{-1/2})$,

(b) $(T-2)\dfrac{\tilde{\sigma}^2}{\sigma^2} \to \chi^2(T-2) + \mathcal{O}(T^0)$,

(c) $\tilde{t}_\beta \to -\dfrac{\int \bar{B}d\bar{B}}{\left[\int \bar{B}^2\right]^{1/2}} + \mathcal{O}(T^{-1/2})$,

(d) $$\frac{\tilde{e}_{T+s}}{\sigma} \to (-1)^{T+s-1}\left[v_{T+1} + \cdots + v_{T+s}\right] + (-1)^{T-1}\frac{s}{\sqrt{T}}\frac{\int \bar{B}d\bar{B}}{\int \bar{B}^2}\bar{B}(1) + \frac{B(1)}{\sqrt{T}} + \mathcal{O}(T^{-1})$$

for $s$ odd; the term $B(1)/\sqrt{T}$ is dropped when $s$ is even.

## 10.5 Conclusions

Assumptions about the initial observation can matter a lot in the AR(1) model with near-unit root. When regressions are fitted without intercept, the stationary and fixed start-up models produce very different sampling properties when the autoregression parameter is near plus or minus one. When regressions are fitted with intercept, the two models have similar sampling properties when the autoregression parameter is near plus one, but very different properties when the parameter is near minus one.

As noted by Jensen (1986), the least-squares statistics in regression without intercept have simple limiting distributions in the stationary model as the autoregression parameter tends to unity. Unfortunately, those limiting distributions do not seem to provide useful approximations for any given stationary model.

## Appendix: Derivations of Propositions

**Proposition 1.** Using the fact that $y_t/\delta \to u_0$, we find

$$\sum y_{t-1}u_t/\delta \to u_0 \sum u_t \tag{17}$$

and

$$\sum y_{t-1}^2/\delta^2 \to Tu_0^2. \tag{18}$$

This, together with (3) proves (a). Then (b) follows from (4) and the fact that

$$\left[\frac{\hat{\beta}-\beta}{\sigma}\right]^2 \sum y_{t-1}^2 \to T(\bar{u})^2.$$

Next, (c) follows from (5), (17), (18) and (b). To prove (d), we use a slight generalisation of (a), proved by a simple Taylor expansion, namely

$$T^{1/2}\left[\frac{\hat{\beta}^s-\beta^s}{s}\right]\frac{\delta}{\sigma} \to \frac{T^{1/2}\bar{u}}{u_0} \sim \text{ Cauchy.} \tag{19}$$

Using (19), we obtain

$$(\hat{\beta}^s - \beta^s)y_T/\sigma \to s\bar{u} \sim N\Big(0, \frac{s^2}{T}\Big)$$

and hence (d).

**Proposition 2.** Similar to the derivation of Proposition 1.

**Proposition 3.** We have, as $\beta \to 1$,

$$y_t/\sigma \to \bar{\mu} + \bar{\delta}u_0 + T^{1/2}w_t,$$

where

$$w_t = T^{-1/2}(u_1 + u_2 + \cdots + u_t), \qquad w_0 = 0. \tag{20}$$

We shall employ the following approximations for large $T$; see Phillips (1987, p. 294):

$$T^{-1}\sum w_{t-1}^2 = \int_0^1 B(r)^2 dr + \mathcal{O}(T^{-1}), \tag{21}$$

$$\sum w_{t-1}(w_t - w_{t-1}) = \int_0^1 B(r)dB(r) + \mathcal{O}(T^{-1/2}). \tag{22}$$

In addition, we note that $w_T = B(1) \equiv N(0,1)$. Then, as $\beta \to 1$,

$$\begin{aligned} T^{-1}\sum y_{t-1}u_t/\sigma &\to \sum w_{t-1}(w_t - w_{t-1}) + T^{-1/2}(\bar{\mu} + \bar{\delta}u_0)w_T \\ &= \int BdB + \mathcal{O}(T^{-1/2}) \end{aligned}$$

and

$$\begin{aligned} T^{-2}\sum y_{t-1}^2/\sigma^2 &\to T^{-1}\sum w_{t-1}^2 + 2T^{-1/2}(\bar{\mu} + \bar{\delta}u_0)\Big(T^{-1}\sum w_{t-1}\Big) \\ &\quad + T^{-1}(\bar{\mu} + \bar{\delta}u_0)^2 \\ &= \int B^2 + \mathcal{O}(T^{-\frac{1}{2}}), \end{aligned}$$

and, hence, using (3)–(6), the proposition follows.

**Proposition 4.** Since $\lambda = 0$, we have from (13), as $\beta \to 1$,

$$\frac{y_t - \bar{y}}{\sigma\sqrt{T}} \to w_t - T^{-1}\sum w_{t-1}$$

where $w_t$ is defined in (20). Hence

$$\begin{aligned} T^{-1}\sum(y_{t-1} - \bar{y})u_t/\sigma &\to \sum w_{t-1}(w_t - w_{t-1}) - w_T\Big(T^{-1}\sum w_{t-1}\Big) \\ &= \int BdB - B(1)\int B + \mathcal{O}(T^{-1/2}), \\ T^{-2}\sum(y_{t-1} - \bar{y})^2/\sigma^2 &\to T^{-1}\sum w_{t-1}^2 - \Big(T^{-1}\sum w_{t-1}\Big)^2 \\ &= \int B^2 - \Big(\int B\Big)^2 + \mathcal{O}(T^{-1}), \end{aligned}$$

where, in addition to (21) and (22), we have used

$$T^{-1}\sum w_{t-1} = \int_0^1 B(r)dr + \mathcal{O}(T^{-1}). \tag{23}$$

The results now follow from (8)–(11).

**Proposition 5.** We define

$$c_t = \left[12T^{-3}\right]^{1/2}\left(t - \frac{T+1}{2}\right) \qquad (t = 1, 2, \ldots, T),$$

so that $\sum c_t = 0$ and $\sum c_t^2 = 1 + \mathcal{O}(T^{-2})$. Then, letting $\gamma = \lambda/\sqrt{12}$, we obtain from (13), as $\beta \to 1$,

$$\frac{y_{t-1} - \bar{y}}{\sigma\sqrt{T}} \to \gamma T c_t + w_{t-1} - \frac{1}{T}\sum w_{t-1}$$

and hence

$$\begin{aligned} T^{-1}\sum \frac{y_{t-1} - \bar{y}}{\sigma\sqrt{T}} u_t \quad &\to \quad \gamma z_1 + T^{-1/2}Q, \\ T^{-2}\sum\left[\frac{y_{t-1} - \bar{y}}{\sigma\sqrt{T}}\right]^2 \quad &\to \quad \gamma^2 + 2\gamma T^{-1/2} z_2 + \mathcal{O}(T^{-1}), \end{aligned}$$

where

$$\begin{aligned} z_1 &= \sum c_t u_t \sim N(0,1) + \mathcal{O}(T^{-2}), \\ z_2 &= T^{-1/2}\sum c_t w_{t-1} \sim N\Big(0, \frac{1}{10}\Big) + \mathcal{O}(T^{-1}), \\ Q &= \sum w_{t-1}(w_t - w_{t-1}) - w_T\Big(T^{-1}\sum w_{t-1}\Big). \end{aligned}$$

Using these results, we obtain, from (8)–(11),

$$\begin{aligned} \gamma T^{3/2}(\tilde{\beta} - \beta) \quad &\to \quad z_1 + T^{-1/2}\frac{Q - 2z_1 z_2}{\gamma} + \mathcal{O}(T^{-1}), \\ \frac{(T-2)\tilde{\sigma}^2}{\sigma^2} \quad &\to \quad \sum(u_t - \bar{u})^2 - z_1^2 - 2T^{-1/2} z_1(Q - z_1 z_2)/\gamma + \mathcal{O}(T^{-1}), \\ \tilde{t}_\beta \quad &\to \quad \left[\frac{\sum(u_t - \bar{u})^2 - z_1^2}{T-2}\right]^{-1/2} z_1 + \mathcal{O}(T^{-1/2}), \\ \frac{\tilde{e}_{T+s}}{\sigma} \quad &\to \quad -(u_{T+1} + \cdots + u_{T+s}) + \frac{s}{\sqrt{T}}(w_T + z_1\sqrt{3}) \\ &+ \quad \frac{s}{\gamma T}\Big[\sqrt{3}(Q - 2z_1 z_2) + z_1(w_T - T^{-1}\sum w_{t-1})\Big] + \mathcal{O}(T^{-3/2}). \end{aligned}$$

The results of Proposition 5 then follow. In fact, we have proved a bit more than Proposition 5. For example, noting that

$$EQ = -\frac{1}{2} + \mathcal{O}(T^{-1}), \qquad Ez_1z_2 = \mathcal{O}(T^{-1}),$$

$$Ez_1w_T = 0, \qquad Ez_1(T^{-1}\sum w_{t-1}) = -\frac{1}{6}\sqrt{3} + \mathcal{O}(T^{-2}),$$

we obtain the following expectations:

$$T^2E(\tilde{\beta} - \beta) \quad \to \quad -\frac{6}{\lambda^2} + \mathcal{O}(T^{-1/2}), \tag{24}$$

$$E\frac{\tilde{\sigma}^2}{\sigma^2} \quad \to \quad 1 + \mathcal{O}(T^{-2}), \tag{25}$$

$$TE\frac{\tilde{e}_{T+s}}{\sigma} \quad \to \quad \frac{-2s}{\lambda} + \mathcal{O}(T^{-1/2}). \tag{26}$$

**Proposition 6.** See text.

**Proposition 7.** From (12), we have

$$(y_t - \bar{y})/\delta \to \bar{k}_t u_0,$$

where

$$\bar{k}_t = \begin{cases} (-1)^t & (T \text{ even}) \\ (-1)^t - 1/T & (T \text{ odd}). \end{cases}$$

Since $\sum \bar{k}_{t-1}^2 = \varphi_T$ and defining $k_t = \varphi_T^{-1/2}\bar{k}_t$, we obtain

$$\sum(y_{t-1} - \bar{y})u_t/\delta \to u_0\varphi_T^{1/2}\sum k_{t-1}u_t$$

and

$$\sum(y_{t-1} - \bar{y})^2/\delta^2 \to u_0^2\varphi_T.$$

Using (8), we then find

$$\varphi_T^{1/2}(\tilde{\beta} - \beta)\delta/\sigma \to \frac{\sum k_{t-1}u_t}{u_0} \sim \text{ Cauchy}.$$

Next, since

$$(\tilde{\beta} - \beta)^2\sum(y_{t-1} - \bar{y})^2/\sigma^2 \to \left(\sum k_{t-1}u_t\right)^2 \sim \chi^2(1),$$

and defining

$$M = I_T - ii'/T, \qquad i = (1, 1, \ldots, 1)',$$

$$k = (k_0, k_1, \ldots, k_{T-1})' \qquad u = (u_1, u_2, \ldots, u_T)',$$

we obtain, using (9),

$$(T-2)\tilde{\sigma}^2/\sigma^2 \to \sum(u_t-\bar{u})^2 - \left(\sum k_{t-1}u_t\right)^2 = u'(M-kk')u \sim \chi^2(T-2),$$

since $Mk = k$. Also, using (10),

$$\tilde{t}_\beta \to \left[\frac{u'(M-kk')u}{T-2}\right]^{-1/2} \sum k_{t-1}u_t \sim \text{ Student } (T-2),$$

since $(M-kk')k = 0$. Finally, using (11),

$$\tilde{e}_{T+s}/\sigma \to z_1 + z_2 + z_3,$$

where $z_1, z_2$ and $z_3$ are independent random variables defined by

$$z_1 = -\sum_{i=0}^{s-1}(-1)^i u_{T+s-i} \sim N(0,s),$$

$$z_2 = \begin{cases} 0 & (s \text{ even}) \\ \bar{u} \sim N(0, 1/T) & (s \text{ odd}), \end{cases}$$

$$z_3 = \begin{cases} T^{-1/2}s\sum k_{t-1}u_t \sim N(0, s^2/T) & (T \text{ even}) \\ -[T(T^2-1)]^{-1/2}s(s+T)\sum k_{t-1}u_t \sim N\left[0, \dfrac{s^2(s+T)^2}{T(T^2-1)}\right] & (T \text{ odd}). \end{cases}$$

The result follows.

**Proposition 8.** Similar to the derivation of Proposition 7.

**Proposition 9.** We have, as $\beta \to 1$, from (12):

$$(y_t - \bar{y})/\sigma \to (-1)^t T^{1/2}\bar{w}_t + (-1)^t(\bar{\delta}u_0 - \frac{1}{2}\lambda) + T^{-1/2}(T^{-1/2}S_T),$$

where

$$\bar{w}_t = T^{-1/2}(v_1 + v_2 + \cdots + v_t).$$

Hence,

$$T^{-1}\sum(y_{t-1}-\bar{y})u_t/\sigma \to -\int \bar{B}d\bar{B} + \mathcal{O}(T^{-1/2})$$

and

$$T^{-2}\sum(y_{t-1}-\bar{y})^2/\sigma^2 \to \int \bar{B}^2 + \mathcal{O}(T^{-1/2}).$$

The results then follow using (8)–(11).

## References


[1] Chipman, J.S., (1979), Efficiency of Least-squares Estimation of Linear Trend when Residuals are Autocorrelated, *Econometrica,* 47, 115–128.

[2] Evans, G.B.A., and N.E. Savin, (1981), Testing for Unit Roots: 1, *Econometrica,* 49, 753–779.

[3] Evans, G.B.A., and N.E. Savin, (1984), Testing for Unit Roots: 2, *Econometrica,* 52, 1241–1269.

[4] Fuller, W.A., (1976), *Introduction to Statistical Time Series*, John Wiley and Sons, New York.

[5] Hoque, A., J.R. Magnus, and B. Pesaran, (1988), The Exact Multi-period Mean-square Forecast Error for the First-order Autoregressive Model, *Journal of Econometrics,* 39, 327–346.

[6] Jensen, J.L., (1986), *Asymptotic Expansions for Sums of Dependent Variables,* Memoirs No. 10, Department of Theoretical Statistics, University of Aarhus, Denmark.

[7] Magnus, J.R., and B. Pesaran, (1989), The Exact Multiperiod Mean-square Forecast Error for the First-order Autoregressive Model with an Intercept, *Journal of Econometrics,* 42, 157–179.

[8] Magnus, J.R., and B. Pesaran, (1991), The Bias of Forecasts from a First-order Autoregression, *Econometric Theory,* 7, 222–235.

[9] Phillips, P.C.B., (1977), Approximations to Some Finite Sample Distributions Associated with a First-order Stochastic Difference Equation, *Econometrica,* 45, 463–485.

[10] Phillips, P.C.B., (1987), Time Series Regression with a Unit Root, *Econometrica,* 55, 277–301.

## References

[1] Chipman, J.S. (1979) Efficiency of Least-Squares Estimation of Linear [illegible]

[illegible]

CHAPTER 11

# Efficiency Comparisons for a System GMM Estimator in Dynamic Panel Data Models

Frank WINDMEIJER[1]

## 11.1 Introduction

The system GMM estimator in dynamic panel data models combines moment conditions for the differenced equation with moment conditions for the model in levels. An initial optimal weight matrix under homoskedasticity and non-serial correlation is not known for this estimation procedure. It is common practice to use the inverse of the moment matrix of the instruments as the initial weight matrix. This paper assesses the potential efficiency loss from the use of this weight matrix using the efficiency bounds as derived by Liu and Neudecker (1997).

A standard practice to estimate the parameters in dynamic panel data models is to take first differences to eliminate the correlated individual specific effects, and estimate the differenced model by generalised method of moments (GMM) using appropriately lagged level variables as instruments. As the information of the instruments for the differenced model decreases as the series become more persistent, Arellano and Bover (1995) and Blundell and Bond (1998) have proposed use of a system GMM estimator that combines the differenced equation with the level equation. The instruments for the level equation are lagged differences of the variables, which are valid when these differences are uncorrelated with the individual effects. Blundell and Bond (1998) show that the system estimator has superior properties in terms of small sample bias and RMSE, especially for persistent series.

The GMM estimator is a two-step estimator. In the first step, an initial positive semidefinite weight matrix is used to obtain consistent estimates of the parameters. Given these consistent estimates, a weight matrix can be constructed that is consistent for the efficient weight matrix, and this weight matrix

[1] I would like to thank Steve Bond for helpful comments. The usual disclaimer applies.

is used for the asymptotically efficient two-step estimates. It is well known, see, for example, Arellano and Bond (1991), that the two-step estimated standard errors have a small sample downward bias in this dynamic panel data setting, and one-step estimates with robust standard errors are often preferred. Although an efficient weight matrix for the differenced model with errors that are homoskedastic and that are not serially correlated is easily derived, this is not the case for the system estimator, combining differences and levels information.

It is common practice to use the inverse of the moment matrix of the instruments as an initial weight matrix. In this paper the potential efficiency loss will be considered in a model with homoskedastic and serially uncorrelated errors. To do this, upper bounds for the efficiency loss will be calculated as derived by Liu and Neudecker (1997) based on the Kantorovich inequality (KI). These upper bounds indicate that the efficiency loss could potentially be quite severe. When the variance of the individual unobserved heterogeneity is small, efficiency can be gained by using a weight matrix that is optimal under the assumption that the variance of the unobserved heterogeneity is equal to zero.

In section 2, an AR(1) dynamic panel data model is considered and a description of the system GMM estimator is given. In section 3 the upper bounds of the efficiency loss are calculated for 3 and 4 time periods respectively. Section 4 concludes.

## 11.2 Model and System GMM Estimator

Consider the AR(1) panel data specification

$$y_{it} = \alpha_0 y_{it-1} + \eta_i + \varepsilon_{it}, \tag{2.1}$$

for $i = 1, \ldots, N$, $t = 2, \ldots, T$, with $N$ large, and $T$ fixed. The error terms follow the error components structure in which

$$E(\eta_i) = 0, \qquad E(\varepsilon_{it}) = 0,$$

$$E\left(\varepsilon_{it}^2\right) = \sigma_\varepsilon^2, \qquad E\left(\eta_i^2\right) = \sigma_\eta^2,$$

$$E(\eta_i \varepsilon_{it}) = 0, \qquad E(\varepsilon_{it}\varepsilon_{is}) = 0, \quad t \neq s.$$

The $y_{it}$ series are assumed stationary with an infinite time horizon and therefore the series can alternatively be written as

$$y_{it} = \frac{\eta_i}{1-\alpha_0} + \sum_{j=0}^{\infty} \alpha_0^j \varepsilon_{it-j}. \tag{2.2}$$

The OLS and within groups estimators of $\alpha_0$ in model (2.1) are biased and inconsistent. A consistent estimator for $\alpha_0$ is the system GMM estimator, as proposed by Arellano and Bover (1995) and Blundell and Bond (1998), utilising the following $(T+1)(T-2)/2$ moment conditions[2]

$$
\begin{aligned}
E\left[\left(\Delta y_{it}-\alpha_0\Delta y_{it-1}\right)\left(y_{it-2},\ldots,y_{i1}\right)\right] &= 0, && (2.3)\\
E\left[\left(y_{it}-\alpha_0 y_{it-1}\right)\Delta y_{it-1}\right] &= 0, && (2.4)
\end{aligned}
$$

for $t=3,\ldots,T$. Moment conditions (2.3) are for the model in first differences, utilising appropriately lagged levels information as instruments, whereas conditions (2.4) are for the model in levels, utilising lagged differences as instruments. As Blundell and Bond (1998) show, the system estimator is considerably more efficient than the traditional GMM estimator utilising the moment conditions of the differenced model only.

Define

$$
Z_i = \begin{bmatrix}
y_{i1} & 0 & 0 & \ldots & 0 & \ldots & 0 & 0 & 0 & \ldots & 0\\
0 & y_{i1} & y_{i2} & \ldots & 0 & \ldots & 0 & 0 & 0 & \ldots & 0\\
0 & 0 & 0 & \ldots & 0 & \ldots & 0 & 0 & 0 & \ldots & 0\\
0 & 0 & 0 & \ldots & y_{i1} & \ldots & y_{iT-2} & 0 & 0 & \ldots & 0\\
0 & 0 & 0 & \ldots & 0 & \ldots & 0 & \Delta y_{i2} & 0 & \ldots & 0\\
0 & 0 & 0 & \ldots & 0 & \ldots & 0 & 0 & \Delta y_{i3} & \ldots & 0\\
0 & 0 & 0 & \ldots & 0 & \ldots & 0 & 0 & 0 & \ldots & 0\\
0 & 0 & 0 & \ldots & 0 & \ldots & 0 & 0 & 0 & \ldots & \Delta y_{T-1}
\end{bmatrix},
$$

$$
v_i = v_i(\alpha_0) = \begin{bmatrix}
\Delta y_{i3}-\alpha_0\Delta y_{i2}\\
\Delta y_{i4}-\alpha_0\Delta y_{i3}\\
\vdots\\
\Delta y_{iT}-\alpha_0\Delta y_{iT-1}\\
y_{i3}-\alpha_0 y_{i2}\\
y_{i4}-\alpha_0 y_{i3}\\
\vdots\\
y_{iT}-\alpha_0 y_{iT-1}
\end{bmatrix},
$$

and

$$
f_i(\alpha_0) = Z_i' v_i.
$$

Moment conditions (2.3) and (2.4) imply that $E\{f_i(\alpha_0)\}=0$. The GMM estimator[3] $\widehat{\alpha}$ for $\alpha_0$ minimises

$$
\left[\frac{1}{N}\sum_{i=1}^{N} f_i(\alpha)\right]' W_N \left[\frac{1}{N}\sum_{i=1}^{N} f_i(\alpha)\right],
$$

with respect to $\alpha$; where $W_N$ is a positive semidefinite weight matrix which satisfies $\text{plim}_{N\to\infty}W_N = W$, with $W$ a positive definite matrix. Regularity

[2]Under homoskedasticity there are additional moment conditions available to improve efficiency, see Ahn and Schmidt (1995). These extra moment conditions are not considered here.

[3]See Hansen (1982), Ogaki (1993). Here, the same notation as in Liu and Neudecker (1997) is used.

conditions are in place such that $\lim_{N\to\infty} N^{-1}\sum_{i=1}^{N} f_i(\alpha) = E\{f(\alpha)\}$ and $N^{-1/2} f_i(\alpha_0) \to N(0,\Psi)$. Let $F(\alpha) = E(\partial f_i(\alpha)/\partial\alpha)$ and $F_0 \equiv F(\alpha_0)$, then $\sqrt{N}(\widehat{\alpha}-\alpha_0)$ has a limiting normal distribution, $\sqrt{N}(\widehat{\alpha}-\alpha_0) \to N(0,V_W)$, where

$$V_W = (F_0'WF_0)^{-1} F_0'W\Psi WF_0 (F_0'WF_0)^{-1}. \tag{2.5}$$

## 11.3 Efficiency Comparisons

As is clear from the expression of the asymptotic variance matrix $V_W$, (2.5), the efficiency of the GMM estimator is affected by the choice of the weight matrix $W_N$. An optimal choice is a weight matrix for which $W = \Psi^{-1}$. The asymptotic variance matrix is then given by $\left(F_0'\Psi^{-1}F_0\right)^{-1}$. For any other $W$, the GMM estimator is less efficient as

$$\left(F_0'\Psi^{-1}F_0\right)^{-1} \leq (F_0'WF_0)^{-1} F_0'W\Psi WF_0 (F_0'WF_0)^{-1}.$$

In panel data models, the efficient estimator is obtained in a two-step procedure. The one-step GMM estimator $\widetilde{\alpha}$ is obtained using an arbitrary weight matrix $W_{N1}$. Let $\widetilde{v}_i = v_i(\widetilde{\alpha})$. The efficient two-step estimator is then based on the weight matrix

$$W_{N2} = \left(\frac{1}{N}\sum_{i=1}^{N} Z_i'\widetilde{v}_i\widetilde{v}_i Z_i\right)^{-1}; \qquad \text{plim } W_{N2} = \Psi^{-1}.$$

Although the efficient estimator is easily obtained, there is a serious problem associated with it as the estimated standard errors of the two-step estimator can be biased downwards quite severely for moderate sample sizes $N$, as has been documented by Arellano and Bond (1991) and Blundell and Bond (1998), who performed Monte Carlo simulations with sample sizes $N = 200$. Therefore, inference based on the two-step estimator can be very unreliable. In contrast, the one-step estimated standard errors based on the asymptotic variance matrix (2.5), using $W_{N2}$ as an estimate for $\Psi$ and substituting $\widetilde{\alpha}$ for $\alpha_0$, are found to be much less biased, and inference, like Wald tests, much more reliable. In practice therefore, one can often only rely on inference based on the less efficient one-step estimator.

For the GMM estimator that only utilises the moment conditions (2.3) for the differenced model, an optimal weight matrix is $\left(N^{-1}\sum_{i=1}^{N} D_i'HD_i\right)^{-1}$, where $D_i$ is the left upper block of $Z_i$ and $H$ is a $(T-2)$ square matrix which has 2's on the main diagonal, $-1$'s on the first subdiagonals and zeros elsewhere. Setting $W_{N1} = \left(N^{-1}\sum_{i=1}^{N} D_i'HD_i\right)^{-1}$ results therefore in an efficient one-step estimator. For the system GMM estimator such an efficient one-step weight matrix is not known, and in practice one uses as an initial weight matrix $W_{N1} = \left(N^{-1}\sum_{i=1}^{N} Z_i'Z_i\right)^{-1}$. To assess the potential loss in efficiency from using this initial weight matrix, the following expression for the upper bound of the efficiency loss has been derived by Liu and Neudecker (1997, p. 350) on the basis of the Kantorovich Inequality (KI):

$$\left(F_0'WF_0\right)^{-1}F_0'W\Psi WF_0\left(F_0'WF_0\right)^{-1}\leq\frac{\left(\lambda_1+\lambda_p\right)^2}{4\lambda_1\lambda_p}\left(F_0'\Psi^{-1}F_0\right)^{-1}, \tag{3.1}$$

where $\lambda_1\geq\cdots\geq\lambda_p$ are the eigenvalues of the matrix $\Psi W$.

For $T=3$, there is one overidentifying moment condition, as the system estimator utilises the following two moment conditions

$$\begin{aligned} E\left[\left(\Delta y_{i3}-\alpha\Delta y_{i2}\right)y_{i1}\right] &= 0, \\ E\left[\left(y_{i3}-\alpha y_{i2}\right)\Delta y_{i2}\right] &= 0, \end{aligned}$$

and $\Psi$ is given by

$$\begin{aligned} \Psi &= \operatorname{plim}\frac{1}{N}\sum_{i=1}^{N}\begin{bmatrix} y_1^2\left(\Delta\varepsilon_{i3}\right)^2 & y_{i1}\Delta y_{i2}\left(\eta_i+\varepsilon_{i3}\right)\Delta\varepsilon_{i3} \\ y_{i1}\Delta y_{i2}\left(\eta_i+\varepsilon_{i3}\right)\Delta\varepsilon_{i3} & \left(\Delta y_{i2}\right)^2\left(\eta_i+\varepsilon_{i3}\right)^2 \end{bmatrix} \\ &= \sigma_\varepsilon^2\begin{bmatrix} 2\sigma_y^2 & -\left(1-\alpha\right)\sigma_y^2 \\ -\left(1-\alpha\right)\sigma_y^2 & 2\dfrac{\sigma_\eta^2+\sigma_\varepsilon^2}{1+\alpha} \end{bmatrix}, \end{aligned}$$

where

$$\sigma_y^2=\frac{\sigma_\eta^2}{\left(1-\alpha\right)^2}+\frac{\sigma_\varepsilon^2}{1-\alpha^2}.$$

Further,

$$\begin{aligned} W_1 &= \left(\operatorname{plim}\frac{1}{N}\sum_{i=1}^{N}Z_i'Z_i\right)^{-1} \\ &= \left(\operatorname{plim}\frac{1}{N}\sum_{i=1}^{N}\begin{bmatrix} y_{i1}^2 & 0 \\ 0 & \left(\Delta y_{i2}\right)^2 \end{bmatrix}\right)^{-1} \\ &= \begin{bmatrix} \sigma_y^2 & 0 \\ 0 & 2\dfrac{\sigma_\varepsilon^2}{1+\alpha} \end{bmatrix}^{-1}, \end{aligned}$$

and the matrix $G=\Psi W_1$ is given by

$$G=\begin{bmatrix} 2\sigma_\varepsilon^2 & -(1-\alpha^2)\sigma_y^2/2 \\ -\left(1-\alpha\right)\sigma_\varepsilon^2 & \sigma_\varepsilon^2+\sigma_\eta^2 \end{bmatrix}.$$

Figure 1 presents the plot of the function $b_{KI}=\left(\lambda_{G1}+\lambda_{Gp}\right)^2/4\lambda_{G1}\lambda_{Gp}$, where the $\lambda_G$'s are the eigenvalues of $G$, for various values of $\alpha_0$ and $\sigma_\eta^2/\sigma_\varepsilon^2$. When $\sigma_\eta^2=\sigma_\varepsilon^2$, $b_{KI}$ is constant for different values of $\alpha$ and equal to $4/3$, indicating that the asymptotic variance of the one-step estimator could potentially be 33% larger than the efficient estimator. When $\sigma_\eta^2/\sigma_\varepsilon^2 < 1$, $b_{KI}$ is declining with $\alpha_0$, whereas it is increasing with $\alpha_0$ when $\sigma_\eta^2/\sigma_\varepsilon^2 > 1$. The value of $b_{KI}$ increases with $\sigma_\eta^2/\sigma_\varepsilon^2$ when $\sigma_\eta^2/\sigma_\varepsilon^2 > 1$.

When $T = 4$, there are 4 overidentifying moment conditions, and the matrices $\Psi$ and $W_1$ are given by

$$\Psi = \sigma_\varepsilon^2 \begin{bmatrix} 2\sigma_y^2 & -\sigma_y^2 & -\delta & -(1-\alpha)\sigma_y^2 & \frac{2-\alpha}{1-\alpha}\sigma_\eta^2 \\ -\sigma_y^2 & 2\sigma_y^2 & 2\delta & \frac{\sigma_\varepsilon^2}{1+\alpha} & -(1-\alpha)\delta \\ -\delta & 2\delta & 2\sigma_y^2 & -\frac{\sigma_\varepsilon^2}{1+\alpha} & -(1-\alpha)\sigma_y^2 \\ -(1-\alpha)\sigma_y^2 & \frac{\sigma_\varepsilon^2}{1+\alpha} & -\frac{\sigma_\varepsilon^2}{1+\alpha} & 2\frac{\sigma_\eta^2+\sigma_\varepsilon^2}{1+\alpha} & -\frac{1-\alpha}{1+\alpha}\sigma_\eta^2 \\ \frac{2-\alpha}{1-\alpha}\sigma_\eta^2 & -(1-\alpha)\delta & -(1-\alpha)\sigma_y^2 & -\frac{1-\alpha}{1+\alpha}\sigma_\eta^2 & 2\frac{\sigma_\eta^2+\sigma_\varepsilon^2}{1+\alpha} \end{bmatrix}$$

$$W_1 = \begin{bmatrix} \sigma_y^2 & 0 & 0 & 0 & 0 \\ 0 & \sigma_y^2 & \delta & 0 & 0 \\ 0 & \delta & \sigma_y^2 & 0 & 0 \\ 0 & 0 & 0 & 2\frac{\sigma_\varepsilon^2}{1+\alpha} & 0 \\ 0 & 0 & 0 & 0 & 2\frac{\sigma_\varepsilon^2}{1+\alpha} \end{bmatrix},$$

where

$$\delta = \sigma_y^2 - \frac{\sigma_\varepsilon^2}{1+\alpha}.$$

Figure 2 presents the efficiency bounds for the one-step system estimator when $T = 4$. The values for $b_{KI}$ are larger than for the $T = 3$ case. When $\sigma_\eta^2/\sigma_\varepsilon^2 = 1$, $b_{KI}$ is no longer constant for different values of $\alpha_0$, and takes values around 3, indicating that the asymptotic variance of the one-step estimator could be 3 times the variance of the efficient estimator. Again, $b_{KI}$ increases with $\sigma_\eta^2/\sigma_\varepsilon^2$ when $\sigma_\eta^2/\sigma_\varepsilon^2 > 1$, and $b_{KI}$ reaches the value 6 when $\sigma_\eta^2/\sigma_\varepsilon^2 = 2.5$.[4]

### 11.3.1 An Optimal Weight Matrix when $\sigma_\eta^2$=0

An optimal weight matrix for the system GMM estimator when $\sigma_\eta^2 = 0$ is given by

$$W_{N,\sigma_\eta^2=0} = \left(\frac{1}{N}\sum_{i=1}^{N} Z_i' A Z_i\right)^{-1},$$

[4]The value of $b_{KI}$ increases with $T$ as the number of moment restrictions increases. For example, when $T = 6$ and $\sigma_\eta^2/\sigma_\varepsilon^2 = 1$, $b_{KI}$ is approximately 14.

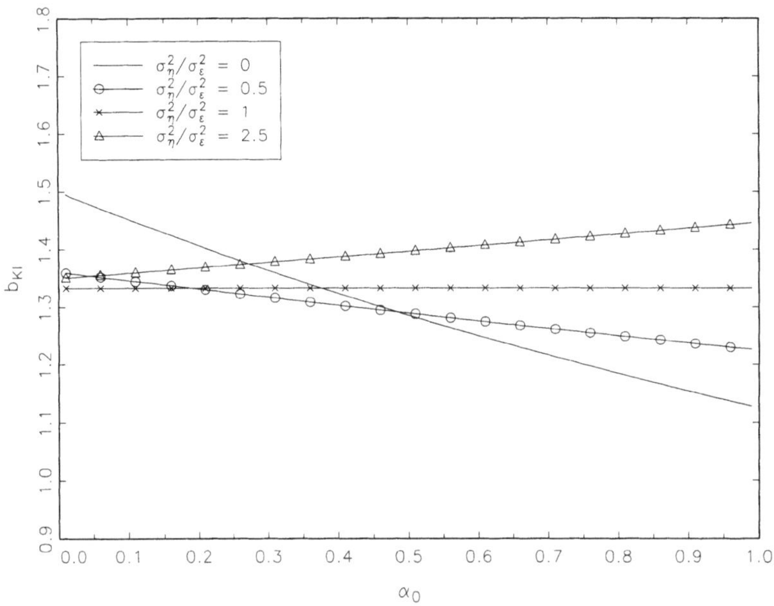


Figure 1: KI efficiency bounds, $T = 3; W_{NI} = \left(\frac{1}{N}\sum_i Z_i'Z_i\right)^{-1}$.

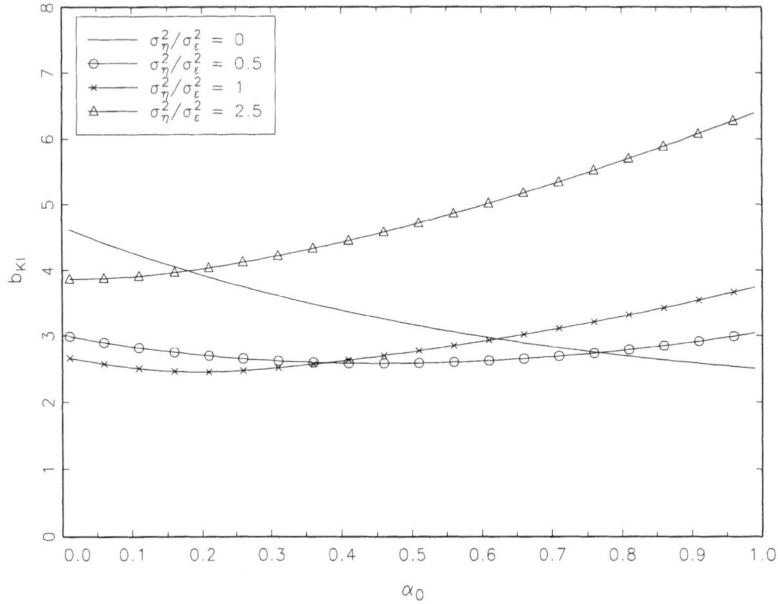


Figure 2: KI efficiency bounds, $T = 4; W_{NI} = \left(\frac{1}{N}\sum_i Z_i'Z_i\right)^{-1}$.

where

$$A = \begin{bmatrix} H & C \\ C' & I_{T-2} \end{bmatrix},$$

with $H$ as defined above, $I_{T-2}$ is the identity matrix of order $(T-2)$, and

$$C = \begin{bmatrix} 1 & 0 & 0 & \dots & 0 \\ -1 & 1 & 0 & & 0 \\ 0 & -1 & 1 & & 0 \\ \vdots & \vdots & \vdots & \ddots & \vdots \\ 0 & 0 & 0 & \dots & 1 \end{bmatrix}.$$

Using this weight matrix instead of $W_{N1} = \left(N^{-1}\sum_{i=1}^{N} Z_i'Z_i\right)^{-1}$ may improve on efficiency when $\sigma_\eta^2$ is small. Figures 3 and 4 display the values for $b_{KI}$ when $W_{N,\sigma_\eta^2=0}$ is used in the one-step estimator, for $T = 3$ and $T = 4$ respectively. Indeed, for small values of $\sigma_\eta^2$ the potential loss in efficiency is seen to be smaller than when $W_{N1}$ is used. However, when $\sigma_\eta^2/\sigma_\varepsilon^2$ is large, the potential efficiency loss gets larger for $W_{N,\sigma_\eta^2=0}$, which is what one would expect.

One way to detect whether use of $W_{N,\sigma_\eta^2=0}$ could be beneficial, without actually calculating the variances of the components, is to calculate the efficiency bounds $b_{KI}$ for the efficiency difference between the one-step and two-step estimators, i.e. calculate $b_{KI}$ from eigenvalues of the matrices $W_{N2}W_{N,\sigma_\eta^2=0}$, and $W_{N2}W_{N1}$. If the former are closer to 1 than the latter, this is an indication that there could be an efficiency gain from using $W_{N,\sigma_\eta^2=0}$ instead of $W_{N1}$.

## 11.4 Discussion

Upper bounds for the efficiency loss of the one-step system GMM estimator for a dynamic AR(1) panel data model as compared to the efficient two-step estimator, show that the efficiency loss could be quite severe when the weight matrix $W_{N1} = \left(N^{-1}\sum_{i=1}^{N} Z_i'Z_i\right)^{-1}$ is used, especially when $T$ gets large. When the variance of the unobserved individual effects, $\sigma_\eta^2$, is small, an efficiency gain can be made by using a weight matrix that is optimal under the assumption that $\sigma_\eta^2 = 0$.

The upper bounds were shown to be quite large, for example when $T = 4$, $W_{N1} = \left(N^{-1}\sum_{i=1}^{N} Z_i'Z_i\right)^{-1}$ and $\sigma_\eta^2/\sigma_\varepsilon^2 = 2.5$, the ratio of the asymptotic variance of the inefficient estimator to that of the efficient estimator can be as large as 6 for high values of $\alpha_0$. In Monte Carlo studies, however, such large differences of the variances are not found, using normal and non-normal data generating processes. This could mean that the KI upper bounds, $b_{KI}$, are too large to be informative for these cases. When the $b_{KI}$ are close to one, there is evidence of an efficient one-step estimator. The opposite statement for large $b_{KI}$ may not be true. Further research is needed to assess whether the $b_{KI}$ are

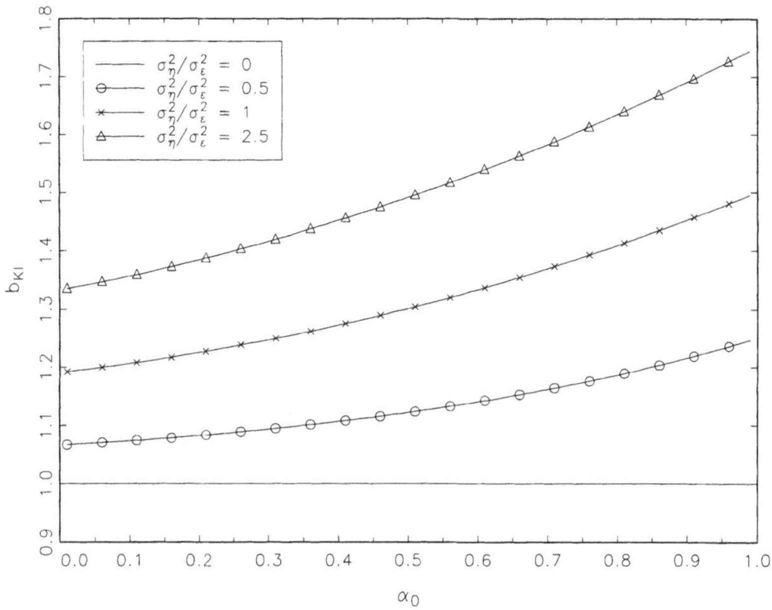


Figure 3: KI efficiency bounds, $T = 3; W_{NI} = W_{N,\sigma_\eta^2=0}$.

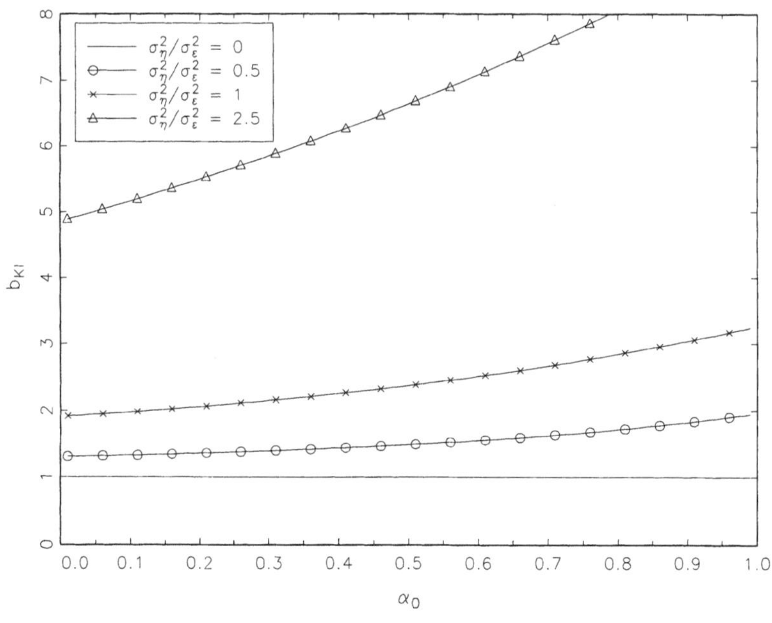


Figure 4: KI efficiency bounds, $T = 4; W_{NI} = W_{N,\sigma_\eta^2=0}$.

informative to rank different one-step estimators on the basis of their relative $KI$-values.

In empirical settings, one can easily compute $b_{KI}$ from the eigenvalues of the matrix $W_{N2}W_{N1}$, where $W_{N2}$ is the two-step efficient weight matrix.

## References

[1] Ahn, S.C., and P. Schmidt, (1995), Efficient Estimation of Models for Dynamic Panel Data, *Journal of Econometrics*, 68, 5–28.

[2] Arellano, M., and S. Bond, (1991), Some Tests of Specification for Panel Data: Monte Carlo Evidence and an Application to Employment Equations, *Review of Economic Studies*, 58, 277–98.

[3] Arellano, M., and O. Bover, (1995), Another Look at the Instrumental-Variable Estimation of Error-Components Models, *Journal of Econometrics*, 68, 29–51.

[4] Blundell, R., and S. Bond, (1998), Initial Conditions and Moment Restrictions in Dynamic Panel Data Models, *Journal of Econometrics*, 87, 115–143

[5] Hansen, L.P., (1982), Large Sample Properties of Generalised Method of Moments Estimators, *Econometrica*, 50, 1029–1054.

[6] Liu, S., and H. Neudecker, (1997), Kantorovich Inequalities and Efficiency Comparisons for Several Classes of Estimators in Linear Models, *Statistica Neerlandica*, 51, 345–355.

[7] Ogaki, M., (1993), Generalised Method of Moments: Econometric Applications, in Maddala, G.S., C.R. Rao, and H.D. Vinod (eds.), *Econometrics, Handbook of Statistics*, 11, North-Holland, Amsterdam.

CHAPTER 12

# The Rank Condition for Forward Looking Models

F.J.H. DON

## 12.1 Introduction

Consider a system of first order linear differential equations

$$\frac{dx}{dt} = Ax - z, \tag{1}$$

with $x$ the $n$-vector of state variables, $z$ an $n$-vector of exogenous variables and $A$ the $n \times n$ transition matrix. The vector $z$ converges to $z_\infty$ for $t \to \infty$. The well-known result of Blanchard and Kahn (1980) and Buiter (1981) is that, for $A$ non-singular, a unique non-explosive solution exists if and only if the number of non-predetermined state variables is equal to the number of unstable roots, i.e. the number of eigenvalues of $A$ with a strictly positive real part.[1] Giavazzi and Wyplosz (1985) extend this result to the case of a singular transition matrix $A$. The treatment in Buiter (1984) also allows for zero roots.

The stated condition is an order condition. As any student of matrix algebra will suspect, it should be supplemented by a rank condition to ensure sufficiency. This has not escaped Blanchard and Kahn (1980) and Buiter (1981, 1984), but they just make the relevant full rank assumption in passing.[2] This contribution intends to provide the intuition behind the rank condition, to derive it for the general case studied by Giavazzi and Wyplosz (1985), and to show how it relates to a rank condition used by Don and Van Stratum (1996) in the analysis of non-linear forward looking models with hysteresis.

[1] For a system of difference equations, the unstable roots of the transition matrix (in discrete time) are the eigenvalues outside the unit circle. This is the case studied by Blanchard and Kahn (1980).

[2] Top of page 1308 in Blanchard and Kahn (1980), top of page 20 in Buiter (1981), around formula (14) on page 670 in Buiter (1984).

## 12.2 The Rank Condition

The need for a rank condition is immediately obvious if we consider the case in which $A$ is diagonal. Then system (1) in fact dissolves into $n$ separate linear differential equations, which yield a unique non-explosive solution provided the predetermined variables are not only matched in number by the stable.[3] roots of $A$, but are also effectively coupled to them in the relevant scalar subsystems. The generalisation to non-diagonal $A$ provides the intuition behind the rank condition: this condition ensures that the stable roots of $A$ correspond to eigenvectors which, when projected into the subspace of predetermined variables, span that space.[4] The geometric formulation also reminds us that just counting predetermined and non-predetermined variables does not necessarily yield the subspace dimensions needed for the order condition. This point is made in section V of Buiter (1981). Here we will ignore that issue and assume that there are no "mixed" state variables, which consist of a linear combination of a predetermined variable and a non-predetermined variable. This assumption can always be satisfied through a proper linear transformation of the vector of state variables (see also Buiter (1984), page 672).

We will now state the rank condition for the case with possibly singular $A$, which is studied by Giavazzi and Wyplosz (1985). Note that a zero root for $A$ in the continuous time system (1) corresponds to a unit root in the discrete time transition matrix. We assume that the order condition is satisfied and that the first $p$ variables in the state vector are predetermined and the next $n-p$ variables are non-predetermined. We follow Giavazzi and Wyplosz (1985) in making the simplifying assumptions that $A$ is diagonalisable and that $z$ is constant.[5] Thus $A = V\Lambda V^{-1}$, with $\Lambda$ a diagonal matrix containing the eigenvalues of $A$ and $V$ the matrix of corresponding eigenvectors. We assume a particular ordering of the eigenvalues in the diagonal matrix $\Lambda$: first the $q$ zero roots, then the other $p-q$ stable roots, and finally the $n-p$ unstable roots. Using the transformation $y = V^{-1}x$, we obtain from (1) the diagonalised system

$$\frac{dy}{dt} = \Lambda y - V^{-1}z. \tag{2}$$

For singular $A$, we have $q > 0$ and existence of a non-explosive solution requires that the first $q$ elements of $V^{-1}z$ vanish (equivalently $AA^{+}z = z$, i.e. $Ax - z = 0$ admits a solution). Then the general solution to (2) is

$$y = \exp(\Lambda t).c + \Lambda^{+}V^{-1}z, \tag{3}$$

with $c$ an $n$-vector of arbitrary constants. Omitting the explosive solutions amounts to putting the last $n-p$ elements of $c$ to zero. The other elements of $c$ should be derived from the initial conditions for the predetermined variables.

[3] All roots with negative or zero real part are considered stable. See also note 6.

[4] For a simple example which does not have "the right root in the right place", see section 3.1 of Buiter (1986).

[5] See note 9 in Giavazzi and Wyplosz (1985) for the case of non-constant $z$.

Define $P$ as the $p \times n$ matrix $[I_p, 0]$. Then we have $c = P'd$ for some $p$-vector $d$, and the initial conditions for the predetermined variables yield

$$Px(0) = PVy(0) = PVP'd + PA^+z. \tag{4}$$

The order condition guarantees that the number of elements in $d$ equals the number of initial conditions $Px(0)$. A mismatch leads to non-uniqueness or non-existence of a solution for $d$, except in special cases that satisfy an equation like (6) below.

The rank condition that should supplement the order condition is that the matrix $PVP'$ is non-singular, i.e. the predetermined variables are related to the stable roots by a full rank submatrix of $V$. This condition is equivalent to the rank conditions identified in note 3 above for Blanchard and Kahn (1980) and Buiter (1981, 1984).

If the rank condition is satisfied, the solution to (1) can be written as

$$x(t) = A^+z + V\exp(\Lambda t)P'(PVP')^{-1}[Px(0) - PA^+z]. \tag{5}$$

This corresponds to formula (1) of Giavazzi and Wyplosz (1985).

Loss of rank in $PVP'$ means that (4) cannot be solved for $d$, unless the initial values of the predetermined variables satisfy a set of linear restrictions given by

$$(PVP')(PVP')^+[Px(0) - PA^+z] = Px(0) - PA^+z. \tag{6}$$

This will hold true only by coincidence; in general failure of the rank condition means that no non-explosive solution exists. The reason then is that some linear combination of predetermined variables cannot be matched with a stable root, hence it must either yield an explosive or a constant component. Explosive components are ruled out *a priori*, and constant components will in general conflict with the steady state condition $Ax = z$.

## 12.3 Hysteresis

The steady state solution[6] is the limit value of (5), which simplifies to

$$x_\infty = A^+z + VQ'QP'(PVP')^{-1}[Px(0) - PA^+z], \tag{7}$$

where $Q$ is the $q \times n$ matrix which partitions as $Q = [I_q, 0]$ .This corresponds to formula (2) of Giavazzi and Wyplosz (1985), who point out that the second term vanishes when $A$ is non-singular ($q = 0$). For singular $A$, the steady state depends on the initial values of the predetermined variables. This property is known as hysteresis.

[6]If $A$ has a pair of purely imaginary roots, the solution will in general show cycles which in the limit neither explode nor dampen. The central value of such a limit cycle will satisfy (7), provided $q$ also counts the purely imaginary roots and $V$ is set up accordingly. The text ignores this type of hysteresis.

Don and Van Stratum (1996) study a numerical algorithm to solve non-linear forward looking models exhibiting hysteresis. Implicit in their formula (4), they use a rank condition which, for the linear case discussed here, amounts to

$$\text{rank}A = \text{rank}A_2 = n - q, \tag{8}$$

where $A = [A_1, A_2]$ is a partitioning, possibly after some reordering of columns, into $q$ columns related to predetermined variables and $n-q$ other columns. We will now show that (8) is implied by the rank condition discussed above.

The first $q$ columns of $V$ correspond to zero roots of $A$, hence $AVQ' = 0$. The first $p$ rows of $VQ'$ are identical to the first $q$ columns of $PVP'$. The rank condition of section 2 states the full rank of the latter matrix, which implies that its first $q$ columns are linearly independent. Hence also the first $p$ rows of $VQ'$ have rank $q$. After a proper reordering of predetermined variables in the state vector, we may assume that the first $q$ rows of $VQ'$ are independent, i.e. $QVQ'$ is non-singular. Then we obtain, using $AVQ' = 0$:

$$AVQ' = [A_1 \quad A_2] \begin{bmatrix} QVQ' \\ W \end{bmatrix} = A_1QVQ' + A_2W = 0, \tag{9}$$

with $W$ an $(n-q) \times q$ matrix collecting the last $n-q$ rows of $VQ'$. From (9) we derive $A_1 = -A_2W(QVQ')^{-1}$ , which implies $\text{rank}A_2 = \text{rank}A$ (because the columns in $A_1$ are all linear combinations of those in $A_2$). This completes the proof of (8).

While Don and Van Stratum (1996) gave no motivation for their rank condition, we see that for the linear case it is a necessary condition for existence of a unique non-explosive solution to the dynamic model.

## 12.4 Concluding Remarks

The extension to transition matrices that are not diagonalisable proceeds through the Jordan form. This is fairly straightforward for the non-singular case, as shown by Blanchard and Kahn (1980). However, if there is a zero root[7] with geometric multiplicity lower than algebraic multiplicity, the solution will have components that are pure polynomials in $t$. In most practical cases, these should probably be treated as unstable. Roots with non-zero real part always generate an exponential factor, which dominates the polynomial and either vanishes or explodes for $t \to \infty$.

This contribution was written in honour of Heinz Neudecker. He taught me the fun of applying matrix algebra to econometric problems, and was my teacher and guide in the field. The matrices $P$ and $Q$ in this paper are simple examples of 0–1 matrices, which Heinz studied with dedication, see e.g. Magnus and Neudecker (1986). I shared some of the fun then, and still enjoy an exercise in matrix algebra when occasionally I run into one. Thank you, Heinz!

[7]A pair of purely imaginary roots with unequal multiplicities generates sine/cosine components with polynomial amplitude.

## References

[1] Blanchard, O.J., and C.M. Kahn, (1980), The Solution of Linear Difference Models under Rational Expectations, *Econometrica,* 48, 1305–1311.

[2] Buiter, W.H., (1981), A Note on the Solution of a Two-point Boundary Value Problem Frequently Encountered in Rational Expectations Models, *NBER technical paper,* no. 12.

[3] Buiter, W.H., (1984), Saddlepoint Problems in Continuous Time Rational Expectations Models: A General Method and Some Macroeconomic Examples, *Econometrica,* 52, 665–680.

[4] Buiter, W.H., (1986), Policy Evaluation and Design for Continuous Time Linear Rational Expectations Models: Some Recent Developments, pp. 84–108 in M.H. Preston and R.E. Quandt (eds.), *Prices, Competition and Equilibrium,* Barnes & Noble Books.

[5] Don, F.J.H., and R.M.G. van Stratum, (1996), Solving Path-Dependent Rational Expectations Models Using the Fair–Taylor Method, pp. 257–269 in M. Gilli (ed.), *Computational Economic Systems,* Kluwer Academic Publishers.

[6] Giavazzi, F., and C. Wyplosz, (1985), The Zero Root Problem: A Note on the Dynamic Determination of the Stationary Equilibrium in Linear Models, *Review of Economic Studies,* 52, 353–357.

[7] Magnus, J.R., and H. Neudecker, (1986), Symmetry, 0-1 Matrices and Jacobians: A Review, *Econometric Theory,* 2, 157–190.

CHAPTER 13

# Notes on the Elementary Properties of Permutation and Reflection Matrices

Richard William FAREBROTHER

## 13.1 Introduction

In this chapter, we discuss some of the elementary properties of permutation and reflection matrices. We define a class of reflection matrices related to the Samuelson–Wise conditions for the stability of a linear difference equation and we discuss a class of permutation matrices associated with functions defining chaotic and sub- chaotic pseudo-random processes.

## 13.2 Definitions

If $P$ is an $n \times n$ matrix satisfying $PP = P$, then it is said to be a projection matrix.

If $R$ is an $n \times n$ matrix satisfying $RR = I_n$, then it is said to be a reflection matrix.

If $Q$ is an $n \times n$ matrix satisfying $Q'Q = I_n$, then it is said to be an orthogonal matrix.

If a projection matrix $P$ is symmetric, then it is known as an orthogonal projection matrix. This terminology is potentially misleading as the identity matrix is the only projection matrix which is also an orthogonal matrix. In this context, it may be pertinent to suggest that $Q$ be renamed a unitary matrix.

If a reflection matrix $R$ is symmetric, then it is known as a Householder matrix. Such matrices are necessarily orthogonal and are widely used in numerical algorithms, see Farebrother (1988) and the references cited therein.

## 13.3 Basic Results

If $P$ is an $n \times n$ projection matrix, then so is $I - P$, whilst $R = I - 2P$ is a reflection matrix. Conversely, if $R$ is a reflection matrix, then $P = \frac{1}{2}(I - R)$

and $I - P = \frac{1}{2}(I + R)$ are both projection matrices.

For example, if

$$R = \begin{bmatrix} 1 & 2a \\ 0 & -1 \end{bmatrix},$$

then $RR = I_2$, and

$$P = \frac{1}{2}(I - R) = \begin{bmatrix} 0 & -a \\ 0 & 1 \end{bmatrix}$$

satisfies the idempotency condition $PP = P$.

Let $P$ be an $n \times n$ projection matrix of rank $q$. Then we may define an $n \times q$ matrix $X$ of rank $q$ whose columns span the columns of $P$. Now, we may express $P$ as the product $P = XZ'$ of $X$ with a $q \times n$ matrix $Z'$ of rank $q$ whose rows span the rows of $P$. Further, $P$ satisfies the condition $PP = P$ so that $XZ'XZ' = XZ'$ and we deduce that $Z'X = I_q$, and hence that we may write $P$ in the form

$$P = X(Y'X)^{-1}Y'$$

and $R = I - 2P$ in the form

$$R = I - 2X(Y'X)^{-1}Y',$$

where $Y$ is some $n \times q$ right transformation of the columns of $Z$ such that the $Q \times Q$ matrix $Y'X$ is nonsingular.

In this context, we note that $R$ has $q$ right eigenvectors of $X$ and $q$ left eigenvectors of $Y'$ corresponding to the $q$ eigenvalues of $-1$; similarly, $R$ has $n-q$ right eigenvectors of $X^{\perp}$ and $n-q$ left eigenvectors of $(Y^{\perp})'$ corresponding to the $n-q$ eigenvalues of $+1$. Further, if the trace of $R$ takes a value of $r$, then $R$ must have $q = (n-r)/2$ negative eigenvalues and $n - q = (n+r)/2$ positive eigenvalues.

For example, since the $3 \times 3$ reflection matrix

$$\frac{1}{2}\begin{bmatrix} 1 & 1 & 1 \\ 2 & 0 & -2 \\ 1 & -1 & 1 \end{bmatrix}$$

has a trace of $r = 1$, we deduce that it necessarily has two eigenvalues of $+1$ and a single eigenvalue of $-1$. Thus we may rewrite this matrix in the form

$$I_3 + \frac{1}{2}\begin{bmatrix} -1 & 1 & 1 \\ 2 & -2 & -2 \\ 1 & -1 & -1 \end{bmatrix} = I_3 - \frac{2}{-4}\begin{bmatrix} -1 \\ 2 \\ 1 \end{bmatrix}\begin{bmatrix} 1 & -1 & -1 \end{bmatrix},$$

from which we deduce that this matrix has a right eigenvector of $[-1\ 2\ 1\ ]'$ and a left eigenvector of $[-1\ 1\ 1\ ]$ corresponding to its negative eigenvalue of $-1$.

## 13.4 Samuelson Transformation Matrices

As an indirect method of identifying a class of reflection matrices, we consider the sequence of linear transformations:

Function $F_0$ carries the vector $[1]$ into the vector $[1]$.

Function $F_1$ carries the vector $[x, y]$ into the vector

$$[(x+y),\quad (x-y)].$$

Function $F_2$ carries the vector $[x^2, xy, y^2]$ into the vector

$$[(x+y)^2,\quad (x+y)(x-y),\quad (x-y)^2].$$

Function $F_3$ carries the vector $[x^3, x^2y,\ xy^2, y^3]$ into the vector

$$[(x+y)^3,\quad (x+y)^2(x-y),\quad (x+y)(x-y)^2,\quad (x-y)^3].$$

Expanding the algebraic expressions in each of the target vectors in terms of those in the source vectors, we immediately obtain the following sequence of matrices:

$$S_0 = [1],$$

$$S_1 = \begin{bmatrix} 1 & 1 \\ 1 & -1 \end{bmatrix},$$

$$S_2 = \begin{bmatrix} 1 & 1 & 1 \\ 2 & 0 & -2 \\ 1 & -1 & 1 \end{bmatrix},$$

$$S_3 = \begin{bmatrix} 1 & 1 & 1 & 1 \\ 3 & 1 & -1 & -3 \\ 3 & -1 & -1 & 3 \\ 1 & -1 & 1 & -1 \end{bmatrix}.$$

Now, there is a simple rule for obtaining each matrix in this series from its predecessor:

The first column of the $2 \times 2$ matrix $S_1$ is obtained from $S_0$ by premultiplying it by the $2 \times 1$ (bidiagonal) matrix $[\ 1\ 1\ ]'$.

The first two columns of the $3 \times 3$ matrix $S_2$ are obtained by premultiplying $S_1$ by the $3 \times 2$ bidiagonal matrix

$$\begin{bmatrix} 1 & 0 \\ 1 & 1 \\ 0 & 1 \end{bmatrix}.$$

The first three columns of the $4 \times 4$ matrix $S_3$ are obtained by premultiplying $S_2$ by the $4 \times 3$ bidiagonal matrix

$$\begin{bmatrix} 1 & 0 & 0 \\ 1 & 1 & 0 \\ 0 & 1 & 1 \\ 0 & 0 & 1 \end{bmatrix}.$$

And, generally, the first $M > 0$ columns of $S_m$ are obtained by premultiplying the $m \times m$ matrix $S_{m-1}$ by the $(M+1) \times m$ bidiagonal matrix with unit elements on both of its diagonals and zeros elsewhere.

Finally, we note that the elements in the final column of $S_m$ may be obtained by alternately attaching plus and minus signs to the corresponding elements in the first column of $S_m$, see Farebrother (1974b) for further details.

This sequence of linear transformations and the associated sequence of matrices arises naturally when one is interested in determining the stability properties of linear difference equations using a procedure developed by Samuelson (1941) and Wise (1956), see Farebrother (1973, 1974a, 1992) and Pollock (1996).

At the recreational level, this sequence of matrices may be regarded as a generalisation of the Arithmetical Triangle, as these matrices define the successive levels of a square pyramid with the Arithmetical Triangle as one of its faces.

## 13.5 Samuelson Reflection Matrices

Returning to the general problem, we apply the transformations defined in Section 4 to their target vectors and find that

Function $F_1$ also carries the vector

$$[(x+y), \quad (x-y)] \qquad \text{into the vector} \qquad [2x, 2y].$$

Function $F_2$ also carries the vector

$$[(x+y)^2, \quad (x+y)(x-y), \quad (x-y)^2$$

into the vector

$$[(2x)^2, \quad (2x)(2y), \quad (2y)^2] = 4[x^2, \quad xy, \quad y^2].$$

Function $F_3$ also carries the vector

$$[(x+y), \quad (x+y)^2(x-y), \quad (x+y)(x-y)^2, \quad (x-y)^3]$$

into the vector

$$8[x^3, \quad x^2y, \quad xy^2, \quad y^3],$$

and so on.

From the definition of the $m$th function in the previous section, we find that the $(m+1)\times(m+1)$ matrix $S_m$ satisfies the condition $S_mS_m = 2^mI_{m+1}$, so that we may define the corresponding reflection matrix by setting $R = S_m^*$, where $S_m^* = 2^{-m/2}S_m$.

Other reflection matrices are readily obtained by forming the Kronecker product of a sequence of scaled Samuelson transformation matrices. In particular, Hadamard matrices of all orders are formed in this way from the $2\times 2$ fundamental Hadamard matrix $H_1 = S_1$. See Hedayat and Wallis (1978) for a detailed discussion of possible applications of this class of matrices.

## 13.6 Givens Rotation Matrices

From the relationship $S_1S_1 = 2I_2$ we have

$$\frac{1}{\sqrt{2}}\begin{bmatrix} 1 & 1 \\ -1 & 1 \end{bmatrix}\frac{1}{\sqrt{2}}\begin{bmatrix} 1 & 1 \\ 1 & -1 \end{bmatrix} = \begin{bmatrix} 1 & 0 \\ 0 & 1 \end{bmatrix}.$$

Reversing the order of the rows in the first matrix and the order of the columns in the second matrix, we may rewrite this expression as

$$\frac{1}{\sqrt{2}}\begin{bmatrix} 1 & -1 \\ 1 & 1 \end{bmatrix}\frac{1}{\sqrt{2}}\begin{bmatrix} 1 & 1 \\ -1 & 1 \end{bmatrix} = \begin{bmatrix} 1 & 0 \\ 0 & 1 \end{bmatrix}.$$

Now, the fact that the first matrix on the left of this expression is the transpose of the second reveals that matrices of the form

$$\begin{bmatrix} a & b \\ -d & a \end{bmatrix}$$

satisfy the condition

$$\begin{bmatrix} a & b \\ -b & a \end{bmatrix}\begin{bmatrix} a & -b \\ b & a \end{bmatrix} = (a^2+b^2)\begin{bmatrix} 1 & 0 \\ 0 & 1 \end{bmatrix}.$$

In particular, if we choose $a$ and $b$ in such a way that $b^2 = 1 - a^2$ then we have the general formula for a Givens rotation matrix. Matrices of this type are widely employed in numerical algorithms, see Farebrother (1988) and the references cited therein.

Further, since the typical matrix in this class may be written as $aI + bJ$ where

$$J = \begin{bmatrix} 0 & 1 \\ -1 & 0 \end{bmatrix}$$

is a $2\times 2$ matrix satisfying $JJ = -I$, we note that the arithmetic of matrices of this form is isomorphic to the arithmetic of (complex) numbers of the form $a + b\sqrt{-1}$.

More generally, we note that the arithmetic of matrices of the form

$$\begin{bmatrix} a & b \\ bd & a \end{bmatrix}$$

is isomorphic to the arithmetic of numbers of the form $a + b\sqrt{d}$ when $a, b$ and $d$ are integers and $d$ is not a square.

Matrices of this more general form satisfy the condition

$$\begin{bmatrix} a & b \\ db & a \end{bmatrix} \begin{bmatrix} 1 & 0 \\ 0 & -d \end{bmatrix} \begin{bmatrix} a & db \\ b & a \end{bmatrix} = (a^2 - db^2) \begin{bmatrix} 1 & 0 \\ 0 & -d \end{bmatrix},$$

which forms the basis of much of the early work on the solution of Pell's equation

$$x^2 - dy^2 = \pm 1$$

in integers. See van der Waerden (1983, pp. 134–154) for details.

In this context we may also note that Farebrother (1988, p. 123) has obtained the general solution of Pythagoras's equation from the corresponding theory of nonsymmetric reflection matrices.

## 13.7 Eigenvalues of Permutation Matrices

If $Q$ is an $n \times n$ matrix whose columns are the columns of the $n \times n$ identity matrix $I_n$ in a different order, then $Q$ is known as a permutation matrix. Furthermore, $Q$ satisfies $Q'Q = I_n$, so that it is a unitary matrix and thus all its eigenvalues have unit modulus. In this section, we outline a simple procedure for obtaining the eigenvalues and eigenvectors of this matrix.

Suppose that the $j$th column of $Q$ is the $f(j)$th column of the $n \times n$ identity matrix. Then the permutation matrix $Q$ carries the $j$th element of an arbitrary $n \times 1$ matrix $x$ into the $f(j)$th element of $y = Qx$.

Now, for each value of $h = 1, 2, \ldots, n$, the sequence of distinct integers $f(h), f^2(h), f^3(h), \ldots$ will eventually return to the value $h$, say at the $m_h$th term, so that $f^{m_h}(h) = h$ for some value of $m_h \geq 1$.

A complete list of all such cycles is readily obtained by choosing successive starting values which have not already appeared in earlier sequences, and continuing in this way until all $n$ values have been used.

Consider a cycle of length $m$. If $z$ is an arbitrary $n \times 1$ matrix with nonzero elements at the $m$ locations indicated by this cycle and zeros elsewhere, then we have $Q^m z = z$, so that the $m$ eigenvalues of $Q$ corresponding to the values in this sequence are all $m$th roots of unity. Further the $m$ eigenvectors corresponding to these $m$ eigenvalues may readily be determined by repeated substitution in the relationship $z = \lambda Q'z$, where $\lambda$ is the relevant eigenvalue of $Q$.

## 13.8 Examples of Permutation Matrices

**Example 1.** Let $n = 2q$ be an even number, and let $A$ and $B$ be $n \times q$ matrices containing the odd and the even numbered columns of $I_n$ respectively. Then

$A$ is an $n \times q$ matrix with unit elements in its $(1,1), (3,2), (5,3), \ldots, (2q-1, q)$ positions and zeros elsewhere. Similarly, $B$ is an $n \times q$ matrix with unit elements in its $(2,1), (4,2), (6,3), \ldots, (2q, q)$ positions and zeros elsewhere. Now the $n \times n$ matrix $Q = [AB]$ is a permutation matrix representation of the function

$$f(j) = \begin{cases} 2j-1, & \text{if } j \leq q; \\ 2j-n, & \text{if } j > q. \end{cases}$$

In particular, if $n = 52$, then this function has two cycles of length one: $1, 1, 1, \ldots$ and $52, 52, 52, \ldots$; one cycle of length two: $18, 35, 18, \ldots$; and six cycles of length eight:

$$\begin{aligned} a &: 2, 3, 5, 9, 17, 33, 14, 27, 2, \ldots \\ b &: 4, 7, 13, 25, 49, 46, 40, 28, 4, \ldots \\ c &: 6, 11, 21, 41, 30, 8, 15, 29, 6, \ldots \\ d &: 10, 19, 37, 22, 43, 34, 16, 31, 10, \ldots \\ e &: 12, 23, 45, 38, 24, 47, 42, 32, 12, \ldots \\ f &: 20, 39, 26, 51, 50, 48, 44, 36, 20, \ldots \end{aligned}$$

This result implicitly establishes that it is possible to restore a pack of playing cards to its original order by a sequence of eight perfect riffles.

**Example 2.** For our second example, we reverse the order of the submatrices and set $Q = [BA]$ which is the matrix equivalent of the function

$$f(j) = \begin{cases} 2j, & \text{if } j \leq q; \\ 2j-1-n, & \text{if } j > q, \end{cases}$$

or, more briefly,

$$f(j) = 2j \bmod (n+1),$$

where the modulus operator indicates that the computed value of $f(j)$ is to be reduced by a multiple of $n+1$ in such a way as to bring the result into the range $0 \leq f(j) \leq n$.

In particular, if $n = 52$ then this process has a single cycle of length 52:

$$\begin{gathered} 1, 2, 4, 8, 16, 32, 11, 22, 44, 35, 17, 34, 15, 30, 7, 14, \\ 28, 3, 6, 12, 24, 48, 43, 33, 13, 26, 52, 51, 49, 45, 37, \\ 23, 46, 39, 25, 50, 47, 41, 29, 5, 10, 20, 40, 27, 1, \ldots \end{gathered}$$

**Example 3.** As a variant of this second example, we suppose that the order of the columns of $A$ has been reversed to obtain the matrix representation of the function

$$f(j) = \begin{cases} n+1-2j, & \text{if } j \leq q; \\ 2j-1-n, & \text{if } j > q. \end{cases}$$

This function is a discretised form of the tent mapping which gives rise to chaotic processes with infinite cycles in the continuous case, but which produces cycles of disappointingly short length when translated into discrete form for implementation in finite arithmetic on a digital computer, see Farebrother (1996) for computational details.

It is somewhat paradoxical to find that the appearance of randomness is only achieved by the careful selection of nonlinear functions suitable for use in this context. For example, the saw-tooth mapping of our second example,

$$f(j) = 2j \bmod (n+1),$$

is often used as the basis of multiplicative congruential generators of pseudo-random numbers. However, not only does the value of the modulus $n+1$ has to be selected with care but the primitive multiplier 2 has to be replaced by a power of 2 close to the square root of $n$ if we wish to obtain a suitably erratic sequence, see Downham and Roberts (1967).

In the particular case when $n = 52$, we have to substitute a value of 8 for the multiplier to obtain the function

$$f(j) = 8j \bmod (53),$$

which produces a more erratic sequence by picking every third element from the earlier sequence:

$$\begin{gathered}1, 8, 11, 35, 15, 14, 6, 48, 13, 51, 37, 39, 47, 5, 40, 2,\\ 16, 22, 17, 30, 28, 12, 43, 26, 49, 23, 25, 41, 10, 27, 4,\\ 32, 44, 34, 7, 3, 24, 33, 52, 45, 46, 50, 29, 20, 1, \ldots\end{gathered}$$

In practice, much larger values of $n+1$ are employed. Indeed the modulus is usually set close to the largest integer that it is possible to store on the computer.

## 13.9 Concluding Remarks

In this paper we have defined a class of reflection matrices related to the Samuelson–Wise conditions for the stability of linear difference equations. We have also discussed a class of permutation matrices associated with a family of discretised nonlinear functions. The first of these classes is a generalisation of the class of Hadamard matrices which has several applications in statistics, whilst the latter is typified by the generators of pseudo-random numbers commonly implemented in statistical computer packages for use in simulation studies.

## References

[1] Downham, D.Y., and Roberts, F.D.K.(1967), Multiplicative Congruential Pseudo-Random Number Generators, *Computer Journal,* 2, 74–77.

[2] Farebrother, R.W., (1973), Simplified Samuelson Conditions for Cubic and Quartic Equations, *The Manchester School,* 41, 396–400. Reprinted in J.C. Wood and R.N. Woods (1988), Vol. 2, 197–203.

[3] Farebrother, R.W., (1974a), Simplified Samuelson Conditions for Quintic Equations, *The Manchester School,* 42, 279–282. Reprinted in J.C. Wood and R.N. Woods (1988), Vol. 2, 276–279.

[4] Farebrother, R.W., (1974b), Recursive Relations for the Samuelson Transformation Coefficients, *International Economic Review,* 15: 805–807. Reprinted in J.C. Wood and R.N. Woods (1988), Vol. 2, 280–282.

[5] Farebrother, R.W.(1987) Independent Conditions for the Stability of a Dynamic Linear Model, *The Manchester School,* 55, 305–309.

[6] Farebrother, R.W., (1988), *Linear Least Squares Computations,* Marcel Dekker Inc., New York.

[7] Farebrother, R.W., (1992), A Note on the Schur and Samuelson Conditions, *The Manchester School,* 60, 79–81.

[8] Farebrother, R.W., (1996), The Role of Chaotic Processes in Econometric Models, *Journal of Statistical Planning and Inference,* 49, 163–176.

[9] Hedayat, A.,, and W. Wallis, (1978), Hadamard Matrices, *Annals of Statistics,* 6, 1184–1234.

[10] Pollock, D.S.G., (1996), Stability Conditions for Linear Stochastic models: A Survey, Paper presented at the Fifth International Workshop on Matrix Methods for Statistics, Shrewsbury.

[11] Samuelson, P.A., (1941), Conditions that the Roots of a Polynomial be Less than Unity in Absolute Value, *Annals of Mathematical Statistics,* 12, 360–364.

[12] Van der Waerden, B.L., (1983), *Geometry and Algebra in Ancient Civilizations,* Springer-Verlag, Berlin.

[13] Wise, J., (1956), Stationarity Conditions for Stochastic Processes of the Autoregressive and Moving Average Type, *Biometrika,* 43: 215–219.

[14] Wood, J.C., and R.N. Woods, (1988), *Paul A. Samuelson: A Critical Assessment,* Routledge, London.

CHAPTER 14

# S-Ancillarity and Strong Exogeneity

H. Peter BOSWIJK[1]

## 14.1 Introduction

This note analyses the differences and similarities between the concepts of S-ancillarity and strong exogeneity. We show that while strong exogeneity of a variable for a parameter implies S-ancillarity of the same variable for the same parameter, the converse is not true. An example illustrates the point.

## 14.2 The Main Result

Let $\{x_t, t = 1, \ldots, T\}$ be an observed $n$-vector time series, and define $X_t = (x_1, \ldots, x_t)', t = 1, \ldots, T$. We represent a statistical model $\mathcal{M}$ of this series by a family of joint densities of $X_T$, indexed by a $k$-dimensional parameter vector $\theta$, with parameter space $\Theta \subseteq \mathcal{R}^k$:

$$\mathcal{M} = \{f(X_T; \theta), \theta \in \Theta\}.$$

Possibly relevant starting values are considered fixed, and are suppressed as conditioning arguments from the densities for notational ease. Let $s(X_T)$ denote a function of the data (i.e., a statistic), and consider the factorisation

$$f(X_T; \theta) = g[X_T \mid s(X_T); \phi] \; h[s(X_T); \lambda],$$

where $g$ and $h$ are conditional and marginal densities, with parameters $\phi \in \Phi$ and $\lambda \in \Lambda$, respectively. The statistic $s(X_T)$ is said to be S-ancillary (cf. Barndorff-Nielsen, 1978) for $\phi$ (or a function $\tau(\phi)$ thereof), if $\phi$ and $\lambda$ are variation free, i.e., if

$$(\phi, \lambda) \in \Phi \times \Lambda.$$

In that case conditional inference on $\phi$ is possible. That is, we may obtain the maximum likelihood estimator $\hat{\phi}$ from the conditional likelihood function

[1]I have benefited from discussions with David Hendry and Søren Johansen.

$L_c(\phi) = g(\cdot \mid \cdot; \phi)$ only, and evaluate its sampling properties under a conditional probability measure with density $g[\cdot \mid s(X_T); \phi_0]$ rather than a joint measure with density $f(\cdot; \theta_0)$.

A particular case is obtained if the statistic $s(\cdot)$ corresponds to the observations on a subset of the variables in $x_t$. Thus, letting $x_t = (y_t', z_t')'$ and defining $Y_t$ and $Z_t$ analogously to $X_t$, we say that $z_t$ is S-ancillary for $\phi$ if

$$f(X_T; \theta) = g(Y_T \mid Z_T; \phi)\ h(Z_T; \lambda), \quad (\phi, \lambda) \in \Phi \times \Lambda. \tag{1}$$

A related concept is that of strong exogeneity, see Engle, Hendry, and Richard (1983). Consider first a sequential factorisation of the joint density of $X_T$:

$$f(X_T; \theta) = \prod_{t=1}^{T} f_x(x_t \mid X_{t-1}; \theta). \tag{2}$$

Next, factorise the density of $x_t = (y_t', z_t')'$ given the past (i.e., $X_{t-1}$) into a conditional density of $y_t$ given $z_t$ and the past, and a marginal density of $z_t$ given the past, with parameter vectors $\theta_1 \in \Theta_1$ and $\theta_2 \in \Theta_2$, respectively:

$$f_x(x_t \mid X_{t-1}; \theta) = f_{y|z}(y_t \mid z_t, X_{t-1}; \theta_1)\ f_z(z_t \mid X_{t-1}; \theta_2). \tag{3}$$

Combining (2) and (3) leads to the following factorisation of the total joint density (and hence the likelihood):

$$L(\theta) = f(X_T; \theta) = \prod_{t=1}^{T} f_{y|z}(y_t \mid z_t, X_{t-1}; \theta_1) \prod_{t=1}^{T} f_z(z_t \mid X_{t-1}; \theta_2) = L_1(\theta_1) L_2(\theta_2). \tag{4}$$

Now $z_t$ is said to be weakly exogenous for $\theta_1$ (or a function $\tau(\theta_1)$ thereof) if $(\theta_1, \theta_2) \in \Theta_1 \times \Theta_2$. In that case the maximum likelihood estimator of $\theta_1$ and likelihood ratio statistics for hypotheses on this parameter may be constructed from the 'conditional' likelihood $L_1(\theta_1)$ only. Note however, that in general $\prod_{t=1}^{T} f_{y|z}(y_t \mid z_t, X_{t-1}; \theta_1)$ need not define a proper density (so that, strictly speaking, $L_1(\theta_1)$ is not a proper likelihood, even though it may be used as one). This implies that weak exogeneity does not warrant conditional inferences on $\theta_1$, that is, we cannot use a conditional probability measure corresponding to $L_1(\theta_1)$ to evaluate the properties of the maximum likelihood estimator $\hat{\theta}_1$, simply because such a measure need not exist.

However, if we add the requirement that $z_t$ is not Granger-caused by $y_t$, i.e., if:

$$f_z(z_t \mid X_{t-1}; \theta_2) = f_z(z_t \mid Z_{t-1}; \theta_2), \tag{5}$$

then obviously $\prod_{t=1}^{T} f_z(z_t \mid X_{t-1}; \theta_2) = h(Z_T; \theta_2)$, which in turn implies

$$\prod_{t=1}^{T} f_{y|z}(y_t \mid z_t, X_{t-1}; \theta_1) = \frac{f(X_T; \theta)}{\prod_{t=1}^{T} f_z(z_t \mid X_{t-1}; \theta_2)} = \frac{f(X_T; \theta)}{h(Z_T; \theta_2)} = g(Y_T \mid Z_T; \theta_1). \tag{6}$$

If this is the case, $z_t$ is said to be strongly exogenous for $\theta_1$. Now $L_1$ and $L_2$ do correspond to proper conditional and marginal densities, so that conditional inference is possible again. In fact, from the last equality of (6) it is obvious that strong exogeneity of $z_t$ for $\theta_1$ implies S-ancillarity of $z_t$ for the same parameter.

On the other hand, S-ancillarity need not imply strong exogeneity. Suppose that we sequentially factorise the conditional and marginal densities in (1) as

$$g(Y_T \mid Z_T; \phi) = \prod_{t=1}^{T} g_y(y_t \mid Y_{t-1}, Z_T; \phi), \qquad h(Z_T; \lambda) = \prod_{t=1}^{T} h_z(z_t \mid Z_{t-1}; \lambda).$$

Next, assume that *lead* terms $z_{t+j}, j > 0$ enter the density of $y_t$ given $Y_{t-1}$ and $Z_T$:

$$g_y(y_t \mid Y_{t-1}, Z_T; \phi) \neq f_{y|z}(y_t \mid Y_{t-1}, Z_t; \theta_2).$$

Then

$$h(Z_T; \lambda) = \frac{f(X_T; \theta)}{g(Y_T \mid Z_T; \phi)} \neq \frac{f(X_T; \theta)}{\prod_{t=1}^{T} f_{y|z}(y_t \mid Y_{t-1}, Z_t; \theta_1)} = \prod_{t=1}^{T} f_z(z_t \mid X_{t-1}; \theta_2), \tag{7}$$

so that

$$f_z(z_t \mid X_{t-1}; \theta_2) \neq h_z(z_t \mid Z_{t-1}; \lambda),$$

i.e., $z_t$ is Granger-caused by $y_t$. Therefore, $z_t$ is not strongly exogenous, neither for $\phi$ nor for $\theta_1$; regardless of whether $\theta_1$ and $\theta_2$ are variation free.

The crucial difference between the two concepts, thus, is that S-ancillarity need not imply Granger non-causality. This is related to the origins and motivations for the development of the two notions: S-ancillarity validates conditional inferences, whereas strong exogeneity does the same, but also allows "*forecasting* $z$ *and then forecasting* $y$ *conditional* on the future $z$'s" (Engle, Hendry and Richard, 1983, p. 286) This is where the Granger non-causality comes in.

What we can prove then, is that S-ancillarity plus Granger non-causality implies strong exogeneity; and of course, the same holds for weak exogeneity plus Granger non-causality. However, this should not be construed to imply that weak exogeneity and S-ancillarity play the same role. As we have noted above, weak exogeneity only allows likelihood calculations to be based on the partial likelihood, whereas S-ancillarity in addition allows for conditional inference, using a probability measure corresponding to the same partial likelihood.

## 14.3 An Example

This example is adapted from Florens and Mouchart (1985, Example 3.2). Let $\{x_t, t = 1, \ldots, T\}$ be a bivariate time series, with $x_t = (y_t, z_t)'$, and suppose that

$$\begin{pmatrix} y_1 \\ z_1 \end{pmatrix} \sim N\left[\begin{pmatrix} 0 \\ 0 \end{pmatrix}, \begin{pmatrix} \sigma^2 & 0 \\ 0 & \tau^2 \end{pmatrix}\right], \tag{8}$$

$$\left( \begin{array}{c} y_t \\ z_t \end{array} \right) \Bigg| X_{t-1} \sim N\left[ \left( \begin{array}{c} 0 \\ (\rho\tau/\sigma)y_{t-1} \end{array} \right), \left( \begin{array}{cc} \sigma^2 & 0 \\ 0 & \tau^2(1-\rho^2) \end{array} \right) \right], \quad t = 2, \ldots, T, \tag{9}$$

with $(\rho, \sigma, \tau) \in (-1, 1) \times \mathcal{R}^+ \times \mathcal{R}^+$. Note that $\tau^2$ is the unconditional variance of $z_t$, and $\rho$ is the correlation coefficient between $y_{t-1}$ and $z_t$. It follows that the distribution of $X_T = (Y_T, Z_T)$ is given by

$$\text{vec}X_T \sim N\left[ \left( \begin{array}{c} 0 \\ 0 \end{array} \right), \left( \begin{array}{cc} \sigma^2 I_T & \rho\sigma\tau L' \\ \rho\sigma\tau L & \tau^2 I_T \end{array} \right) \right], \tag{10}$$

where $L$ is a $T \times T$ backward shifting matrix:

$$L = \begin{bmatrix} 0 & 0 & \cdots & \cdots & 0 \\ 1 & 0 & \ddots & & \vdots \\ 0 & 1 & \ddots & \ddots & \vdots \\ \vdots & \ddots & \ddots & \ddots & 0 \\ 0 & \cdots & 0 & 1 & 0 \end{bmatrix}.$$

This implies that the conditional and marginal distributions are

$$Y_T \mid Z_T \sim N\left[ \frac{\rho\sigma}{\tau} L' Z_T, \sigma^2 \left( I_T - \rho^2 L' L \right) \right], \qquad Z_T \sim N(0, \tau^2 I_T).$$

Because $L'$ is a forward shifting matrix, i.e., $L' Z_T = (z_2, \ldots, z_T, 0)'$, the conditional distribution can be expressed as

$$y_t \mid Z_T \sim N\left[ \frac{\rho\sigma}{\tau} z_{t+1}, \sigma^2 \left( 1 - \rho^2 \right) \right], \quad t = 1, \ldots, T-1, \tag{11}$$

$$y_T \mid Z_T \sim N(0, \sigma^2). \tag{12}$$

Because the conditioning set $Z_T$ does not contain $z_{T+1}$, the distribution of the last observation differs from the others.

From (11)–(12), we see that the parameters of the distribution of $Y_T \mid Z_T$ are $[(\rho\sigma/\tau), \sigma^2(1-\rho^2), \sigma^2]$, which are in one-to-one correspondence to $(\rho, \sigma, \tau)$. Thus the parameter of the conditional distribution are not variation independent of $\tau^2$, the parameter of the marginal distribution of $Z_T$, and S-ancillarity of $z_t$ cannot be established (for any parameter). However, $\sigma^2$ enters the distribution only via $y_t$, and because its contribution to the likelihood function will be asymptotically negligible, one could argue that $z_t$ is 'asymptotically' or 'approximately' S-ancillary for $[(\rho\sigma/\tau), \sigma^2(1-\rho^2)]$, because these parameters are variation independent of $\tau^2$. Alternatively, the problem is simplified by adding the restriction $\sigma^2 = \tau^2 = 1$, i.e., to replace the original parameter space for $(\rho, \sigma, \tau)$ by $(-1, 1) \times \{1\} \times \{1\}$: then the marginal distribution of $Z_T$ contains no unknown parameters anymore, so that it is actually an ancillary statistic (this is the case considered originally by Florens and Mouchart, 1985).

To check weak and strong exogeneity of $z_t$, we need the distribution of $y_t$ given $z_t$ and $X_{t-1}$, which is

$$y_t \mid z_t, X_{t-1} \sim N(0, \sigma^2).$$

From (9) we see that the parameters of the distribution of $z_t$ given the past are $[(\rho\tau/\sigma), \tau^2(1-\rho^2)]$, at least for $t = 2, \ldots, T$. Since these two parameters are variation independent of $\sigma^2$, weak exogeneity of $z_t$ is established, but only for the parameter $\sigma^2$. Furthermore, we see immediately from (9) that $z_t$ is Granger caused by $y_t$, so that $z_t$ is not strongly exogenous for any parameter. This result holds regardless of the restriction $\sigma^2 = \tau^2 = 1$, which was shown above to be sufficient for ancillarity of $z_t$. In summary, this example illustrates that variables may be ancillary but not strongly exogenous for a parameter of interest.

## References

[1] Barndorff-Nielsen, O.E., (1978), *Information and Exponential Families in Statistical Theory,* New York, John Wiley.

[2] Engle, R.F., D.F. Hendry, and J.-F. Richard, (1983), Exogeneity, *Econometrica,* 51, 277–304.

[3] Florens, J.-P., and M. Mouchart, (1985), Conditioning in Dynamic Models, *Journal of Time Series Analysis,* 6, 15–34.

CHAPTER 15

# Asymptotic Inference Based on Eigenprojections of Covariance and Correlation Matrices

Tõnu KOLLO[1]

## 15.1 Introduction

In this chapter, we aim to examine problems of testing which are based on asymptotic distributions of eigenprojections of the sample covariance matrix $S_n$ and the sample correlation matrix $R_n$. The topic was introduced by Tyler (1981, 1983) who designed certain test statistics using the asymptotic normality of the eigenprojections of the sample covariance matrix in the case of normally and elliptically distributed populations. Some of his results were carried over to the case of the sample correlation matrix by Kollo (1984). For rotationally invariant distributions, the asymptotic properties of estimation and hypothesis testing on the basis of eigenprojections of the sample covariance matrix were discussed by Fan and Fang (1990).

The necessity of using eigenprojections in problems of testing arises whenever the statistic of interest is designed via eigenvalues and eigenvectors of a symmetric matrix, where some eigenvalues can be repeated. In this case, eigenvectors are no longer uniquely determined, and we have to use eigenprojections which correspond to the repeated eigenvalues.

When all of the eigenvalues of a symmetric matrix $M$ are non-repeated, the asymptotic distributions of the eigenvalues and eigenvectors of an estimator $\hat{M}$ of $M$ have been presented under general conditions (existence of the fourth order moments of the population distribution) in Kollo and Neudecker (1993, 1997).

In this chapter, we construct asymptotic chi-square test-statistics and we examine the properties of the tests. In Section 2, some basic notions and nota-

[1]The author is thankful to Professor Heinz Neudecker for inspiring discussions on matrix differentiation for many years, which have also stimulated the writing of the present paper. The author is thankful to the Estonian Research Foundation, which has supported the study through grant Nr. 3013.

tions are introduced, and, in Section 3, the asymptotic normal laws for eigenprojections are presented. The derivation of the results is based on the matrix derivative, introduced by Neudecker (1969). In Section 4, asymptotic chi-square test-statistics are designed and their properties are examined.

## 15.2 Preliminaries

Let $\lambda$ be an eigenvalue with multiplicity $r$ of the $p \times p$ matrix $A$. Then there exist $r$ linearly independent eigenvectors, corresponding to $\lambda$. The space of all linear combinations of the eigenvectors is called eigenspace; and the space corresponding to $\lambda$ and may be denoted by $\mathcal{V}(\lambda)$. The dimension of $\mathcal{V}(\lambda)$ equals to the multiplicity of $\lambda$. If $\lambda$ and $\mu$ are two different eigenvalues of a symmetric matrix $A$, then $\mathcal{V}(\lambda)$ and $\mathcal{V}(\mu)$ are orthogonal subspaces in $\mathcal{R}^p$. If $\lambda_1, \ldots, \lambda_k$ are all different eigenvalues of $A$ with multiplicities $m_1, \ldots, m_k$, then the space $\mathcal{R}^p$ can be presented as a direct sum of subspaces $\mathcal{V}(\lambda_i)$

$$\mathcal{R}^p = \bigoplus_{i=1}^{k} \mathcal{V}(\lambda_i)$$

and therefore any vector $x \in \mathcal{R}^p$ can be presented, in a unique way, as a sum

$$x = \sum_{i=1}^{k} x_i, \tag{2.1}$$

where $x_i \in \mathcal{V}(\lambda_i)$.

The eigenprojection $P_{\lambda_i}$ of the matrix $A$ which corresponds to $\lambda_i$ is the $p \times p$ matrix which transforms the space $\mathcal{R}^p$ onto the space $\mathcal{V}(\lambda_i)$. An arbitrary vector $x \in \mathcal{R}^p$ is transformed by the projector $P_{\lambda_i}$, to the vector $x_i$:

$$P_{\lambda_i} x = x_i,$$

where $x_i$ is the $i$th term in the sum (2.1). If $v$ is a subset of the set of all different eigenvalues of $A$, i.e.

$$v \subset \{\lambda_1, \ldots, \lambda_k\},$$

then the eigenprojection $P_v$ which corresponds to the eigenvalues $\lambda_i \in v$ has the form

$$P_v = \sum_{\lambda_i \in v} P_{\lambda_i}.$$

The eigenprojection $P_{\lambda_i}$ of a symmetric matrix $A$ can be presented through unit-length eigenvectors $y_i$, i.e.

$$P_{\lambda_i} = \sum_{j=1}^{m_i} y_j y_j'.$$

The eigenprojections $P_{\lambda_i}$ enable us to present a symmetric matrix $A$ through its spectral decomposition:

$$A = \sum_{i=1}^{k} \lambda_i P_{\lambda_i}.$$

Basic properties of eigenprojections of symmetric matrices are given by the following equalities:

$$\begin{aligned} P_{\lambda_i} P_{\lambda_i} &= P_{\lambda_i}; \\ P_{\lambda_i} P_{\lambda_j} &= 0, \qquad i \neq j; \\ \sum_{i=1}^{k} P_{\lambda_i} &= I. \end{aligned}$$

For eigenvalues and eigenprojections of a symmetric matrix see Kato (1977), Chapter II or Watson (1983), Appendix B, if necessary.

A generalised inverse of a $p \times q$ matrix $A$ is any $q \times p$ matrix $G$ satisfying $AGA = A$. Any generalised inverse of $A$ is denoted by $A^-$. The Moore–Penrose inverse $A^+$ of $A$ is unique and its entries are continuous functions of the entries of $A$. The Moore–Penrose inverse of a $p \times p$ symmetric matrix $M$ can be presented in the form

$$M^+ = \sum_{\substack{i=1 \\ \lambda_i \neq 0}}^{k} = \frac{1}{\lambda_i} P_{\lambda_i}.$$

For $p \times q$ matrix $A$, let $\mathcal{M}(A)$ be the manifold of $A$:

$$\mathcal{M}(A) = \{x : x \in \mathcal{R}^p, x = Ay \quad \text{for some} \quad y \in \mathcal{R}^q\},$$

and $\mathcal{N}(A)$ its null space

$$\mathcal{N}(A) = \{y : y \in \mathcal{R}^q, Ay = 0\}.$$

Let $X = (x_1, \ldots, x_n)$ be a sample from a $p$-dimensional population $x \sim P_x$ with the first four moments being finite and with $Ex = \mu$, $Dx = \Sigma$. In special cases, the population distribution $P_x$ will be considered as multivariate normal or elliptical. For an elliptical population $x \sim EC_p(\mu, V, g)$, we assume that $x$ is a continuous random vector with the density function

$$f_x(x) = |V|^{-1/2} g\{(x - \mu)' V^{-l} (x - \mu)\},$$

where $g(\cdot)$ is some non-negative function, and the positive definite parameter matrix $V$ is related with the covariance matrix of $x$:

$$Dx = -2\phi'(0)V,$$

where the function $\phi(\cdot)$ determines the characteristic function

$$\psi_x(t) = E(e^{it'x}) = e^{it'\mu} \phi(t'Vt), \quad t \in \mathcal{R}^p.$$

In this case, we write $x \sim EC_p(\mu, V, \phi)$. The kurtosis parameter $\kappa$, which appears later in formulae, is defined by the equality

$$\kappa = \frac{\phi''(0)}{\{\phi'(0)\}^2} - 1. \tag{2.2}$$

The reader is referred to Fang and Zhang (1990) for details of elliptical distributions. Let $\bar{x}$ and $S_n$ be the usual unbiased estimators of $\mu$ and $\Sigma$ respectively. Let us denote the correlation matrix of $x$ by $P$ and its estimator by $R_n$:

$$P = \Sigma_d^{-1/2} \Sigma \Sigma_d^{-1/2}, \tag{2.3}$$

where $\Sigma_d^{-1/2}$ is the diagonal matrix with diagonal elements $\sigma_{11}^{-1/2}, \ldots, \sigma_{pp}^{-1/2}$ The subscript $n$ in the quantities $S_n$, $R_n$ etc. will be omitted unless specifically required. For example, the sample analogue of (2.3) is

$$R = S_d^{-1/2} S S_d^{-1/2}. \tag{2.4}$$

When the discussions of covariance and correlation matrices are analogues to each other, we shall use the single notation $M$ for either matrix and $M_n$ for its estimator ($S_n$ or $R_n$). Moments of a random $p$ vector $x$ are defined by the following equalities:

$$M_k(x) = E[(x \otimes x')^{\otimes m} \otimes x], \quad k = 2m+1, m = 0, 1, \ldots \tag{2.5}$$

$$M_k(x) = E[(x \otimes x')^{\otimes m}], \quad k = 2m, m = 1, 2, \ldots \tag{2.6}$$

and central moments

$$\bar{M}_k(x) = E[\{(x-\mu) \otimes (x-\mu)'\}^{\otimes m} \otimes (x-\mu)], \quad k = 2m+1, \ m = 0, 1, \ldots \tag{2.7}$$

$$\bar{M}_k(x) \doteq E[\{(x-\mu) \otimes (x-\mu)'\}^{\otimes m}], \quad k = 2m, \ m = 1, 2, \ldots \tag{2.8}$$

Here the Kronecker power $A^{\otimes k}$ denotes the $k$-times Kronecker product of the matrix $A$ with itself. To shorten the notation, the argument $(x)$ will be omitted from notations for moments if there is no need to highlight it, so $M_k$ and $\bar{M}_k$ will be used instead of $M_k(x)$ and $\bar{M}_k(x)$. From matrix algebra, the Kronecker product, the commutation matrix, the vec-operator and matrix derivatives will be used extensively. An interested reader is referred to Magnus and Neudecker (1991). Here we shall give the usual definition of the derivative—see Neudecker (1969)—together with a few formulae which we shall use repeatedly later. If $Y : r \times s$ is a function of $X : p \times q$, then the matrix derivative $dX/dY$ is an $rs \times pq$ matrix

$$\frac{dY}{dX} = \frac{d}{(d\text{vec}X)'} \otimes \text{vec}Y \tag{2.9}$$

where

$$\frac{d}{(d\text{vec}X)'} = \left( \frac{\partial}{\partial x_{11}}, \ldots, \frac{\partial}{\partial x_{p1}}, \frac{\partial}{\partial x_{12}}, \ldots, \frac{\partial}{\partial x_{p2}}, \ldots, \frac{\partial}{\partial x_{1q}}, \ldots, \frac{\partial}{\partial x_{pq}} \right).$$

The most useful formulae of Kronecker product and vec-operator are the following two equalities:

$$(AB) \otimes (CD) = (A \otimes C)(B \otimes D); \tag{2.10}$$

$$\mathrm{vec}(ABC) = (C' \otimes A)\mathrm{vec}B. \tag{2.11}$$

The commutation matrix $K_{pq}$ is used to commute terms in Kronecker products; for $A : p \times q$ and $B : r \times s$

$$A \otimes B = K_{rp}(B \otimes A)K_{qs}. \tag{2.12}$$

## 15.3 Asymptotic Distribution of Eigenprojections

Let $\xrightarrow{\mathcal{D}}$ and $\xrightarrow{\mathcal{P}}$ denote the convergence in distribution and probability respectively. In accordance with Anderson (1984, p. 121), we state the following lemma:

**Lemma 3.1.** Let $U_n$ be an $m$ component random vector and $b$ a fixed $m$ vector. Assume $U_n, \xrightarrow{\mathcal{P}} b$ and

$$\sqrt{n}(U_n - b) \xrightarrow{\mathcal{D}} N(0, T) \tag{3.1}$$

as $n \to \infty$. If $h(x) : \mathcal{R}^m \longrightarrow \mathcal{R}^q$ is a differentiable function at $x = b$ with the non-zero differentials of the coordinates $h_i(x)$ at $x = b$, then

$$\sqrt{n}\{h(U_n) - h(b)\} \xrightarrow{\mathcal{D}} N(0, \xi_h T \xi_h') \tag{3.2}$$

as $n \to \infty$ and $\xi_h = dh(x)/dx|_{x=b}$ is the matrix derivative (2.9).

The next lemma gives us asymptotic normal distribution of the sample covariance matrix (Parring, 1979).

**Lemma 3.2.** Let $X = (x_1, \ldots, x_n)$ be a sample of size $n$, $Ex_i = \mu_i$, $Dx_i = \Sigma$, $M_4(x_i) < \infty$. Then, if $n \to \infty$

$$\sqrt{n}\mathrm{vec}(S_n - \Sigma) \xrightarrow{\mathcal{D}} N(0, \Pi), \tag{3.3}$$

where

$$\Pi = \bar{M}_4 - \mathrm{vec}(\Sigma)\{\mathrm{vec}(\Sigma)\}' \tag{3.4}$$

and central moments $\bar{M}_k$ are defined by (2.7) and (2.8).

It is known (cf. Anderson (1993)) that, for elliptical family $x \sim EC_p(\mu, V, \phi)$, we have $\bar{M}_3 = 0$ and

$$\Pi_E = (1 + \kappa)(I_{p^2} + K_{pp})(\Sigma \otimes \Sigma) + \kappa\mathrm{vec}(\Sigma)\{\mathrm{vec}(\Sigma)\}', \tag{3.5}$$

where $\kappa$ is the kurtosis parameter, defined by (2.2). For a normally distributed sample, $x_i \sim N(\mu, \Sigma)$, we have $\kappa = 0$ and

$$\Pi_N = (I_{p^2} + K_{pp})(\Sigma \otimes \Sigma). \tag{3.6}$$

In the following, we shall use the sequence $\{\mathrm{vec} S_n\}$ in the role of the sequence $\{U_n\}$ in Lemma 3.1. In constructing test-statistics later on, we need the asymptotic distributions of certain eigenprojections. If we denote the eigenvalues of $M$ by $\mu_1 \geq \cdots \geq \mu_p$ and set

$$w = \{\mu_i, \ldots, \mu_{i+m-1}\},$$

assuming $\mu_{i-1} \neq \mu_i, \mu_{i+m-1} \neq \mu_{i+m}$, and use $\lambda$ for different eigenvalues of $M$, then, as a direct conclusion from Lemmas 3.1 and 3.2, we have the following statement:

**Theorem 3.1.** (Kollo, 1984). Let $X = (x_1, \ldots, x_n)$ be a sample of size $n$, $Ex_i = \mu$, $Dx_i = \Sigma$, $M_4(x_i) < \infty$ and let $\tilde{P}_w$ be the estimator of the eigenprojection $P_w$ associated with the roots $\lambda_i \in w$ of the covariance matrix $\Sigma$. Then, as $n \to \infty$,

$$\sqrt{n}\mathrm{vec}(\tilde{P}_w - P_w) \xrightarrow{\mathcal{D}} N(0, \xi_w \Pi \xi_w'), \tag{3.7}$$

where $\Pi$ is given by relation (3.4) and

$$\xi_w = \sum_{\lambda_i \in w} \sum_{\lambda_j \notin w} \frac{1}{\lambda_i - \lambda_j} [(P_{\lambda_i} \otimes P_{\lambda_j}) + (P_{\lambda_j} \otimes P_{\lambda_i})]. \tag{3.8}$$

**Corollary 3.1.1.** For the normal sample, $x_i \sim N(\mu, \Sigma)$, the covariance matrix $\tau_w$ of the asymptotic distribution for the eigenprojection $P_w$ is given by

$$\tau_w = (I_{p^2} + K_{pp}) \sum_{\lambda_i \in w} \sum_{\lambda_j \notin w} \frac{\lambda_i \lambda_j}{(\lambda_i - \lambda_j)^2} [(P_{\lambda_i} \otimes P_{\lambda_j}) + (P_{\lambda_j} \otimes P_{\lambda_i})]. \tag{3.9}$$

**Proof.** Using equality (3.6) we get from Theorem 3.1:

$$\begin{aligned} \tau_w &= \xi_w \Pi_N \xi_w' = \sum_{\lambda_i \in w} \sum_{\lambda_j \notin w} \frac{1}{(\lambda_i - \lambda_j)^2} [(P_{\lambda_i} \otimes P_{\lambda_j}) + (P_{\lambda_j} \otimes P_{\lambda_i})] \\ &\quad \times (I_{p^2} + K_{pp})(\Sigma \otimes \Sigma)[(P_{\lambda_i} \otimes P_{\lambda_j}) + (P_{\lambda_j} \otimes P_{\lambda_i})] \\ &= (I_{p^2} + K_{pp}) \sum_{\lambda_i \in w} \sum_{\lambda_j \notin w} \frac{\lambda_i \lambda_j}{(\lambda_i - \lambda_j)^2} [(P_{\lambda_i} \otimes P_{\lambda_j}) + (P_{\lambda_j} \otimes P_{\lambda_i})]. \end{aligned}$$

**Corollary 3.1.2.** For an elliptically distributed sample, the covariance matrix $\tau_w$ of the asymptotic distribution for the eigenprojector $P_w$ is given by

$$\tau_w = (1 + \kappa)(I_{p^2} + K_{pp}) \sum_{\lambda_i \in w} \sum_{\lambda_j \notin w} \frac{\lambda_i \lambda_j}{(\lambda_i - \lambda_j)^2} [(P_{\lambda_i} \otimes P_{\lambda_j}) + (P_{\lambda_j} \otimes P_{\lambda_i})]. \tag{3.10}$$

**Proof.** Using equality (3.5) and Theorem 3.1, we get

$$\tau_w = \xi_w \Pi_N \xi_w' = \sum_{\lambda_i \in w} \sum_{\lambda_j \notin w} \frac{1}{\lambda_i - \lambda_j} [(P_{\lambda_i} \otimes P_{\lambda_j}) + (P_{\lambda_j} \otimes P_{\lambda_i})]$$

$$\times \ [(1+\kappa)(I_{p^2}+K_{pp})(\Sigma\otimes\Sigma)+\kappa\text{vec}\Sigma\text{vec}'\Sigma]$$
$$\sum_{\lambda_i\in w}\sum_{\lambda_j\notin w}\frac{1}{\lambda_i-\lambda_j}[(P_{\lambda_i}\otimes P_{\lambda_j})+(P_{\lambda_j}\otimes P_{\lambda_i})]$$
$$= \ (1+\kappa)(I_{p^2}+K_{pp})\sum_{\lambda_i\in w}\sum_{\lambda_j\notin w}\frac{\lambda_i\lambda_j}{(\lambda_i-\lambda_j)^2}[(P_{\lambda_i}\otimes P_{\lambda_j})+(P_{\lambda_j}\otimes P_{\lambda_i})].$$

The second term vanishes because, from relation (2.11),

$$(P_{\lambda_i}\otimes P_{\lambda_j})\text{vec}\Sigma=\text{vec}(P_{\lambda_j}\Sigma P_{\lambda_i})=\text{vec}(P_{\lambda_j}\lambda_i P_{\lambda_i})=\lambda_i\text{vec}(P_{\lambda_j}P_{\lambda_i})=0.$$

For the correlation matrix, the asymptotic behaviour of eigenprojections is described by

**Theorem 3.2.** (Kollo, 1991). Let $X=(x_1,\ldots,x_n)$ be a sample of size $n$, $Ex_i=\mu$, $Dx_i=\Sigma$, $M_4(x_i)<\infty$, and let $\tilde{P}_w$ be the estimator of the eigenprojection $P_w$ associated with the roots $\lambda_i\in w$ of the population correlation matrix $P$. Then, as $n\to\infty$,

$$\sqrt{n}\text{vec}(\tilde{P}_w-P_w)\xrightarrow{\mathcal{D}}N(0,\eta_w\Pi\eta_w'),$$

where $\Pi$ is given by relation (3.4), and

$$\eta_w \ = \ \left\{\left(\Sigma_d^{-1/2}\otimes\Sigma_d^{-1/2}\right)-\frac{1}{2}(K_{pp})_d[(I_p\otimes\Sigma_d^{-1}P)+(\Sigma_d^{-1}P\otimes I_p)]\right\}$$
$$\times\sum_{\lambda_i\in w}\sum_{\lambda_j\notin w}\frac{1}{\lambda_i-\lambda_j}[(P_{\lambda_i}\otimes P_{\lambda_j})+(P_{\lambda_j}\otimes P_{\lambda_i})]. \qquad (3.11)$$

The expression in curly brackets in (3.11) is the derivative $dR_n/dS_n|_{S_n}=\Sigma$, which can be found in Neudecker and Wesselmann (1990), for instance. In the correlation matrix case, the forms of the the matrix $\eta_w$ are not essentially any simpler for normal and elliptical samples.

## 15.4 Testing $H_0$ by the Chi-Square Test

Let $\mu_1\geq\cdots\geq\mu_p>0$ be the eigenvalues of $M$, and let $A$ be a real $p\times r$ matrix with rank $r$. Assume that $\mu_{i-1}\neq\mu_i$, $\mu_{i+m-1}\neq\mu_{i+m}$ and consider the following hypothesis for given $r<m$:

*$H_0$: the columns of $A$ lie in the subspace generated by the set of eigenvectors of $M$ associated with the eigenvalues $\mu_i,\ldots,\mu_{i+m-1}$.* (4.1)

The problem of testing the hypothesis arises, for example, in principal component analysis. Then the population and sample principal component vectors are eigenvectors of $M$ and $M_n$ respectively. The vector of principal components $z$ is given by

$$z=W'x,$$

where $W = (w_1, \ldots, w_p)$ is a matrix of eigenvectors of $M$ and $w_i$ is an eigenvector of $M$ corresponding to the eigenvalue $\mu_i$ $(i = 1, \ldots, p)$.

Traditionally, we regard the principal components as the independent variables, where the first few components (corresponding to the largest eigenvalues $\mu_1, \mu_2, \ldots$) account for most of the variability of the original variables and are regarded as "real". The last few principal components are regarded as "noise", or as variables describing the errors of measurement, and they should be neglected. Suppose it were possible to check, by the sphericity test, that last $k$ eigenvalues of $M$ are small and equal. Then we can discard the original variable $X_i$, if vector $(0 \ldots 0\ 1\ 0 \ldots 0)'$ (with 1 in $i$th position) lies in the space spanned by the last $k$ principal components. Using the eigenprojection $P_w$, the null hypothesis (4.1) can be expressed simply as

$$H_0 : P_w A = A. \tag{4.2}$$

For testing $H_0$, we consider the statistic

$$T_w(A) = A - \tilde{P}_w A = (I_p - \tilde{P}_w)A. \tag{4.3}$$

We can easily obtain the asymptotic distribution for the statistic $T_w(A)$ from the asymptotic distribution of the eigenprojection $\tilde{P}_w$. Using the assumptions of Theorems 3.1 and 3.2, Kollo (1984) showed that, as $n \to \infty$,

$$\sqrt{n}\mathrm{vec}[(I_p - \tilde{P}_w)A - (I_p - P_w)A] \xrightarrow{\mathcal{D}} N(0, \xi_w \Pi \xi_w'), \tag{4.4}$$

where

$$\xi_w = -(A' \otimes I_p) \left. \frac{d\tilde{P}_w}{dS_n} \right|_{S_n = \Sigma}.$$

This means that

$$\xi_w = \sum_{\lambda_i \in w} \sum_{\lambda_j \notin w} \frac{1}{\lambda_i - \lambda_j} [(A'P_{\lambda_i} \otimes P_{\lambda_j}) + (A'P_{\lambda_j} \otimes P_{\lambda_i})] \tag{4.5}$$

for eigenprojections of the covariance matrix $S_n$, and that

$$\begin{aligned} \xi_w &= \left\{ \left( \Sigma_d^{-1/2} \otimes \Sigma_d^{-1/2} \right) - \frac{1}{2}(K_{pp})_d[(I_p \otimes \Sigma_d^{-1}P) + (\Sigma_d^{-1}P \otimes I_p)] \right\} \\ &\quad \times \sum_{\lambda_i \in w} \sum_{\lambda_j \notin w} \frac{1}{\lambda_i - \lambda_j} [(A'P_{\lambda_i} \otimes P_{\lambda_j}) + (A'P_{\lambda_j} \otimes P_{\lambda_i})] \end{aligned} \tag{4.6}$$

for eigenprojections of the sample correlation matrix $R_n$. Suppose that $H_0$ is valid. Then

$$P_{\lambda_j} = 0$$

for all $\lambda_j \notin w$ and we have the next result.

**Theorem 4.1.** Let $X = (x_1, \dots, x_n)$ be a sample of size $n$, $Ex_i = \mu$, $Dx_i = \Sigma$, $M_4(x_i) < \infty$, and suppose the null hypothesis (4.2) holds. Then, as $n \to \infty$,

$$\sqrt{n}\text{vec}T_w(A) \xrightarrow{\mathcal{D}} N(0, \gamma_w \Pi \gamma_w'),$$

where $\Pi$ is given by relation (3.4) and

$$\gamma_\omega = \left. \frac{dT_\omega(A)}{dS_n} \right|_{S_n=\Sigma}.$$

If $P_w$ is an eigenprojection of $\Sigma$, then

$$\gamma_w = \sum_{\lambda_i \in w} \sum_{\lambda_j \notin w} \frac{1}{\lambda_i - \lambda_j} (A'P_{\lambda_i} \otimes P_{\lambda_j}); \tag{4.7}$$

whereas, if $P_w$ is an eigenprojection of the correlation matrix $P$, then

$$\begin{aligned} \gamma_w &= \left\{ \left( \Sigma_d^{-1/2} \otimes \Sigma_d^{-1/2} \right) - \frac{1}{2} (K_{pp})_d [(I_p \otimes \Sigma_d^{-1} P) + (\Sigma_d^{-1} P \otimes I_p)] \right\} \\ &\quad \times \sum_{\lambda_i \in w} \sum_{\lambda_j \notin w} \frac{1}{\lambda_i - \lambda_j} (A'P_{\lambda_i} \otimes P_{\lambda_j}). \end{aligned} \tag{4.8}$$

**Corollary 4.1.1.** (Tyler, 1981). For the normal sample $x_i \sim N(\mu, \Sigma)$, the asymptotic covariance matrix $\omega_w^E$ of the statistic $T_w(A)$ is given by

$$\omega_w^N = \sum_{\lambda_i \in w} \sum_{\lambda_j \notin w} \frac{\lambda_i \lambda_j}{(\lambda_i - \lambda_j)^2} (A'P_{\lambda_i} A \otimes P_{\lambda_j})$$

if the null hypothesis holds.

A similar statement holds also for an elliptical population.

**Corollary 4.1.2.** (Tyler, 1981). For the elliptically distributed sample, the asymptotic covariance matrix $\omega_w^E$ of the statistic $T_w(A)$ is, under the null hypothesis (4.2), given by

$$\omega_w^E = (1 + \kappa) \sum_{\lambda_i \in w} \sum_{\lambda_j \notin w} \frac{\lambda_i \lambda_j}{(\lambda_i - \lambda_j)^2} (A'P_{\lambda_i} A \otimes P_{\lambda_j}),$$

where the parameter $\kappa$. is defined by relation (2.2).

**Proof.** Using relations (2.10)–(2.12) and expression of the asymptotic covariance matrix (3.5), we obtain

$$\begin{aligned} \omega_w^E &= \sum_{\lambda_i \in w} \sum_{\lambda_j \notin w} \frac{1}{\lambda_i - \lambda_j} (A'P_{\lambda_i} \otimes P_{\lambda_j}) \Pi_E \sum_{\lambda_i \in w} \sum_{\lambda_j \notin w} \frac{1}{\lambda_i - \lambda_j} (P_{\lambda_i} A \otimes P_{\lambda_j}) \\ &= (1 + \kappa) \sum_{\lambda_i \in w} \sum_{\lambda_j \notin w} \frac{\lambda_i \lambda_j}{(\lambda_i - \lambda_j)^2} (A'P_{\lambda_i} A \otimes P_{\lambda_j}) \end{aligned}$$

$$+ \kappa \sum_{\lambda_i \in w} \sum_{\lambda_j \notin w} \frac{1}{\lambda_i - \lambda_j} (A'P_{\lambda_i} \otimes P_{\lambda_j}) \text{vec}\Sigma \text{vec}'\Sigma \sum_{\lambda_i \in w} \sum_{\lambda_j \notin w} \frac{1}{\lambda_i - \lambda_j} (P_{\lambda_i} A \otimes P_{\lambda_j}).$$

The second term on the right hand side vanishes because

$$(A'P_{\lambda_i} \otimes P_{\lambda_i})\text{vec}\Sigma = \text{vec}(P_{\lambda_j}\Sigma P_{\lambda_i}A) = \lambda_i \text{vec}(P_{\lambda_j}P_{\lambda_i}A) = 0,$$

if $i \neq j$.

By means of $T_w(A)$, Tyler (1981) introduced the statistic

$$T_n(A) = n\{\text{vec}T_w(A)\}'\hat{\Sigma}_w^-(A)\text{vec}T_w(A), \tag{4.9}$$

where $\hat{\Sigma}_w^-(A)$ is a generalised inverse of the sample estimator $\hat{\Sigma}_w(A)$ of the asymptotic covariance matrix of $T_w(A)$. The properties of the statistic $T_n(A)$ are derived from the following lemmas (Moore, 1977).

The central and non-central chi-squared distributions with $k$ degrees of freedom are denoted by $\chi_k^2$ and $\chi_k^2(\Delta)$ where $\Delta$ is the non-centrality parameter.

**Lemma 4.1.** Suppose that

$$y_n \xrightarrow{\mathcal{D}} N(0, \Sigma)$$

with $\text{rank}(\Sigma) = k$, and that $\{\Sigma_n\}$ is a sequence of estimators of $\Sigma$ such that

$$\Sigma_n \xrightarrow{\mathcal{P}} \Sigma$$

and that, for each $n$, the vector $y_n \in \mathcal{M}(\Sigma_n)$. Then

(i) $T_n = y_n'\Sigma_n^- y_n$ is invariant under choice of $\Sigma_n^-$;

(ii) $T_n \xrightarrow{\mathcal{D}} \chi_k^2$, if $\mu = 0$;

(iii) $T_n \xrightarrow{\mathcal{D}} \chi_k^2(\mu'\Sigma\mu)$ in general.

**Lemma 4.2.** Suppose that

$$y_n \xrightarrow{\mathcal{D}} N(0, \Sigma)$$

with $\text{rank}(\Sigma) = k$, and that $B$ is a generalised inverse of $\Sigma$. If $B_n$ are random matrices such that

$$B_n \xrightarrow{\mathcal{P}} B$$

then

(i) $y_n'B_n y_n \xrightarrow{\mathcal{D}} \chi_k^2$, if $\mu = 0$;

(ii) in general $y_n'B_n y_n \xrightarrow{\mathcal{D}} \chi_k^2(\mu'\Sigma^-\mu)$, if $\mu \in \mathcal{M}(\Sigma)$.

If we denote

$$U = \begin{cases} I_{p^2}, & \text{if } P_w \text{ is an eigenprojection of } \Sigma, \\ (\Sigma_d^{-1/2} \otimes \Sigma_d^{-1/2}) - \frac{1}{2}(K_{pp})_d[(I_p \otimes \Sigma_d^{-1}P) + (\Sigma_d^{-1}P \otimes I_p)], & \\ \quad \text{if } P_w \text{ is an eigenprojection of } P, & \end{cases} \tag{4.10}$$

then relations (4.7) and (4.8) can be combined as

$$\gamma_w = U \sum_{\lambda_i \in w} \sum_{\lambda_j \notin w} \frac{1}{\lambda_i - \lambda_j} (A' P_{\lambda_i} \otimes P_{\lambda_j}). \tag{4.11}$$

An estimator of $\Sigma_w(A)$ can be obtained by defining

$$\hat{\Sigma}_w(A) = \hat{\gamma}_w \hat{\Pi} \hat{\gamma}_w', \tag{4.12}$$

where

$$\hat{\gamma}_w = \hat{U} \sum_{l_i \in w} \sum_{l_j \notin w} \frac{1}{l_i - l_j} (A' \tilde{P}_{l_i} \otimes \tilde{P}_{l_j}), \tag{4.13}$$

and $\hat{U}$ is obtained from relation (4.10) by replacing the matrices $\Sigma$ and $P$ with their estimates $S$ and $R$ throughout. For the different eigenvalues of $M_n$, which correspond to the eigenvalues $\mu_i, \ldots, \mu_{i+m-1}$ of $M$, we use the notation $l_i \in w$ and, for the remaining ones, $l_j \in w$; estimators of eigenprojections $P_{\lambda_i}$ are denoted by $\tilde{P}_{l_i}$.

**Theorem 4.2.**

(i) The statistic $T_n(A)$, defined by (4.9), is invariant under postmultiplication of $A$ by an $r \times r$ nonsingular matrix $B$.

(ii) In the covariance matrix case, if $\text{rank}(\tilde{P}_w A) = r$, then, under $H_0$, the statistic $T_w(A)$ does not depend on the choice of generalised inverse of $\hat{\Sigma}_w(A)$, $\text{rank}\{\hat{\Sigma}(A)\} = r(p-m)$, and

$$\mathcal{M}\{\hat{\Sigma}_w(A)\} = \mathcal{M}\{I_r \otimes (I_p - \tilde{P}_w)\}.$$

(iii) In the correlation matrix case under $H_0$, the statistic $T_n(A)$ is the same, with probability 1, for all choices of generalised inverse of $\hat{\Sigma}_w(A)$, at least asymptotically.

**Proof.**
(i) from the definition of $T_n(A)$

$$T_n(AB) = n\{\text{vec}T_w(AB)\}' \hat{\Sigma}_w^-(AB) \text{vec}T_w(AB).$$

Using relations (4.12) and (4.13)

$$\hat{\Sigma}_w^-(AB) = [(B' \otimes I_p)\hat{\Sigma}_w(A)(B \otimes I_p)]^- = (B^{-1} \otimes I_p)\hat{\Sigma}_w^-(A)(B'^{-1} \otimes I_p).$$

This last relation is in consequence of the following identity for conformable matrices $A$, $B$, $C$, where $A$, $C$ are nonsingular and $B$ is arbitrary:

$$(ABC)(C^{-1}B^-A^{-1})(ABC) = ABB^-BC = ABC.$$

Using equality (2.11),

$$\text{vec}T_w(AB) = \text{vec}[(I_p - \tilde{P}_w)AB] = (B' \otimes I_p)\text{vec}T_w(A).$$

Altogether

$$\begin{aligned}
&T_n(AB) \\
&= n[(B'\otimes I_p)\mathrm{vec}T_w(A)]'(B^{-1}\otimes I_p)\hat{\Sigma}_w^-(A)(B'^{-1}\otimes I_p)(B'\otimes I_p)\mathrm{vec}T_w(A) \\
&= n[\mathrm{vec}T_w(A)]'(B\otimes I_p)(B^{-1}\otimes I_p)\hat{\Sigma}_w^-(A)(B'^{-1}\otimes I_p)(B'\otimes I_p)\mathrm{vec}T_w(A) \\
&= T_n(A).
\end{aligned}$$

The first statement is proved.

(ii) To show that $T_n(A)$ is invariant under different choices of generalised inverse, we have to check, according to Lemma 4.1, whether $\mathrm{vec}T_w(A) \in \mathcal{M}\{\hat{\Sigma}_w^-(A)\}$. Let $x = \mathrm{vec}G$, where $G$ is a $p \times r$ matrix. Then

$$\begin{aligned}
\hat{\Sigma}_w(A)\mathrm{vec}G &= \hat{\gamma}_w\hat{\Pi}\sum_{l_i\in w}\sum_{l_j\notin w}\frac{1}{l_i-l_j}(\tilde{P}_{l_i}A\otimes\tilde{P}_{l_j})\mathrm{vec}G \\
&= \hat{\gamma}_w\hat{\Pi}\sum_{l_i\in w}\sum_{l_j\notin w}\frac{1}{l_i-l_j}\mathrm{vec}(\tilde{P}_{l_i}GA'\otimes\tilde{P}_{l_j}).
\end{aligned}$$

This expression is zero if and only if

$$\sum_{l_i\in w}\sum_{l_j\notin w}\frac{1}{l_i-l_j}\mathrm{vec}(\tilde{P}_{l_i}GA'\otimes\tilde{P}_{l_j}) = 0, \tag{4.14}$$

because $\hat{\Pi}$ has full rank. We now show that relation (4.14) holds if and only if $\tilde{P}_{l_i}G = 0$ for all $l_j \in w$. Suppose, conversely, that there exists $l_{j0} \in w$, $\tilde{P}_{l_{j0}}G \neq 0$. Then $\mathrm{vec}(\tilde{P}_{l_{j0}}GA'\otimes\tilde{P}_w) \neq 0$, because $\mathrm{rank}(\hat{P}_wA) = r$. The last implication is justified by the following argument. Suppose

$$\tilde{P}_{l_{j0}}GA'\otimes\tilde{P}_w = 0.$$

From our assumptions, $\mathrm{rank}(A'\hat{P}_w) = r$ and $\mathrm{rank}(P_{l_{j0}}G) \geq 1$. Now we use the following property of rank of matrices (Muirhead, 1982, p. 582)[13]:

If $A : p \times q$ and $B : q \times r$ such that $AB = 0$, then

$$\mathrm{rank}(B) < q - \mathrm{rank}(A).$$

In our case, this leads to the inequality

$$r < r - \mathrm{rank}(\tilde{P}_{l_{j0}}G),$$

which is impossible. This means that $\tilde{P}_{l_{j0}}GA'\otimes\tilde{P}_w \neq 0$. Then

$$\sum_{l_i\in w}\frac{1}{l_i-l_{j0}}\mathrm{vec}(\tilde{P}_{l_{j0}}GA'\otimes\tilde{P}_{l_i}) \neq 0.$$

For different values of $l_j \notin w$, the projections $\tilde{P}_{l_j}G$ belong to different orthogonal subspaces and

$$\sum_{l_i \in w} \sum_{l_j \notin w} \frac{1}{l_i - l_j} \text{vec}(\tilde{P}_{l_i} GA' \otimes \tilde{P}_{l_j}) \neq 0.$$

too, in contradiction to (4.14).

The converse, $\tilde{P}_{l_j}G = 0$ for all $l_j \notin w$, implies directly relation (4.14). Hence

$$\mathcal{M}\{\hat{\Sigma}_w(A)\} = \{\text{vec}G : (I_p - \tilde{P}_w)G = 0\}.$$

This means that columns $g_i$ of $G$ must belong to the subspace spanned by the eigenvectors associated with $l_i \in w$. Therefore, if $G = T_w(A)$,

$$\text{vec}\{\tilde{P}_{l_j} T_w(A) A' \tilde{P}_{l_i}\} = \text{vec}[\tilde{P}_{l_j}(I_p - \tilde{P}_w)AA'\tilde{P}_{l_i}] = \text{vec}(\tilde{P}_{l_j} AA' \tilde{P}_{l_i}) \neq 0$$

and hence $\text{vec}T_w(A) \in \mathcal{M}\{\hat{\Sigma}_w(A)\}$. From Tyler's (1981) Corollary 5.1, we get $\text{rank}\{\hat{\Sigma}_w(A)\} = r(p - m)$ and

$$\mathcal{M}\{\hat{\Sigma}_w(A)\} = \mathcal{M}\{I_r \otimes (I_p - \tilde{P}_w)\}.$$

It remains to prove the last statement.

(iii) In the correlation matrix case, $\text{vec}G \in \mathcal{M}\{\hat{\Sigma}_w(A)\}$ if $(I_p - \tilde{P}_w)G = 0$, as in the previous case, but the null space may be wider because of the singularity of the asymptotic covariance matrix for $R$. Analytically, $\text{vec}G \in \mathcal{M}\{\hat{\Sigma}_w(A)\}$, if and only if

$$\sum_{l_i \in w} \sum_{l_j \notin w} \frac{1}{l_i - l_j} \{\text{vec}(S_d^{-1/2} \tilde{P}_{l_j} GA' \tilde{P}_{l_i} S_d^{-1/2}) - \frac{1}{2}(K_{pp})_d \text{vec}[l_i(\tilde{P}_{l_j} GA' \tilde{P}_{l_i} S_d^{-1}) + l_j(S_d^{-1} \tilde{P}_{l_j} GA' \tilde{P}_{l_i})]\} = 0.$$

The solution of this equation depends on $A$ and, in general, we cannot show that $\text{vec}T_w(A) \in \mathcal{M}\{\hat{\Sigma}_w(A)\}$. It remains to show that the statistic $T_n(A)$ is the same with probability 1 for all choices of generalised inverse of $\hat{\Sigma}_w(A)$, at least asymptotically. From convergence (4.4)

$$\sqrt{n}\text{vec}[T_w(A) - (I_p - P_w)A] \xrightarrow{\mathcal{D}} N\{0, \Sigma_w(A)\}.$$

If $H_0$ is valid, then

$$\sqrt{n}\text{vec}T_w(A) \xrightarrow{\mathcal{D}} y \sim N\{0, \Sigma_w(A)\}.$$

According to Moore (1977) (Theorem 2), the quadratic form $y'\Sigma_w^-(A)y$ is the same for all choices of generalised inverse with probability 1.

Suppose we take the Moore–Penrose inverse $\hat{\Sigma}_w^+(A)$ as a generalised inverse for $\hat{\Sigma}_w(A)$. Then

$$T_n(A) = n\{\text{vec}T_w(A)\}'\hat{\Sigma}_w^+(A)\text{vec}T_w(A).$$

This expression can be simplified by noting that, for a symmetric matrix $B$,

$$Bx = 0 \iff B^{+}x = 0.$$

We show that

$$\hat{\Sigma}_w(A)\mathrm{vec}(\tilde{P}_w A) = 0.$$

Using relations (4.12),(4.13) and (2.11), we get

$$\begin{aligned} \hat{\Sigma}_w(A)\mathrm{vec}(\tilde{P}_w A) &= \hat{\gamma}_w \hat{\Pi}\hat{U} \sum_{l_i \in w} \sum_{l_j \notin w} \frac{1}{l_i - l_j} (\tilde{P}_{l_i} A \otimes \tilde{P}_{l_j}) \mathrm{vec}(\tilde{P}_w A) \\ &= \hat{\gamma}_w \hat{\Pi}\hat{U} \sum_{l_i \in w} \sum_{l_j \notin w} \frac{1}{l_i - l_j} \mathrm{vec}(\tilde{P}_{l_j} \tilde{P}_w A A' \otimes \tilde{P}_{l_i}) = 0, \end{aligned}$$

because of $\tilde{P}_{l_j}\tilde{P}_w = 0$, when $l_j \notin w$. Thus

$$T_n(A) = n(\mathrm{vec}A)'\hat{\Sigma}_w^{+}(A)(A)\mathrm{vec}A.$$

Owing to Theorem 4.2, the assumptions of Tyler's (1981) Theorem 5.3 are satisfied in the covariance matrix case, and the following result holds:

**Theorem 4.3.** Under the null hypothesis

$$H_0 : P_w A = A,$$

where $P_w$ is an eigenprojection of the covariance matrix $\Sigma$, the statistic

$$T_n(A) \xrightarrow{\mathcal{D}} \chi^2_{r(p-m)}$$

If $P_w A \neq A$, then, for any fixed $x$,

$$P[T_n(A) > x] \longrightarrow 1, \qquad n \longrightarrow \infty.$$

For the correlation matrix we have a similar statement.

**Theorem 4.4.** Under the null hypothesis

$$H_0 : P_w A = A,$$

where $P_w$ is an eigenprojection of the correlation matrix $P$, the statistic

$$T_n(A) \xrightarrow{\mathcal{D}} \chi^2_{\nu(A)},$$

where $\nu(A) = \mathrm{rank}\{(\Sigma_w(A)\}$.

We can use the asymptotic distribution in Theorem 4.3 for testing $H_0$:

Suppose first that $\mathrm{rank}(\tilde{P}_w A) = r$. If, for a significance level $\alpha > 0$, we have the statistic $T_n(A) < h^2_{r(p-m),1-\alpha}$, then we accept $H_0$. However, if $T_n(A) \geq$

$h^2_{r(p-m),1-\alpha}$, then we reject $H_0$. Moreover, if $\text{rank}(\tilde{P}_w A) < r$, then, by convention, we reject $H_0$.

In the above, $h^2_{r(p-m),1-\alpha}$ is the $(1-\alpha)$ percentile of a $\chi^2_{r(p-m)}$ distribution. Using the asymptotic chi-square distribution in Theorem 4.4, we get similar rules as in the previous case:

If $T_n(A) < h^2_{\nu(A),1-\alpha}$, then we accept $H_0$. However, if $T_n(A) \geq h^2_{\nu(A),1-\alpha}$, then we reject $H_0$.

But, in this case, we have the additional problem of calculating $\text{rank}\{\hat{\Sigma}_w(A)\}$ for every $A$; and the statistic $T_n(A)$ is only asymptotically invariant under different choices of the generalised inverse of $\hat{\Sigma}_w(A)$.

## References

[1] Anderson, T.W., (1984), *An Introduction to Multivariate Statistical Analysis, Second Edition,* Wiley, New York.

[2] Anderson, T.W., (1993), Non-Normal Multivariate Distributions: Inference Based on Elliptically Contoured Distributions, in *Multivariate Analysis: Future Directions,* C.R. Rao (ed.), North-Holland, Amsterdam.

[3] Fan, J., and K.-T. Fang, (1990), Maximum Likelihood Characterisation of Distributions, in *Statistical Inference in Elliptically Contoured and Related Distributions,* K.-T. Fang and T.W. Anderson (eds.), Allerton Press, New York, 421–427.

[4] Fang, K.-T., and Y.-T. Zhang, (1990), *Generalised Multivariate Analysis,* Springer-Verlag, Berlin.

[5] Kato, T., (1976), *Perturbation Theory for Linear Operators, Second Edition,* Springer, Berlin.

[6] Kollo, T., (1984), Asymptotic Distributions of Eigenprojections of Covariance and Correlation Matrices for Testing Hypothesis, *Acta et Commentationes Universitatis Tartuensis,* 685, 3–13.

[7] Kollo, T., (1991), *Matrix Derivatives in Multivariate Statistics* (in Russian), Tartu University Press, Tartu.

[8] Kollo, T., and H. Neudecker, (1993), Asymptotics of Eigenvalues and Unit-length Eigenvectors of Sample Variance and Correlation Matrices, *Journal of Multivariate Analysis,* 47, 283–300.

[9] Kollo, T., and H. Neudecker, (1997), The Derivative of an Orthogonal Matrix of Eigenvectors of a Symmetric Matrix, *Linear Algebra and its Applications,* 264, 489–493.

[10] Magnus, J., and H. Neudecker, (1991), *Matrix Differential Calculus with Applications in Statistics and Econometrics,* Second Edition, Wiley, New York.

[11] Moore, D.S., (1972), Chi-square Tests, in *Studies in Statistics,* MAA Studies in Mathematics, R.S. Hogg (ed.), 19, 66–106.

[12] Moore, D.S., (1977), Generalised Inverses, Wald's Method, and the Construction of Chi-squared Tests of Fit, *Journal of the American Statistical Association,* 72, 131–137.

[13] Muirhead, R., (1982), *Aspects of Multivariate Statistical Theory,* Wiley, New York.

[14] Neudecker, H., (1969), Some Theorems on Matrix Differentiation with Special Reference to Kronecker Matrix Products, *Journal of the American Statistical Association,* 64, 953–963.

[15] Neudecker, H., and A.M. Wesselmann, (1990), The Asymptotic Variance Matrix of the Sample Correlation Matrix, *Linear Algebra and its Applications,* 127, 589–599.

[16] Parring, A.-M., (1979), Calculation of Asymptotic Characteristics for the Sample Functions, (in Russian) *Acta et Commentationes Universitatis Tartuensis,* 492, 86–90.

[17] Tyler, D., (1981), Asymptotic Inference for Eigenvectors. *Annals of Statistics,* 9, 725–736.

[18] Tyler, D., (1983), A Class of Asymptotic Tests for Principal Component Vectors, *Annals of Statistics,* 11, 1243–1250.

[19] Watson, G.S., (1983), *Statistics on Spheres,* Wiley, New York.

CHAPTER 16

# On a Fisher–Cornish Type Expansion of Wishart Matrices

D.G. NEL and P.C.N. GROENEWALD

## 16.1 Introduction

This chapter considers the limiting distributions of $U = \sqrt{n}(n^{-1}A - \Sigma)$ where $A \sim W(\Sigma, n)$ or $A \sim W(\Sigma, n, \Omega)$, $\Omega = 0(1)$ are respectively extended to asymptotic approximations to the distributions of $U$. The Fisher–Cornish approximation to a chi–square distribution is shown to be a special case of this approximation up to the order terms involved. An asymptotic approximation to the distribution of $U = \sqrt{n}\{n^{-1}A - \Sigma(I+\theta)\}$ is also derived when $A \sim W(\Sigma, n, \Omega)$ and $\Omega = 0(n)$.

Anderson (1958) (Theorem 4.2.4) proved that if $A : p \times p$ has a Wishart distribution denoted as $A \sim W(\Sigma, n)$ and if $B = n^{-1/2}(A - n\Sigma)$, then the limiting distribution of $B$, as $n \to \infty$, is normal with mean 0 and covariances.

$$E(b_{ij}b_{k1}) = \sigma_{ik}\sigma_{jl} + \sigma_{il}\sigma_{jk}. \tag{1.1}$$

Fujikoshi (1968) gave a limiting distribution for the noncentral Wishart case: If $A \sim W(\Sigma, n, \Omega)$ then

(i) if $\Omega = 0(n) = n\theta$ say, the limiting distribution of

$$U = \sqrt{n}\{n^{-1}A - \Sigma(I+\theta)\}$$

is a $p(p+1)/2$ variate normal with mean 0 and some covariance matrix.

(ii) if $\Omega = 0(1)$, the limiting distribution of $U = \sqrt{n}(n^{-1}A - \Sigma)$ is a $p(p+1)/2$ variate normal with mean 0 and some covariance matrix.

The properties of this limiting normal distribution and the structure of the covariance matrices involved were not described. Nel (1978) gave a derivation of the symmetric normal distribution, which is briefly discussed in section 2.

More detailed results including marginal and conditional distributions can be found in Nel (1977).

Nel (1978) derived an asymptotic approximation up to order $m^{-1/2}$ for the distribution of $U = \sqrt{m}(m^{-1}A - \Sigma)$, where $A \sim W(\Sigma, n)$ and $m = n - (p+1)/3$ in terms of symmetric multivariate normal variates. In section 3 this approximation is improved, the asymptotic density function of $U$ is derived and the approximation is also extended to the case of noncentral Wishart distributions.

This asymptotic density function has been applied by Groenewald and van der Merwe (1979) and Groenewald and de Waal (1979) to find asymptotic distributions for the multiple regression coefficient $\hat{\beta} = A_{12}A_{22}^{-1}$ for the cases where $A$ are central and noncentral Wishart and also for the Stein estimator in multiple regression, but has not been described in detail yet.

## 16.2 The Symmetric Multivariate Normal Distribution

If $X : p \times p$ is symmetric, vec$X$ : $p^2 \times 1$ denotes the column vector formed when the rows of $X$ are stacked as columns in order of appearance, into a single column vector. Similarly vecs$X$ : $\frac{1}{2}p(p+1) \times 1$ denotes the column vector formed when the rows of $X$ up to and including the main diagonal are stacked as columns, in order of appearance into a single column vector. See Koopmans (1950), Nel (1980) and Magnus and Neudecker (1991).

Browne (1974) defined transition matrices $K_p : p^2 \times \frac{1}{2}p(p+1)$ and $K_p^- : \frac{1}{2}p(p+1) \times p^2$ (the Moore–Penrose inverse of $K_p$) for symmetric matrices such that

$$\text{vecs}X = K_p'\text{vec}X \quad \text{and} \quad \text{vec}X = K_p^{-'}\text{vecs}X.$$

Using the properties of these matrices, Nel (1978) proved that

$$|K_p'(A \otimes B)K_p| = 2^{-p^2} \prod_{i \leq j} (\lambda_i \delta_j + \lambda_j \delta_i),$$

where $\lambda_i$ and $\delta_i, i = 1, \ldots, p$ are the latent roots of $A = A' : p \times p$ and $B = B' : p \times p$ respectively, where $AB = BA$. In particular

$$|K_p'(A \otimes 2A)K_p| = 2^p |A|^{p+1}.$$

Let $X : p \times p$ be a symmetric variable matrix and $M$, $\Sigma$ and $R$ be constant symmetric matrices such that $\Sigma R = R\Sigma$. If the vector vecs$X$ : $\frac{1}{2}p(p+1) \times 1$, formed from $X$, is distributed $N\{\text{vecs}M, K_p'(\Sigma \otimes R)K_p\}$, then $X$ is said to have a symmetric multivariate normal distribution with mean $M$ and covariance matrix $K_p'(\Sigma \otimes R)K_p$, which we denoted as

$$X = X' \sim N\{M, K_p'(\Sigma \otimes R)K_p\}.$$

A typical element of the covariance matrix $K_p'(\Sigma \otimes R)K_p$ is

$$\text{Cov}(X_{ij}, X_{k1}) = \frac{1}{4}(\sigma_{ik}r_{jl} + \sigma_{jk}r_{il} + \sigma_{il}r_{jk} + \sigma_{jl}r_{ik}).$$

A typical element of the covariance matrix $K_p'(\Sigma \otimes 2\Sigma)K_p$ is given by

$$\mathrm{Cov}(X_{ij}, X_{k1}) = \sigma_{ik}\sigma_{jl} + \sigma_{il}\sigma_{jk}, \qquad \text{as in (1.1).}$$

The characteristic function of the diagonal elements of $X$ and two times the lower triangular, off–diagonal elements of $X$ can be written as

$$\phi_X(T) = \mathrm{etr}(i\, TM - \frac{1}{2}\Sigma TRT), \qquad T = T' : p \times p,$$

where trA denotes the trace of A : $p \times p$ and etr(A) = exp(trA).

If $X = X' \sim N\{M, K_p'(\Sigma \otimes 2\Sigma)K_p\}$, the probability density function of $X$ is given by

$$f(X) = \frac{\mathrm{etr}\{-\frac{1}{4}\Sigma^{-1}(X-M)\Sigma^{-1}(X-M)\}}{(2\pi)^{p(p+1)/4}2^{p/2}|\Sigma|^{(p+1)/2}}, \qquad -\infty < X < \infty. \tag{2.1}$$

Some expected values of functions of $X$ are given in Nel (1977) and Nel (1978).

## 16.3 Asymptotic Approximations for Wishart Matrices

### 16.3.1 Central Wishart : $A \sim W(\Sigma, n)$

If $A \sim W(\Sigma, n)$, $S = A/n$ and $U = \sqrt{n}(S - \Sigma)$, then the characteristic function of $U$, which is given by Sugiura (1973), is

$$\begin{aligned}\phi_U(T) &= \mathrm{etr}(-T\Sigma T\Sigma)[1 + \frac{4}{3\sqrt{n}}\mathrm{tr}(iT\Sigma)^3 + \frac{2}{n}\{\mathrm{tr}(iT\Sigma)^4 \\ &\quad + \frac{8}{9}\mathrm{tr}^2(iT\Sigma)^3\} + 0(n^{-3/2})].\end{aligned} \tag{3.1}$$

We approximate $U$ by

$$B_n(V) = V + \frac{1}{\sqrt{n}}f(V) + \frac{1}{n}g(V), \tag{3.2}$$

where $V = V' \sim N\{0, K_p'(\Sigma \otimes 2\Sigma)K_p\}$, and then find $f(V)$ and $g(V)$ such that $\phi_{B_n(V)}(T)$ yields $\phi_U(T)$ up to order terms $n^{-1}$.

The characteristic function of $B_n(V)$, which is given by Nel (1978), is

$$\begin{aligned}\phi_{B_n(V)}(T) &= E\Big[\mathrm{etr}iT\Big\{V + \frac{1}{\sqrt{n}}f(V) + \frac{1}{n}g(V)\Big\}\Big] \\ &= E(\mathrm{etr}iTV)\Big(1 + \frac{1}{\sqrt{n}}\{\mathrm{tr}\ iTf(V)\} + \frac{1}{2n}[\mathrm{tr}^2\{iTf(V)\} \\ &\quad + 2\mathrm{tr}\{iTg(V)\}]\Big) + 0(n^{-3/2}),\end{aligned} \tag{3.3}$$

where $V = V' \sim N\{0, K_p'(\Sigma \otimes 2\Sigma)K_p\}$

$$\begin{aligned}\phi_{B_n(V)}(T) &= \operatorname{etr}(-T\Sigma T\Sigma)\Big(1 + \frac{1}{\sqrt{n}}\mathrm{Etr}\{iTf(W)\} + \frac{1}{2n}[E\mathrm{tr}^2\{iTf(W)\} \\ &\quad + 2\mathrm{Etr}\{iTg(W)\}]\Big) + 0(n^{-3/2}),\end{aligned}$$

where $W = W' \sim N\{2i\Sigma T\Sigma, K_p'(\Sigma \otimes 2\Sigma)K_p\}$ and EtrA denotes the expected value of trA.

To find $f(W)$ and $g(W)$, the following expected values are of interest

$$E\{\mathrm{tr}(\Sigma^{-1}W)\} = 2\mathrm{tr}(iT\Sigma), \tag{3.4}$$

$$E\{\mathrm{tr}(iTW)\} = 2\mathrm{tr}(iT\Sigma)^2, \tag{3.5}$$

$$E\{\mathrm{tr}(iTW\Sigma^{-1}W)\} = 4\mathrm{tr}(iT\Sigma)^3 + (p+1)\mathrm{tr}(iT\Sigma) \tag{3.6}$$

$$\begin{aligned}E\{\mathrm{tr}^2(iTW\Sigma^{-1}W)\} &= 16\mathrm{tr}^2(iT\Sigma)^3 + 32\mathrm{tr}(iT\Sigma)^4 \\ &\quad + 8(p+1)\mathrm{tr}(iT\Sigma)\mathrm{tr}(iT\Sigma)^3 \\ &\quad + 2(p+2)\mathrm{tr}(iT\Sigma)^2 \\ &\quad + (p^2+2p+3)\mathrm{tr}^2(iT\Sigma),\end{aligned} \tag{3.7}$$

$$\begin{aligned}E\{\mathrm{tr}(iTW\Sigma^{-1}W\Sigma^{-1}W)\} &= 8\mathrm{tr}(iT\Sigma)^4 + 2\mathrm{tr}^2(iT\Sigma) \\ &\quad + 2(2p+3)\mathrm{tr}(iT\Sigma)^2.\end{aligned} \tag{3.8}$$

From (3.6) follows that:

$$\frac{1}{3}E\{\mathrm{tr}(iTW\Sigma^{-1}W)\} = (p+\frac{1}{3})\mathrm{tr}(iT\Sigma) + \frac{4}{3}\mathrm{tr}(iT\Sigma)^3$$

and thus, comparing with (3.1) and (3.3),

$$f(W) = \frac{1}{3}\{W\Sigma^{-1}W - (p+1)\Sigma\}.$$

From (3.7) and (3.3) it is clear that $g(W)$ must be such that

$$E\mathrm{tr}\{iTg(W)\} = \frac{1}{18}\{2\mathrm{tr}(iT\Sigma)^4 - \mathrm{tr}^2(iT\Sigma) - (p+2)\mathrm{tr}(iT\Sigma)^2\}.$$

Using (3.4), (3.5) and (3.8), it follows that

$$g(W) = \frac{1}{36}\{W\Sigma^{-1}W\Sigma^{-1}W - 3\Sigma\mathrm{tr}\Sigma^{-1}W - (4p+7)W\}.$$

Thus $S$ can be approximated by:

$$\begin{aligned}A_n(V) &= \Sigma + \frac{1}{\sqrt{n}}V + \frac{1}{3n}\{V\Sigma^{-1}V - (p+1)\Sigma\} \\ &\quad + \frac{1}{36n^{3/2}}\{V\Sigma^{-1}V\Sigma^{-1}V - 3\Sigma\mathrm{tr}\Sigma^{-1}V - (4p+7)V\},\end{aligned} \tag{3.9}$$

where $V = V' \sim N\{0, K_p'(\Sigma \otimes 2\Sigma)K_p\}$.

It is interesting to compare the first two exact moments of $S$ with the approximated moments, i.e. the moments of $A_n(V)$.

*Moments of* $S$ ( de Waal and Nel (1978))

$$E(S) = \Sigma \qquad E(S^2) = \Sigma^2 + \frac{1}{n}(\Sigma^2 + \Sigma \mathrm{tr}\Sigma).$$

*The moments of* $A_n(V)$

$$E\{A_n(V)\} = \Sigma \qquad E\{A_n(V)\}^2 = \Sigma^2 + \frac{1}{n}(\Sigma^2 + \Sigma\mathrm{tr}\Sigma) + 0(n^{-2}).$$

If $p = 1$ and $\sigma^2 = 1$ it is possible to improve the approximation by further order terms. Now $S$ denotes a $\chi_n^2/n$ variable and it can be approximated by:

$$\begin{aligned} A_n(V) &= 1 + \frac{1}{\sqrt{n}}V + \frac{1}{3n}(V^2 - 2) + \frac{1}{36n^{3/2}}(V^3 - 14V) \\ &- \frac{1}{810n^2}(3V^4 + 14V^2 - 64) + \frac{1}{38880n^{5/2}}(9V^5 + 512V^3 - 1732V), \end{aligned}$$

where $V \sim N(0, 2)$.

Note that this approximation to the distribution of $S$ (obtained from (3.9) by setting $p = 1, \sigma^2 = 1$) in terms of $V \sim N(0, 2)$ is identical (up to the order terms considered) to the asymptotic expansion of Fisher and Cornish (1960). Thus (3.9) can be considered as the multivariate analogue of the Fisher and Cornish expansion of $\chi_n^2$.

If $R_n = S - A_n(V)$, then

$$\begin{aligned} E(R_n^2) &= \frac{2}{n}(\Sigma^2 + \Sigma\mathrm{tr}\Sigma) + \frac{1}{36n^{3/2}}(5p^2 + 67p + 86)\Sigma^2 \\ &+ (5p^2 + 21p + 28)\Sigma\mathrm{tr}\Sigma + 0(n^{-4}), \end{aligned}$$

so $\lim_{n\to\infty} E(R_n^2) = 0$ and $R_n$ converges in quadratic mean to a zero matrix.

### 16.3.2 The Probability Density Function (p.d.f.) of $U$

The exact p.d.f. of $U = \sqrt{n}\,(S - \Sigma)$ is given by:

$$f(U) = \frac{|\sqrt{n}U + n\Sigma|^{(n-p-1)/2}\mathrm{etr}\{-\frac{1}{2}\Sigma^{-1}(\sqrt{n}U + n\Sigma)\}}{n^{-p(p+1)/4}2^{np/2}|\Sigma|^{n/2}\Gamma_p(n/2)}.$$

Expanding the p.d.f. we get

$$\begin{aligned} f(U) &= \frac{n^{p(2n-p-1)/4}e^{-np/2}\mathrm{etr}(-\frac{1}{4}\Sigma^{-1}U\Sigma^{-1}U)}{2^{np/2}\Gamma_p(n/2)|\Sigma|^{(p+1)/2}} \\ &\times \left[1 + \frac{1}{\sqrt{n}}g_1(U) + \frac{1}{n}\{g_2(U) + \frac{p}{24}(2p^2 + 3p - 1)\} + R_1(U)\right], \end{aligned}$$

where $R_1(U)$ is a remainder term, and

$$\Gamma_p(1/2n) = \pi^{p(p-1)4} \prod_{i=1}^{p} \Gamma\{(n+1-i)/2\}$$

Using the Euler–Bernoulli expansion of $\log \Gamma(x+h)$ ( Anderson (1958) p.204)

$$\frac{n^{p(2n-p-1)/4} e^{np/2}}{2^{np/2}\Gamma_p(n/2)} = \frac{1}{(2\pi)^{p(p+1)/4} 2^{p/2}} \Big\{1 - \frac{p}{24n}(2p^2+3p-1) + 0(n^{-2})\Big\},$$

it follows that

$$f(U) = \Phi(U)\Big\{\frac{1}{\sqrt{n}} g_1(U) + \frac{1}{n} g_2(U) + R_2(U)\Big\}, \tag{3.10}$$

where $\Phi(U)$ is given by (2.1) with $m = 0$ and

$$\begin{aligned} g_1(U) &= \frac{1}{6}\{\mathrm{tr}(\Sigma^{-1}U)^3 - 3(p+1)\mathrm{tr}(\Sigma^{-1}U)\}, \\ g_2(U) &= \frac{1}{72}\{\mathrm{tr}^2(\Sigma^{-1}U)^3 - 9\mathrm{tr}(\Sigma^{-1}U)^4 - 6(p+1)\mathrm{tr}(\Sigma^{-1}U)\mathrm{tr}(\Sigma^{-1}U)^3 \\ &\quad + 9(p+1)^2\mathrm{tr}^2(\Sigma^{-1}U) + 18(p+1)\mathrm{tr}(\Sigma^{-1}U)^2 - 3p(2p^2+3p-1)\}. \end{aligned}$$

The p.d.f. of $U$ can also be obtained from the characteristic function (3.1) by using the inverse Fourier transform

$$f(U) = (2\pi)^{-p(p+1)/2} 2^{p(p-1)/2} \int_{\infty}^{\infty} \mathrm{etr}(-iUT)\phi_u(T)dT,$$

where the integration is taken over all the $p(p+1)/2$ different elements of $T$. This gives

$$\begin{aligned} f(U) &= (2\pi)^{p(p+1)/2} 2^{-p/2} \mathrm{etr}\Big(-\frac{1}{4}U\Sigma^{-1}U\Sigma^{-1}\Big) \\ &\quad \int_{\infty}^{\infty} \mathrm{etr}\Big\{-\frac{1}{4}(W + i\Sigma^{-1}U\Sigma^{-1})\Sigma(W + i\Sigma^{-1}U\Sigma^{-1})\Big\} h(W)dW, \end{aligned}$$

where $T = \frac{1}{2}W$ and

$$h(W) = 1 + \frac{1}{6\sqrt{n}}\mathrm{tr}(iW\Sigma)^3 + \frac{1}{8n}\Big\{\mathrm{tr}(iW\Sigma)^4 + \frac{1}{9}\mathrm{tr}^2(iW\Sigma)^3\Big\} + 0_p(n^{-3/2}).$$

Thus

$$f(U) = \frac{\mathrm{etr}(-\frac{1}{4}U\Sigma^{-1}U\Sigma^{-1})}{(2\pi)^{p(p+1)/2} 2^{p/2} |\Sigma|^{p+1)/2}} E\{h(W)\}, \tag{3.11}$$

where $W = W' \sim N\{-i\Sigma^{-1}U\Sigma^{-1}, K_p'(2\Sigma^{-1} \otimes \Sigma^{-1})K_p\}$. Now

$$E\{\mathrm{tr}(iW\Sigma)^3\} = -3(p+1)\mathrm{tr}\Sigma^{-1}U + \mathrm{tr}(\Sigma^{-1}U)^3,$$

$$\begin{aligned}
E\{\mathrm{tr}(iW\Sigma)^4\} &= (2p^3+5p^2+5p) - 2(2p+3)\mathrm{tr}(\Sigma^{-1}U)^2 \\
&\quad - 2\mathrm{tr}^2(\Sigma^{-1}U) + \mathrm{tr}(\Sigma^{-1}U)^4, \\
E\{\mathrm{tr}^2(iW\Sigma)^3\} &= \mathrm{tr}^2(\Sigma^{-1}U)^3 - 18\mathrm{tr}(\Sigma^{-1}U)^4 - 6(p+1)\mathrm{tr}(\Sigma^{-1}U)\mathrm{tr}(\Sigma^{-1}U)^3 \\
&\quad - 18(3p+4)\mathrm{tr}(\Sigma^{-1}U)^2 + 9(p^2+2p+3)\mathrm{tr}^2(\Sigma^{-1}U) \\
&\quad - 6p(4p^2+9p+7).
\end{aligned}$$

Substitution of these values into (3.11) also yields (3.10).

These results imply that the error term involved in approximating $U$ by $B_n(V)$ in (3.2) is of order $0(n^{-3/2})$, and thus the error term in (3.9) is of order $0(n^{-2})$ since $A_n(V) = \Sigma + (n^{-1/2}B_n(V)$.

### 16.3.3 Noncentral Wishart

These results can be extended to noncentral Wishart matrices. If $A \sim W(\Sigma, n, \Omega)$ where $\Omega = 0(1)$, then the characteristic function of $U = \sqrt{n}(S - \Sigma)$, where $S = n^{-1}A$ is given by

$$\Phi_U(T) = \mathrm{etr}(-\sqrt{ni}\,T\Sigma)\Big|I - \frac{2i}{\sqrt{n}}T\Sigma\Big|^{-n/2}\mathrm{etr}\Big\{\frac{i}{\sqrt{n}}\Omega T\Sigma(I - \frac{2i}{\sqrt{n}}T\Sigma)^{-1}\Big\},$$

which can be expanded asymptotically as

$$\begin{aligned}
\Phi_U(T) &= \mathrm{etr}(-T\Sigma T\Sigma)\Big[1 + \frac{1}{\sqrt{n}}\Big\{\frac{4}{3}\mathrm{tr}(iT\Sigma)^3 + \mathrm{tr}(\Omega iT\Sigma)\Big\}\Big] \\
&\quad + \frac{1}{n}\Big\{\frac{8}{9}\mathrm{tr}^2(iT\Sigma)^3 + 2\mathrm{tr}(iT\Sigma)^4 + 2\mathrm{tr}\Omega(iT\Sigma)^2 + \frac{1}{2}\mathrm{tr}^2(\Omega iT\Sigma) \\
&\quad + \frac{4}{3}\mathrm{tr}(iT\Sigma)^3\mathrm{tr}(\Omega iT\Sigma)\Big\} + 0(n^{-3/2}). \qquad (3.12)
\end{aligned}$$

Suppose $B_n(V) = V + n^{-1/2}f(V) + n^{-1}g(V)$ , where $V = V' \sim N\{0, K_p'(\Sigma \otimes 2\Sigma)K_p\}$. By using the same approach as in Section 3.1, it follows that

$$f(V) = \frac{1}{3}\{V\Sigma^{-1}V - (p+1)\Sigma + 3\Sigma\Omega\}.$$

Thus $S$ can be approximated by

$$A_n(V) = \Sigma + \frac{1}{\sqrt{n}}V + \frac{1}{3n}\{V\Sigma^{-1}V - (p+1)\Sigma + 3\Sigma\Omega\}. \qquad (3.13)$$

It was not possible to find an expression for $g(V)$ which will yield the coefficient of $n^{-1}$ in (3.12). If $A \sim W(\Sigma, n, \Omega)$ where $\Omega = 0(n) = n\theta$, then the characteristic function of $U = \sqrt{n}\{S - \Sigma(I+\theta)\}$, where $S = n^{-1}A$, is given by

$$\phi_U(T) = \mathrm{etr}\{-i\sqrt{n}T\Sigma(I+\theta)\}\Big|I - \frac{2i}{\sqrt{n}}T\Sigma\Big|^{-n/2}\mathrm{etr}\Big\{i\sqrt{n}\,\theta T\Sigma\Big(I - \frac{2i}{\sqrt{n}}T\Sigma\Big)^{-1}\Big\},$$

which can be expanded asymptotically as

$$\phi_U(T) = \mathrm{etr}\{-T\Sigma T(\Sigma + 2\theta\Sigma)\}\Big[1 + \frac{4}{3\sqrt{n}}\mathrm{tr}\{(I+3\theta)(iT\Sigma)^3\} + 0(n^{-1})\Big]. \quad (3.14)$$

Suppose $B_n(V) = V + n^{-1/2}f(V)$, where $V = V' \sim N[0, K_p'\{2\Sigma \otimes (\Sigma + 2\Sigma\theta)\}K_p]$, and where it is assumed that $\Sigma\theta$ is symmetric. Then the characteristic function of $B_n(V)$ is given by

$$\begin{aligned}\phi_{B_n(V)}(T) &= E(\mathrm{etr}iTV)\Big[1 + \frac{1}{\sqrt{n}}\{\mathrm{tr}iTf(V)\} + 0_p(n^{-1})\Big] \\ &= \mathrm{etr}\{-T\Sigma T(\Sigma + 2\theta\Sigma)\}\Big[1 + \frac{1}{\sqrt{n}}E\{\mathrm{tr}iTf(W)\} + 0_p(n^{-1})\Big],\end{aligned}$$

where

$$W = W' \sim N[2i\Sigma T(\Sigma + 2\Sigma\theta), K_p'\{\Sigma \otimes 2\Sigma(I+2\theta)\}K_p]. \quad (3.15)$$

Now

$$\begin{aligned}&E\Big\{\frac{1}{3}W(\Sigma + 2\Sigma\theta)^{-1}W(I+2\theta)^{-1}(I+3\theta)\Big\} \\ &= \frac{1}{6}\{(p+2)\Sigma(I+2\theta)^{-1}(I+3\theta) + \Sigma(I+3\theta)\mathrm{tr}(I+2\theta)^{-1}\} \\ &+ \frac{4}{3}\Sigma(iT\Sigma)^2(I+3\theta).\end{aligned}$$

Comparing this result with (3.14) and (3.15), it is clear that for this case

$$\begin{aligned}f(V) &= \frac{1}{3}\Big\{V(\Sigma + 2\Sigma\theta)^{-1}V(I+2\theta)^{-1}(I+3\theta) \\ &\quad - \frac{1}{2}\Sigma(p+2)(I+2\theta)^{-1} + I\mathrm{tr}(I+2\theta)^{-1})(I+3\theta)\Big\}.\end{aligned}$$

Thus $S$ can be approximated by

$$\begin{aligned}A_n(V) &= \Sigma(I+\theta) + \frac{1}{\sqrt{n}}V + \frac{1}{3n}\Big\{V(\Sigma + 2\Sigma\theta)^{-1}V(I+2\theta)^{-1}(I+3\theta) \\ &\quad - \frac{1}{2}\Sigma(p+2) + I\mathrm{tr}(I+2\theta)^{-1})(I+3\theta)\Big\}.\end{aligned}$$

Another possible approach to an asymptotic approximation of a noncentral Wishart matrix is to approximate the noncentral distribution first with a central Wishart distribution and then to use the method of Section 3.1 to approximate the central distribution.

Steyn and Roux (1972) showed, by using the method of moments on the first two moments of the noncentral Wishart, and ignoring terms of order higher than $0(n^{-1})$, that $A$ is approximately distributed as a central Wishart $W\{\Sigma((I + n^{-1}\Omega), n\}$.

Thus, as in Section 3.3, where $\Omega = 0(1)$ if $A \sim W(\Sigma, n, \Omega)$ is approximated by $W(\Sigma(I + n^{-1}\Omega),\ n)$, then, from Section 3.1, it follows, with $\Sigma(I + n^{-1}\Omega)$ replacing $\Sigma$ and writing $(I + n^{-1}\Omega)^{-1} = I - n^{-1}\Omega + 0(n^{-2})$, that $S$ can be approximated by

$$\begin{aligned} A_n(V) &= \Sigma + \frac{1}{\sqrt{n}} V + \frac{1}{3n}\{V\Sigma^{-1}V - (p+1)\Sigma + 3\Sigma\Omega\} \\ &+ \frac{1}{36n^{3/2}}\{V\Sigma^{-1}V\Sigma^{-1}V - 3\Sigma\mathrm{tr}(\Sigma^{-1}V) - (4p+7)V\}. \end{aligned} \tag{3.16}$$

The order terms to $n^{-1}$ are the same as in (3.13) while the last order term is the same as in the central case (3.9), but here

$$V = V' \sim N\Big[0, K_p'\Big\{2\Sigma\Big(I + \frac{1}{n}\Omega\Big) \otimes \Sigma\Big(I + \frac{1}{n}\Omega\Big)\Big\}K_p\Big].$$

In the case when $\Omega = n\theta$, $A \sim W(\Sigma, n, \Omega)$ can be approximated by $W\{\Sigma(I + \theta), n\}$. From Section 3.1, it follows, with $\Sigma(I + \theta)$ replacing $\Sigma$, that $S$ can be approximated by

$$\begin{aligned} A_n(V) &= \Sigma(I+\theta) + \frac{1}{\sqrt{n}} V + \frac{1}{3n}\{V(\Sigma + \Sigma\theta)^{-1}V - (p+1)\Sigma(I+\theta)\} \\ &+ \frac{1}{36n^{3/2}}\{V(\Sigma + \Sigma\theta)^{-1}V(\Sigma + \Sigma\theta)^{-1}V(\Sigma + \Sigma\theta)^{-1}V \\ &- 3\Sigma(I+\theta)\mathrm{tr}(\Sigma + \Sigma\theta)V - \ (4p+7)V\}, \end{aligned} \tag{3.17}$$

where

$$V = V' \sim N[0, K_p'\{2\Sigma(I+\theta) \otimes \Sigma(I+\theta)\}K_p].$$

The advantage of a relatively simple extra order term, when using this approximation, is offset by the fact that it is a double approximation and that terms appearing in (3.16) and (3.17) of order higher than $0(n^{-1})$ are merely approximations of the appropriate order terms. The relative efficiencies of the different approximations have not been examined.

## References


[1] Anderson, T.W., (1958), *An Introduction to Multivariate Statistical Analysis,* John Wiley and Sons, New York,

[2] Browne, M.W., (1974), Generalised Least Squares Estimators in the Analysis of Covariance Structures, *South African Statistical Journal,* 8, 1–24.

[3] De Waal, D.J., and D.G. Nel, (1973), On Some Expectations with Respect to Wishart Matrices, *South African Statistical Journal,* 7, 61–67.

[4] Fisher, R.A., and E.A. Cornish, (1960), The Percentile Points of Distributions Having Known Cumulants, *Technometrics,* 2, 209–225.

[5] Fujikoshi, Y., (1968), Asymptotic Expansion of the Distribution of the Generalised Variance in the Noncentral case. *J. Sc. Hiroshima Univ., Ser. A–I,* 32, 293–299.

[6] Groenewald, P.C.N., and A.J van der Merwe, (1979), Asymptotic Distributions for the Stein Estimator in Multiple Regression, *South African Statistical Journal,* 13, 139–152.

[7] Groenewald, P.C.N., and D.J. de Waal, (1979), An Asymptotic Distribution for the Coefficient in a Multiple Time Series, *South African Statistical Journal,* 13, 15–28.

[8] Koopmans, T.C., (1950), *Statistical Inference in Dynamic Economic Models,* Cowles Commission for Research in Economics, Monograph No. 10, John Wiley, New York.

[9] Magnus, J.R., and H. Neudecker, (1991), *Matrix Differential Calculus with Applications in Statistics and Econometrics, 2nd Edition*, John Wiley, Chichester.

[10] Nel, H.M., (1977), On Distributions and Moments Associated with Matrix Normal Distributions, Technical Report No. 24, Department of Statistics, University of the Orange Free State,

[11] Nel, D.G., (1978), On the Symmetric Multivariate Normal Distribution and the Asymptotic Expansion of a Wishart Matrix, *South African Statistical Journal,* 12, 145–159.

[12] Nel, D.G., (1980), On Matrix Differentiation in Statistics, *South African Statistical Journal,* 14, 137–193.

[13] Steyn, H.S., and Roux, J.J., (1972), Approximation for the Noncentral Wishart distribution, *South African Statistical Journal,* 6, 165–173.

[14] Sugiura, N., (1973), Derivatives of the Characteristic Root of a Hermitian Matrix with Applications in Multivariate Analysis, *Communications in Statistics,* 1, 393–417.

CHAPTER 17

# Scaled and Adjusted Restricted Tests in Multi-Sample Analysis of Moment Structures

Albert SATORRA[1]

## 17.1 Introduction

Moment structure analysis is widely used in behavioural, social and economic studies to analyse structural relations between variables, some of which may be latent (i.e., unobservable); see, for example, Bollen (1989), Yuan and Bentler (1997), and references therein. Commercial computer programs to carry out such analysis, for a general class of structural equation models, are available, for example, LISREL of Jöreskog and Sörbom (1994) and EQS of Bentler (1995). In multi-sample analysis, data from several samples are combined into one analysis, making it possible, among other features, to test for across-group invariance of specific model parameters. One issue which is central in moment structure analysis is the goodness-of-fit test of the model and the test of restrictions on parameters.

Asymptotic distribution-free (ADF) methods which do not require distribution assumptions on the observable variables have been developed, see Browne (1984), Muthén (1989) and Bentler, Lee and Weng (1987). The ADF methods, however, involve fourth-order sample moments, thus they may lack robustness to small and medium-sized samples. In the case of non-normal data, an alternative to the ADF approach is to use a normal-theory estimation method in conjunction with asymptotic robust standard errors and test statistics, see Satorra (1992). Asymptotic robust test statistics may still lack robustness to small and medium-sized samples.

As an alternative to the asymptotic robust test statistics, Satorra and Bentler (1988a, b, 1994) developed a family of corrected normal-theory test statistics which are easy to implement in practice, and which have been shown to out-

[1]The author is grateful to Peter Bentler and Bengt Muthén for their very useful comments. Research supported by the Spanish DGES grant PB96-0300.

perform the asymptotic robust test statistics in small and medium-sized samples, See, for example, Chou, Bentler and Satorra (1991), Hu, Bentler and Kano (1992) and Curran, West and Finch (1996). Extension of Satorra–Bentler (SB)'s corrections to goodness-of-fit test statistics in the case of the analysis of augmented moment structures, multi-samples and categorical data, have been discussed respectively by Satorra (1992) and Muthén (1993). In a recent personal communication, Bengt Muthén has kindly pointed to this author the need to extend SB's corrections to score and difference test statistics.

The purpose of this chapter is to extend SB's corrections to score, difference and Wald-test statistics. This is undertaken in the general context of multi-sample analysis. We rely heavily on the notation and results of Satorra (1989), Bentler (1995) and Satorra and Bentler (1988a, b).

The chapter is structured as follows. In Section 2 we review the general theory of normal-theory GLS analysis of multi-sample moment structures, placing special emphasis on goodness-of-fit test statistics. The extension of the scaling and adjusted corrections to the restricted test is undertaken in Section 3. Finally, Section 4 illustrates, by means of a Monte Carlo study, the finite sample size performance of competing score-test statistics developed in Section 3.

## 17.2 Goodness-of-fit tests

Let $\sigma$ and $s$ be $p$-dimensional vectors of population and sample moments respectively, with

$$\sqrt{n}(s - \sigma^0) \xrightarrow{L} \mathcal{N}(0, \Gamma), \tag{1}$$

where $\sigma^0$ is the probability limit of $s$, as sample size $n \to +\infty$, $\Gamma$ is the asymptotic variance matrix of $s\sqrt{n}$, and $\xrightarrow{L}$ denotes convergence in distribution as $n \to \infty$. Consider a moment structure for $\sigma$, $\sigma = \sigma(\theta)$, where $\sigma(.)$ is a twice-continuously differentiable vector-valued function of a $q$-dimensional parameter vector $\theta \in \Theta$, a compact subset of $R^q$. Let $\hat{\theta}$ be an estimator of $\theta$ that satisfies

$$\sqrt{n}(\hat{\theta} - \theta_0) = O_p(1), \tag{2}$$

with $\theta_0$ in the interior of $\Theta$. The usual notation of stochastic orders of magnitude is used. Thus, $O_p(1)$ and $o_p(1)$ denote, respectively, "bounded" and "tends to zero in probability", as $n \to +\infty$. Usually, (2) is fulfilled whenever $\sigma = \sigma(\theta)$ is locally identified (viz., the function $\sigma(\theta)$ is injective in a neighbourhood of the true parameter value). Denote $\hat{\sigma} = \sigma(\hat{\theta})$ and $\sigma_0 = \sigma(\theta_0)$. It is assumed that $\sqrt{n}\{\hat{\theta} - g(s)\} = o_p(1)$, where $g(.)$ is a $q$-dimensional continuously differentiable vector-valued function of $s$, with $g(.)$ Fisher-consistent for $\theta$, i.e. $g\{\sigma(\theta)\} = \theta$, $\theta \in \Theta$. The theory discussed involves the use of the Jacobian matrices $G = (\partial/\partial s')g(s)$ and $\Delta = (\partial/\partial\theta')\sigma(\theta)$, which will be assumed to be regular at $\sigma_0 = \sigma(\theta_0)$ and $\theta_0$ respectively. Note that Fisher-consistency of $g(s)$ implies that $G\Delta = I_q$. In the developments that follow, $\Delta$ and $G$ denote the corresponding matrices evaluated at $\theta_0$ and $\sigma_0$ respectively.

The above mentioned properties for $\hat{\theta}$ are satisfied by most of the estimators commonly in use in moment structure analysis. In particular, they are satisfied by the ML and GLS estimators discussed below. They are also satisfied in the case of instrumental variable estimators, see, for example, Satorra and Bentler (1991).

A Wald-type statistic for testing the goodness-of-fit of the model can be developed. Since

$$\sqrt{n}(s-\hat{\sigma}) = \sqrt{n}(s-\sigma^0) + \sqrt{n}(\sigma^0-\sigma_0) - \sqrt{n}(\hat{\sigma}-\sigma_0),$$

a Taylor series expansion of $\sqrt{n}(\hat{\sigma}-\sigma_0)$ at $\sigma_0$, gives

$$\sqrt{n}(s-\hat{\sigma}) = \sqrt{n}(s-\sigma^0) + \sqrt{n}(\sigma^0-\sigma_0) - \Delta\sqrt{n}(\hat{\theta}-\theta_0) + o_p(1),$$

which combined with (1) yields

$$\sqrt{n}(s-\hat{\sigma}) \xrightarrow{L} \mathcal{N}(\tau, H), \tag{3}$$

where

$$\tau = \lim_{n\to+\infty} \sqrt{n}(\sigma^0-\sigma_0) \tag{4}$$

and

$$H = (I-\Delta G)\Gamma(I-\Delta G)'. \tag{5}$$

We further assume

$$\tau = \lim_{n\to+\infty} \sqrt{n}(\sigma^0-\sigma_0) \quad < \quad +\infty,$$

which is the usual assumption of a sequence of local alternatives (needed to ensure finite asymptotic distribution of test statistics when the model is not exactly true).

Let $\Delta_\perp$ be an orthogonal complement of $\Delta$, i.e., $\Delta_\perp$ is a $p\times(p-q)$ matrix of full-column rank such that $\Delta_\perp'\Delta = 0$. From (3), we have

$$\sqrt{n}\Delta_\perp'(s-\hat{\sigma}) \xrightarrow{L} \mathcal{N}(\Delta_\perp'\tau, \Delta_\perp'\Gamma\Delta_\perp), \tag{6}$$

since $\Delta_\perp' H\Delta_\perp = \Delta_\perp'\Gamma\Delta_\perp$, as $\Delta_\perp'(I-\Delta G) = \Delta_\perp'$; hence, a residual-based goodness-of-fit test for the model is defined as the Wald-test statistic

$$GF = n(s-\hat{\sigma})'\hat{\Delta}_\perp(\hat{\Delta}_\perp'\hat{\Gamma}\hat{\Delta}_\perp)^+\hat{\Delta}_\perp'(s-\hat{\sigma}), \tag{7}$$

where $\hat{\Delta}_\perp$ and $\hat{\Gamma}$ are consistent estimators of $\Delta_\perp$ and $\Gamma$, respectively. The matrix $\hat{\Delta}_\perp$ will be taken to be an orthogonal complement of $\hat{\Delta}$, the matrix $\Delta$ evaluated at the estimated value $\hat{\theta}$. Clearly, from the standard theory of Wald-test statistics—see, for example, Moore (1977)—$GF$ is asymptotically chi-square distributed with degrees of freedom $r = \operatorname{rank}\Delta_\perp'\Gamma\Delta_\perp$ and non-centrality parameter $\lambda = \tau'\Delta_\perp(\Delta_\perp'\Gamma\Delta_\perp)^+\Delta_\perp'\tau$. When $\hat{\Gamma}$ is non-singular, we have

$$GF = n(s-\hat{\sigma})'\left\{\hat{\Gamma}^{-1} - \hat{\Gamma}^{-1}\hat{\Delta}(\hat{\Delta}'\hat{\Gamma}^{-1}\hat{\Delta})^{-1}\hat{\Delta}'\hat{\Gamma}^{-1}\right\}(s-\hat{\sigma}), \tag{8}$$

When $\hat{\Gamma}$ is the (distribution-free) consistent estimator of $\Gamma$ in (10) below, then $GF$ will be called the asymptotic robust goodness-of-fit test statistic, since it is an asymptotic chi-square statistic regardless of the distribution of $z$. In the context of single-sample covariance structure analysis, this statistic was first introduced by Browne (1984). Its performance was studied by Yuan and Bentler (1998), and found to be adequate only in very large samples.

Consider now multi-sample data from $G$ population or groups; that is to say, let $\{z_{gi}\}_{i=1}^{n_g}$, $g = 1, \ldots, G$, where it is assumed that, for each group $g$, the $z_{gi}$ are independent and identically distributed $p_g$-dimensional vectors of observable variables, with finite fourth-order moments. Define $\Sigma_g = E z_{gi} z'_{gi}$, and let the multi-sample vectors $\sigma$ and $s$ of population and sample moments, respectively, be given as $\sigma = (\sigma'_1, \ldots, \sigma'_1)'$ and $s = (s'_1, \ldots, s'_1)'$, where $\sigma'_g = \operatorname{vec} \Sigma_g$ and $s_g = \operatorname{vec} S_g$, with

$$S_g = \frac{1}{n_g} \sum_{i=1}^{n_g} z_{gi} z'_{gi}.$$

Here "vec" denotes the column-wise vectorisation operator See Magnus and Neudecker, (1999), for full details on this operator. Clearly, when there is independence across samples, the asymptotic variance matrix of $s\sqrt{n}$ is of the form

$$\Gamma = \text{block diag}\left(\frac{n}{n_1}\Gamma_1, \ldots, \frac{n}{n_G}\Gamma_G\right), \tag{9}$$

where $\Gamma_g$ is the asymptotic variance of $s_g\sqrt{n_g}$. We further assume that the matrices $S_g$ and $\Sigma_g$ are positive definite, and that $n_g/n \to f_g > 0$, as $n \to +\infty$ $(g = 1, \ldots, G)$; in this case, a distribution-free consistent estimator of $\Gamma$ is

$$\hat{\Gamma} = \text{block diag}\left(\frac{n}{n_1}\hat{\Gamma}_1, \ldots, \frac{n}{n_G}\hat{\Gamma}_G\right), \tag{10}$$

where

$$\hat{\Gamma}_g = \frac{1}{n_g - 1} \sum_{i=1}^{n_g} (d_{gi} - s_g)(d_{gi} - s_g)', \tag{11}$$

with $d_{gi} = \text{vec } z_{gi} z'_{gi}$.

We consider the generalised least-square (GLS) estimation method with normality as a working assumption; viz., the estimator $\hat{\theta}$ is defined as the minimiser over the parameter space $\Theta$ of the fitting function

$$F_{GLS}(\theta) = (s - \sigma)'\hat{V}(s - \sigma),$$

with

$$\hat{V} = \text{block diag}\left(\frac{n_1}{n}\hat{V}_1, \ldots, \frac{n_G}{n}\hat{V}_G\right) \tag{12}$$

and

$$\hat{V}_g = \frac{1}{2}({S_g}^{-1} \otimes {S_g}^{-1}).$$

Clearly,

$$F_{GLS}(\theta) = \frac{1}{2}\sum_{g=1}^{G} \frac{n_g}{n} \text{tr}\left\{(S_g - \Sigma_g)S_g^{-1}\right\}^2 .$$

For single-sample covariance structure analysis, this method of estimation was first proposed by Browne (1974). When in the minimisation process the matrices $S_g$ in (12) are replaced iteratively by the fitted $\hat{\Sigma}_g$, the so-called normal-theory reweighted GLS estimates are obtained.

Differentiation with respect to $\theta$ yields the following first-order condition for the estimator $\hat{\theta}$ :

$$(s - \hat{\sigma})' \hat{V} \hat{\Delta} = 0, \qquad \sum_{g=1}^{G} \frac{n_g}{n} (s_g - \hat{\sigma}_g)' \hat{V}_g \hat{\Delta}_g = 0. \tag{13}$$

A related estimation method which is asymptotically equivalent to normal-theory GLS, and which produces estimators numerically equal to the re-weighted GLS estimators, is pseudo-ML estimation; viz., $\hat{\theta}$ is the minimiser over $\Theta$ of

$$F_{ML}(\theta) = \sum_{g=1}^{G} \frac{n_g}{n} F_{ML}{}^{(g)}(\theta),$$

with

$$F_{ML}{}^{(g)}(\theta) = \ln|\Sigma_g| + \text{tr}\, S\Sigma_g{}^{-1} - \ln|S_g| - p_g^{\star},$$

and $p_g^{\star}$ denoting the dimension of the $z_{gi}$.

A goodness-of-fit test statistic associated with GLS estimation is defined as $GF_{GLS}^{\star} = nF_{GLS}(s, \hat{\sigma})$, which, using (13), can be written as

$$GF_{GLS}^{\star} = (s - \hat{\sigma})' \left\{\hat{V} - \hat{V}\hat{\Delta}(\hat{\Delta}'\hat{V}\hat{\Delta})^{-1}\hat{\Delta}'\hat{V}\right\} (s - \hat{\sigma})$$

The goodness-of-fit test associated with pseudo ML estimation is in turn defined as $GF_{ML} = nF_{ML}(\hat{\theta})$, which corresponds to the log-likelihood ratio test statistic for testing the model $\sigma = \sigma(\theta)$, against the general alternative in which the $\Sigma_g$ are unrestricted. Under normality, the statistics $GF$, $GF_{GLS}^{\star}$ and $GF_{ML}$ are asymptotically equivalent (i.e., they have the same asymptotic distribution as that stated above for $GF$). Under non-normality of the $z_g$, $GF_{GLS}^{\star}$ and $GF_{ML}$ are asymptotically equivalent, though not necessarily asymptotic chi-square. See Satorra (1989) for more details on the above goodness-of-fit test statistics, with discussion of their equivalence.

Under general distribution of the $z_g$, when the model holds exactly, $GF_{GLS}^{\star}$ and $GF_{ML}$ are asymptotically distributed as a mixture of chi-square distributions of 1 degree of freedom (df); that is

$$T \xrightarrow{L} \sum_{j=1}^{r} \alpha_j \chi_j^2, \tag{14}$$

as $n \to \infty$, where $T$ denotes either of the statistics $GF^{\star}_{GLS}$ or $GF_{ML}$. Here the $\chi^2_j$ are independent chi-square variables of 1 df, the $\alpha_j$ are the non-null eigenvalues of the matrix $U\Gamma$, with

$$U = V - V\Delta J^{-1}\Delta' V,$$

$J = \Delta' V \Delta = \sum_{g=1}^{G} f_g \Delta_g' V_g \Delta_g$ and $r = \mathrm{rank}(\Delta_\perp' \Gamma \Delta_\perp)$, see Satorra and Bentler (1986). If the $z_g$ are normally distributed, then the $\alpha_j$ are equal to 1 and thus one obtains an asymptotic exact chi-square distribution for the goodness-of-fit test statistics. This result motivated the scaling correction to the goodness-of-fit test statistic proposed by Bentler (1995) and Satorra and Bentler (1988a, b), the specialisation of which to $GF^{\star}_{GLS}$ and $GF_{ML}$ is as follows. Let

$$\begin{aligned} c = \frac{1}{r}\mathrm{tr}\, U\Gamma &= \frac{1}{r}\mathrm{tr}\left\{V\Gamma - J^{-1}\Delta' V\Gamma V\Delta\right\} \\ &= \frac{1}{r}\sum_{g=1}^{G}\mathrm{tr}\left\{V_g\Gamma_g - f_g J^{-1}\Delta_g' V_g \Gamma_g V_g \Delta_g\right\} \\ &= \frac{1}{r}\mathrm{tr}\left\{\left(\Delta_\perp' V^{-1}\Delta_\perp\right)^{-1}\left(\Delta_\perp'\Gamma\Delta_\perp\right)\right\} \\ &= \frac{1}{r}\mathrm{tr}\left\{\left(\sum_{g=1}^{G} f_g^{-1} H_g' V_g^{-1} H_g\right)^{-1}\left(\sum_{g=1}^{G} f_g^{-1} H_g'\Gamma_g H_g\right)\right\}, \end{aligned} \tag{15}$$

where $\Delta_\perp' = [H_1', \ldots, H_G']$ is partitioned conformably with the partition given above of $V$ as a block-diagonal matrix. From the expression (10) of a consistent estimator of $\Gamma$, a consistent estimator of $c$ is

$$\hat{c} = \sum_{g=1}^{G}\sum_{i=1}^{n_g}\frac{n}{n_g(n_g-1)}(d_{gi}-s_g)'\left\{H_g\left(\sum_{g=1}^{G}\frac{n}{n_g}H_g'V_g^{-1}H_g\right)^{-1}H_g'\right\}(d_{gi}-s_g); \tag{16}$$

thus, the scaled (adjusted for mean) goodness-of-fit test statistic is

$$\bar{T} = T/\hat{c}, \tag{17}$$

where $T$ denote either of the test statistics $GF^{\star}_{GLS}$ or $GF_{ML}$.

We now discuss the adjusted test statistic (adjusted for mean and variance) introduced in structural equation models by Bentler (1995) and Satorra and Bentler (1988a, b); viz.,

$$\bar{\bar{T}} = \frac{d}{\mathrm{tr}\,(U\Gamma)}T,$$

where $d$ is the integer closest to

$$d' = \frac{(\mathrm{tr}\, U\Gamma)^2}{\mathrm{tr}\,(U\Gamma)^2},$$

and probability levels of $\bar{\bar{T}}$ computed using a chi-square distribution of $d'$ degrees of freedom. The statistic $\bar{\bar{T}}$ is a Satterthwaite type test statistic, see Satterthwaite, (1941). When tables from a chi-square distribution with fractional degrees of freedom are available, then $d'$ is used instead of $d$.

For the sake of completeness, we also review the asymptotic variance matrix of estimators. In the current set-up, the estimator $\hat{\theta}$ is asymptotically normal with asymptotic variance matrix, see, Satorra and Neudecker (1994), for example,

$$\text{avar}(\hat{\vartheta}) = n^{-1}J^{-1}\Delta'V\Gamma V\Delta J^{-1}, \tag{18}$$

where $\Delta'V\Gamma V\Delta = \sum_{g=1}^{G} f_g\Delta_g'V_g\Gamma_g V_g\Delta_g$. Obviously, standard errors of parameter estimators are produced by replacing population matrices for consistent estimators in the above formulae, and replacing the $f_g$ by $n_g/n$. Clearly, when

$$\Delta'V\Gamma V\Delta = \Delta'V\Delta, \tag{19}$$

then $\text{avar}(\hat{\vartheta}) = \text{avar}^\star(\hat{\theta}) = n^{-1}J^{-1}$. Note that a sufficient condition for (19) to hold is

$$V_g\Gamma_g V_g = V_g, \quad g = 1, \ldots, G, \tag{20}$$

which is verified when the $z_g$ are normally distributed.

## 17.3 Restricted tests

In this section we focus on testing a specific set of restrictions on the parameters of the model, when the alternative hypothesis is that the restrictions do not hold (rather than the general alternative that the $\Sigma_g$ are unrestricted).

Let us parametrise the fitted model $\sigma = \sigma(\theta)$ as $\sigma = \sigma^\star(\delta)$ with the constraint $a(\delta) = a_0$, where $\delta$ is a $(q+m)$-dimensional vector of parameters, $a_0$ is an $m \times 1$ vector of constants, and $\sigma^\star(.)$ and $a(.)$ are twice-continuously differentiable vector-valued functions of $\delta \in \Theta_1$, a compact subset of $R^{q+m}$. Our interest is in testing $H_0 : a(\delta) = a_0$ against $H_1 : a(\delta) \neq a_0$. In multi-sample analysis, a typical example arises when testing across-sample invariance of model parameters (see the example in the illustration below).

Define the Jacobian matrices

$$\Pi = (\partial/\partial\delta')\sigma^\star(\delta) \quad \text{and} \quad A = (\partial/\partial\delta')a(\delta),$$

which we assume to be regular at the value of $\delta$ associated with $\theta_0$, say $\delta_0$, with $A$ of full row rank. Let

$$P = \Pi'V\Pi = \sum_{g=1}^{G} f_g\Pi_g'V_g\Pi_g$$

and

$$h' = (s - \hat{\sigma})'\hat{V}\hat{\Pi} = \sum_{g=1}^{G} \frac{n_g}{n}(s_g - \hat{\sigma}_g)'\hat{V}_g\hat{\Pi}_g.$$

Generalised score and Wald-test statistics for restricted tests were investigated by Satorra (1989, Section 5); viz., the generalised score-test statistic (here called robust score-test statistic) is

$$Sr = nh'\hat{P}^{-1}\hat{A}'\left(\hat{A}\hat{P}^{-1}\hat{\Pi}'\hat{V}\hat{\Gamma}\hat{V}\hat{\Pi}\hat{P}^{-1}\hat{A}'\right)^{-1}\hat{A}\hat{P}^{-1}h,$$

where $\hat{A}$, $\hat{\Pi}$ and $\hat{P}$ are the matrices $A$, $\Pi$ and $P$ evaluated at $\hat{\theta}$, and $\hat{\Gamma}$ is the (distribution-free) consistent estimator of $\Gamma$ defined in (10).

Clearly, when

$$\Pi'V\Gamma V\Pi = \Pi'V\Pi,$$

i.e., when

$$\sum_{g=1}^{G}\frac{n_g}{n}\Pi_g'V_g\Gamma_g V_g\Pi_g = \sum_{g=1}^{G}\frac{n_g}{n}\Pi_g'V_g\Pi_g,$$

then $Sr$ is asymptotically equivalent to the (non-robust, asymptotically) score-test statistic

$$S = nh'\hat{P}^{-1}\hat{A}'\left(\hat{A}\hat{P}^{-1}\hat{A}'\right)^{-1}\hat{A}\hat{P}^{-1}h.$$

The statistic $S$ is asymptotic chi-square under the condition that the estimation method is asymptotically optimal, see Satorra (1989). Specifically, when $V$ has the normal-theory form given above and the $z_{gi}$ are normally distributed, then $S$ corresponds to the usual score-test statistic. In the case of one-dimensional restrictions, $S$ is the "Modification Index" of Jöreskog and Sörbom (1994) and Sörbom (1989).

When the fitted model is the less restricted one, that is when the model analysed is $\sigma = \sigma^\star(\delta)$, then the generalised Wald-test statistic for $H_0 : a(\delta) = a_0$ against $H_0 : a(\delta) \neq a_0$ is, see Satorra, (1989, Section 5),

$$Wr = n\hat{a}'\left(\hat{A}\hat{P}^{-1}\hat{\Pi}'\hat{V}\hat{\Gamma}\hat{V}\hat{\Pi}\hat{P}^{-1}\hat{A}'\right)^{-1}\hat{a},$$

where the matrices are now evaluated at the estimated value $\hat{\delta}$ under the unrestricted model and $\hat{a} = a(\hat{\delta})$. When $\Pi'V\Gamma V\Pi = J$, then $Wr$ is asymptotically equivalent to the (non-robust, asymptotically) Wald-test statistic

$$W = n\hat{a}'\left(\hat{A}\hat{P}^{-1}\hat{A}'\right)^{-1}\hat{a}.$$

Under the assumption of normality of the $z_{gi}$, when we use ML or normal-theory GLS estimation methods then W is the typical Wald-test statistic.

When both the restricted and unrestricted models are analysed using GLS or ML estimation methods, the restricted test of $H_0 : a(\delta) = a_0$ can be carried out using the difference-test statistic

$$DGF = GF_0 - GF_1, \tag{21}$$

where $GF_0$ and $GF_1$ denote the goodness-of-fit test statistic associated with the restricted and unrestricted models respectively. Under an arbitrary distribution of the observable variables, the three test statistics $S$, $W$ and $DGF$ are asymptotically equivalent—see Satorra (1989, Theorem 4.1)—though not necessarily chi-square distributed; when the estimation method is asymptotically optimal, then all the statistics $S$, $W$, $DGF$, $Sr$ and $Wr$ are asymptotically equivalent, with the same asymptotic chi-square distribution, see Satorra, (1989).

We will now obtain the expression of $Sr$ and $S$ for the particular case in which $a(\delta) = a_0$ corresponds to equating parameters to specific values; in this case, without loss of generality, we have $A = [0_{m\times q}, I_m]$ and $h' = [0_{1\times q}, h_2']$; consequently,

$$h'\hat{P}^{-1}\hat{A}' = h_2'\hat{P}^{22} \quad \text{and} \quad \left(\hat{A}\hat{P}^{-1}\hat{A}'\right)^{-1} = \left(\hat{P}^{22}\right)^{-1} = \hat{P}_{22} - \hat{P}_{21}\hat{P}_{11}^{-1}\hat{P}_{12},$$

where we used

$$\hat{P} = \begin{pmatrix} \hat{P}_{11} & \hat{P}_{12} \\ \hat{P}_{21} & \hat{P}_{22} \end{pmatrix} \quad \text{and} \quad \hat{P}^{-1} = \begin{pmatrix} \hat{P}^{11} & \hat{P}^{12} \\ \hat{P}^{21} & \hat{P}^{22} \end{pmatrix},$$

(with the partitions corresponding to $h' = [0_{1\times q}, h_2'])$, and standard formulae for the inverse of partitioned matrices, see, for example, Magnus and Neudecker (1999, p. 11). Consequently,

$$\begin{aligned} Sr &= nh_2'\hat{P}^{22}\left\{\begin{pmatrix} \hat{P}^{21}, & \hat{P}^{22} \end{pmatrix}\hat{\Pi}'V\Gamma V\hat{\Pi}\begin{pmatrix} \hat{P}^{12} \\ \hat{P}^{22} \end{pmatrix}\right\}^{-1}\hat{P}^{22}h_2 \\ &= nh_2'\left\{\begin{pmatrix} -\hat{P}_{21}\hat{P}_{11}^{-1}, & I_m \end{pmatrix}\hat{\Pi}'V\Gamma V\hat{\Pi}\begin{pmatrix} -\hat{P}_{11}^{-1}\hat{P}_{12} \\ I_m \end{pmatrix}\right\}^{-1} h_2, \end{aligned}$$

where we used $\hat{P}^{21} = -\hat{P}^{22}\hat{P}_{21}\hat{P}_{11}^{-1}$ and $(\hat{P}^{21})' = \hat{P}^{12}$. Similarly, we can express $S$ as

$$\begin{aligned} S &= nh_2'\hat{P}^{22}(\hat{P}^{22})^{-1}\hat{P}^{22}h_2 \\ &= nh_2'\left(\hat{P}_{22} - \hat{P}_{21}\hat{P}_{11}^{-1}\hat{P}_{12}\right)^{-1} h_2. \end{aligned}$$

Note that $h_2$ corresponds to the components of $h$ that are not equal to zero when evaluated at the restricted parameter value $\hat{\theta}$. Note also that $\hat{P}_{11}^{-1}$ is the "information" matrix associated with the restricted model, and it does not change with the restrictions being tested.

The SB scaled versions of $S$, $W$ and $DGF$ will thus be obtained by dividing the corresponding statistic by a consistent estimate of $c$ in (15), where, as in Satorra (1989, p. 146), $U$ is now given by

$$U = V\Pi P^{-1}A'\left(AP^{-1}A'\right)^{-1}AP^{-1}\Pi'V, \tag{22}$$

with $m$ (the number of rows of $A$) replacing $r$; that is, the scaled statistic is

$$Ts = T/\hat{c},$$

where $T$ denotes any of the statistics $S$, $W$ and $DGF$, $\hat{c} = m^{-1}\mathrm{tr}\,\hat{U}\hat{\Gamma}$ and $\hat{U}$ is the matrix $U$ of (22) with the matrices $\Pi$, $P$ and $A$ evaluated at estimated values, and $\hat{V}$ replaces $V$.

We will now obtain an explicit expression for $c$ in the case of restricted tests and multi-samples. Clearly,

$$\begin{aligned} c = \frac{1}{m}\mathrm{tr} U\Gamma &= \frac{1}{m}\mathrm{tr}\left(AP^{-1}\Pi' V\Gamma V\Pi P^{-1}A'\right)\left(AP^{-1}A'\right)^{-1} \\ &= \frac{1}{m}\mathrm{tr}\left(V\Gamma V\right)\left\{\Pi P^{-1}A'\left(AP^{-1}A'\right)^{-1}AP^{-1}\Pi'\right\} \\ &= \frac{1}{m}\sum_{g=1}^{G} f_g \mathrm{tr}\left(V_g\Gamma_g V_g\right)\left\{\Pi_g P^{-1}A'\left(AP^{-1}A'\right)^{-1}AP^{-1}\Pi_g'\right\}, \end{aligned} \tag{23}$$

where $\Pi_g = (\partial/\partial\delta')\sigma_g^{\star}(\delta)$. Note that this scaling factor does not necessarily coincide with the scaling factor associated with $GF_0$, nor with that associated with $GF_1$; i.e., the SB scaled $DGF$ is not the difference of the SB scaled goodness-of-fit tests $GF_1$ and $GF_0$.

We now show that when $m = 1$, then the scaled score statistic $Ss$ coincides with the robust score statistic $Sr$. When $m = 1$, the scaling factor $c$ defined above is

$$c = \frac{AP^{-1}\Pi' V\Gamma V\Pi P^{-1}A'}{AP^{-1}A'}, \tag{24}$$

and the robust and non-robust score statistics can be be expressed as

$$Sr = n\,\frac{h'\hat{P}^{-1}\hat{A}'\hat{A}\hat{P}^{-1}h}{\hat{A}\hat{P}^{-1}\hat{\Pi}'\hat{V}\hat{\Gamma}\hat{V}\hat{\Pi}\hat{P}^{-1}\hat{A}'} \tag{25}$$

and

$$S = n\,\frac{h'\hat{P}^{-1}\hat{A}'\hat{A}\hat{P}^{-1}h}{\hat{A}\hat{P}^{-1}\hat{A}'}. \tag{26}$$

Consequently, we have

$$Sr = S/\hat{c}, \tag{27}$$

where

$$\hat{c} = \frac{\hat{A}\hat{P}^{-1}\hat{\Pi}'\hat{V}\hat{\Gamma}\hat{V}\hat{\Pi}\hat{P}^{-1}\hat{A}'}{\hat{A}\hat{P}^{-1}\hat{A}'}, \tag{28}$$

is a consistent estimator of the SB scaling factor $c$ of (23). Moreover, if in addition to $m = 1$ we have that $a(\delta) = a_0$ restricts a parameter to a specific value, then $\hat{c}$ of (28) is the ratio of two standard errors of the restricted parameter: the asymptotic robust standard error (computed from (18), unrestricted model), and the NT standard error (computed from $J^{-1}$, unrestricted model).

We now provide an explicit expression for $c$ when $A = [0_{m\times q}, I_m]$. In this case we have

$$c = \frac{1}{m}\mathrm{tr}\, V\Gamma V\Pi\left\{P^{-1}A'\left(AP^{-1}A'\right)^{-1}AP^{-1}\right\}\Pi'$$

$$= \frac{1}{m}\text{tr}\, V\Gamma V\Pi \left\{ \begin{pmatrix} \hat{P}^{21} & \hat{P}^{22} \end{pmatrix} \left(\hat{P}^{22}\right)^{-1} \begin{pmatrix} \hat{P}^{12} \\ \hat{P}^{22} \end{pmatrix} \right\} \Pi'$$

$$= \frac{1}{m}\text{tr}\, V\Gamma V\Pi M \Pi',$$

where

$$M = \begin{pmatrix} P^{12}\left(P^{22}\right)^{-1}P^{21} & P^{12} \\ P^{21} & P^{22} \end{pmatrix} = \begin{pmatrix} P_{11}^{-1}P_{12}P^{22}P_{21}P_{11}^{-1} & -P_{11}^{-1}P_{12}P^{22} \\ -P^{22}P_{21}P_{11}^{-1} & P^{22} \end{pmatrix}.$$

Finally, the specialisation of the SB adjusted (for mean and variance) goodness-of-fit test to $S$, $W$ and $DGF$ test statistics is as follows. We define

$$Ta = (d/\text{tr}\, \hat{U}\hat{\Gamma})\, T,$$

where $T$ denotes any of the statistics $S$, $W$ or $DGF$, and $d$ is the integer closest to

$$d' = \frac{(\text{tr}\, \hat{U}\hat{\Gamma})^2}{\text{tr}\, (\hat{U}\hat{\Gamma})^2},$$

with $\hat{U}$ being the matrix of (22) evaluated at estimated values, and $\hat{\Gamma}$ given in (10). The null distribution of the adjusted test statistic $Ta$ is taken to be a chi-square distribution with $d$ degrees of freedom. Here $d'$ also replaces $d$ when tables of a chi-square distribution with fractional degrees of freedom are available.

## 17.4 Illustration

To illustrate the performance of the above statistics in finite samples, a specific model context is used to simulate the competing score-test statistics introduced in the previous section for various sample sizes. The performance of the test statistics is evaluated by comparing Monte Carlo rejection frequencies with the corresponding nominal values.

We use a very simple model context. We consider a regression equation

$$y_{gi}^{\star} = \beta x_{gi} + v_{gi}, \quad i = 1, \ldots, n_g, \tag{29}$$

where for case $i$ in group $g$ $(g = 1, 2)$, $y_{gi}^{\star}$ and $x_{gi}$ are the values of the response and explanatory variables, respectively, $v_{gi}$ is the value of the disturbance term, and $\beta$ is the regression coefficient. The model assumes that $x_{gi}$ is unobservable, but there are two observable variables $x_{1gi}^{\star}$ and $x_{2gi}^{\star}$ related to $x_{gi}$ by the following measurement-error equations

$$x_{1gi}^{\star} = x_{gi} + u_{1gi}, \qquad x_{2gi}^{\star} = x_{gi} + u_{2gi}, \tag{30}$$

where $u_{1gi}$ and $u_{2gi}$ are mutually independent and also independent of $v_{gi}$ and $x_{gi}$. It is assumed that the observations are independent and identically distributed within each group. Equations (29) and (30) with the associated assumptions yield an identified model. See Fuller (1987) for a comprehensive overview of measurement-error models in regression analysis. Inference is usually carried out in this type of model under the assumption that the observable variables are normally distributed.

Write the model of (29) and (30) as

$$z_{gi} = \Lambda \xi_{gi}, \quad i = 1, 2, \ldots, n, \tag{31}$$

where

$$z_{gi} = \begin{pmatrix} y^{\star}_{gi} \\ x^{\star}_{1gi} \\ x^{\star}_{2gi} \end{pmatrix}, \quad \xi_{gi} = \begin{pmatrix} x_{gi} \\ v_{gi} \\ u_{1gi} \\ u_{2gi} \end{pmatrix}$$

and

$$\Lambda = \begin{pmatrix} \beta & 1 & 0 & 0 \\ 1 & 0 & 1 & 0 \\ 1 & 0 & 0 & 1 \end{pmatrix}. \tag{32}$$

Define

$$\Phi = E\xi_{gi}\xi'_{gi} = \begin{pmatrix} \sigma_{xx} & 0 & 0 & 0 \\ 0 & \sigma_{vv} & 0 & 0 \\ 0 & 0 & \sigma_{uu} & 0 \\ 0 & 0 & 0 & \sigma_{uu} \end{pmatrix}, \tag{33}$$

and the parameter vector $\theta = (\sigma_{vv}, \sigma_{xx}, \sigma_{uu}, \beta)'$. Under this set-up, we obtain the moment structure

$$\Sigma = \Lambda\Phi\Lambda' = \Sigma(\theta), \tag{34}$$

where $\Sigma(.)$, $\Lambda(.)$ and $\Phi(.)$ are (twice-continuously differentiable) matrix-valued functions of $\theta$, as deduced from (32), (33) and (34). Note that the model restricts the variances of $u_1$ and $u_2$ by equality.

The simulation of two-sample data from the above model and computation of score-test statistics was replicated 1000 times, for various combinations of two-sample sizes as shown in Table 1. In all the replications we used $\theta_0 = (1, 1, 0.3, 2)'$. The distributions of $v$ and $x$ were independent conveniently scaled zero mean and unit variance chi-squared of 1 df (i.e., a highly non-normal distribution); the distribution of $u_1$ and $u_2$ were set to be normal mutually, independent, and independent of $v$ and $x$. The normal-theory GLS estimation method described in Section 2 was used. The model imposed across-group invariance of model parameters. In each replication, we computed the various versions of the score-test statistic of Section 3, for the null hypothesis of across-group invariance of all model parameters. Clearly, in our Monte Carlo set-up, the null hypothesis holds true, with the asymptotic distribution of $Sr$ being chi-square with $m = 4$ degrees of freedom. Note that in our Monte Carlo set-up, severe non-normality of random constituents of the model requires the use of robust and/or corrected versions of the score-test statistic.

We note that the normal-theory chi-square goodness-of-fit $GF_1$ of the unrestricted model (i.e. the model that does not restrict parameters across-groups), is an asymptotic chi-square statistic despite non-normality of the data (this

Table 1: Monte Carlo results.

| nominal significance levels: | 1% | 5% | 10% | 20% |
|---|---|---|---|---|
| $n_1 = 50$ and $n_2 = 60$ | | | | |
| Sr | 2.20 | 9.90 | 16.60 | 28.30 |
| Ss | 1.00 | 5.60 | 11.00 | 24.10 |
| Sa | 0.80 | 3.00 | 7.40 | 18.90 |
| S | 14.60 | 33.40 | 45.20 | 59.80 |
| $n_1 = 100$ and $n_2 = 120$ | | | | |
| Sr | 1.40 | 7.60 | 14.90 | 24.70 |
| Ss | 1.40 | 5.40 | 11.00 | 20.70 |
| Sa | 0.60 | 3.00 | 6.30 | 18.10 |
| S | 22.00 | 40.20 | 49.20 | 60.60 |
| $n_1 = 200$ and $n_2 = 240$ | | | | |
| Sr | 1.00 | 4.20 | 10.10 | 22.40 |
| Ss | 1.60 | 6.60 | 11.30 | 20.40 |
| Sa | 0.90 | 3.40 | 8.50 | 18.40 |
| S | 25.70 | 41.50 | 49.90 | 60.30 |
| $n_1 = 600$ and $n_2 = 700$ | | | | |
| Sr | 0.80 | 5.40 | 10.00 | 20.30 |
| Ss | 2.60 | 7.70 | 11.60 | 20.80 |
| Sa | 1.10 | 4.50 | 9.50 | 17.20 |
| S | 29.00 | 45.90 | 54.50 | 66.70 |

follows from the asymptotic robustness theory for multi-sample as in Satorra, (1992)); in contrast, the normal-theory chi-square goodness of fit $GF_0$, of the restricted model (i.e., the model that imposes parameter invariance across-samples), is not necessarily an asymptotic chi-square statistic (since variances of non-normal constituents of the model are restricted by equality across-groups; see, Satorra (1992). Consequently, since $DGF$ $(= GF_1 - GF_0)$ is asymptotically equal to $S$—see Satorra, (1989)—the scaling factor $c$ for $S$ does not need to coincide with the scaling factor $c$ associated with $GF_1$ (which is 1, as deduced from the mentioned theory of asymptotic robustness), nor with the scaling factor $c$ associated to $GF_0$.

As shown in Table 1, in our specific model context, in the case of small samples, the SB scaled statistic, $Ss$, seems to outperform the alternative robust test statistic $Sr$. As expected from theory, in the case of large samples, $Sr$ outperforms the alternative test statistics. The adjusted test statistic, $Sa$, shows acceptable performance in various sample sizes. Clearly, the normal-theory score-test statistic $S$ performs very poorly for all sample sizes considered in this

Monte Carlo study.

## References


[1] Bentler, P.M., (1995), *EQS Structural Equations Program Manual,* Multivariate Software Inc., Encino, California.

[2] Bentler, P. M., S.-Y. Lee and J. Weng, (1987), Multiple Population Covariance Structure Analysis under Arbitrary Distribution Theory, *Communications in Statistics–Theory,* 16, 1951–1964.

[3] Bollen, K.A., (1989), *Structural Equations with Latent Variables,* John Wiley, New York.

[4] Browne, M.W., (1974), Generalised Least Squares Estimators in the Analysis of Covariance Structures, *South African Statistical Journal,* 8, 1–24.

[5] Browne, M.W., (1984), Asymptotically Distribution-free Methods for the Analysis of Covariance Structures, *British Journal of Mathematical and Statistical Psychology,* 37, 62–83.

[6] Chou, C.-P., P.M. Bentler and A. Satorra, (1991), Scaled Test Statistics and Robust Standard Errors for Non-normal Data in Covariance Structure Analysis: A Monte Carlo Study, *British Journal of Mathematical and Statistical Psychology,* 44, 347–357.

[7] Curran, P. J., S. G. West and J. F. Finch, (1996), The Robustness of Test Statistics to Nonnormality and Specification Error in Confirmatory Factor Analysis. *Psychological Methods,* 1, 16–29.

[8] Fuller, W.A., (1987), *Measurement Error Models,* John Wiley, New York.

[9] Hu, L., P. M. Bentler and Y. Kano, (1992), Can Test Statistics in Covariance Structure Analysis be Trusted? *Psychological Bulletin,* 112, 351–362.

[10] Jöreskog, K., and D. Sörbom, (1994), *8 User's Reference Guide,* Scientific Software Inc., Mooresville, Indiana.

[11] Magnus J., and H. Neudecker, (1999), *Matrix Differential Calculus with Applications in Statistics and Econometrics,* John Wiley, New York.

[12] Moore, D.S., (1977), Generalised Inverses, Wald's Method, and the Construction of Chi-squared Tests of Fit, *Journal of the American Statistical Association,* 72, 131–137.

[13] Muthén, B., (1989), Multiple Group Structural Modelling with Non-normal Continuous Variables, *British Journal of Mathematical and Statistical Psychology,* 42, 55–62.

[14] Muthén, B., (1993), Goodness of Fit Test with Categorical and Other Non-normal Variables, in *Testing Structural Equation Models,* K.A. Bollen and J. Scott Long (eds.), Sage Publications Inc., Newbury Park.

[15] Satorra, A., (1989), Alternative Test Criteria in Covariance Structure Analysis: A Unified Approach, *Psychometrika,* 54, 131–151.

[16] Satorra, A., (1992), Asymptotic Robust Inferences in the Analysis of Mean and Covariance Structures, *Sociological Methodology,* 22, 249–278.

[17] Satorra, A., and P.M. Bentler, (1986), Some Robustness Properties of Goodness of Fit Statistics in Covariance Structure Analysis, *1986 ASA Proceedings of the Business and Economic Statistics Section,* 549–554, American Statistical Association, Alexandria (Virginia).

[18] Satorra, A., and P.M. Bentler, (1988a), Scaling Corrections for Chi-square Statistics in Covariance Structure Analysis, *ASA 1988 Proceedings of the Business and Economic Statistics Section,* 308–313, American Statistical Association: Alexandria (Virginia).

[19] Satorra, A., and P.M. Bentler, (1988b), Scaling Corrections for Statistics in Covariance Structure Analysis, *UCLA Statistics Series # 2,* University of California, Los Angeles.

[20] Satorra, A., and P.M. Bentler, (1991), Goodness-of-Fit Test under IV Estimation; Asymptotic Robustness of a NT Test Statistic, in R. Gutierrez, and M.J. Valderrama, (eds.), *Applied Stochastic Models and Data Analysis,* World Scientific, London.

[21] Satorra, A., and H. Neudecker, (1994), On the Asymptotic Optimality of Alternative Minimum-Distance Estimators in Linear Latent-Variable Models, *Econometric Theory,* 10, 867–883.

[22] Satorra, A., and P. M. Bentler, (1994), Corrections to Test Statistics and Standard Errors in Covariance Structure Analysis, pp. 399–419 in A. von Eye and C.C. Clogg (eds.), *Latent Variables Analysis: Applications for Developmental Research,* Sage Publications, Thousand Oaks CA.

[23] Satterthwaite, F.E., (1941), Synthesis of Variance, *Psychometrika,* 6, 309–316.

[24] Sörbom, D., (1989), Model Modification, *Psychometrika,* 54, 371–384.

[25] Yuan, K.-H., and P.M. Bentler, (1997), Mean and Covariance Structure Analysis: Theoretical and Practical Improvements, *Journal of The American Statistical Association,* 92, 767–774.

[26] Yuan, K.-H., and P.M. Bentler, (1998), Normal Theory Based Test Statistics in Structural Equation Modelling, *British Journal of Mathematical and Statistical Psychology,* 51, 289–309.

CHAPTER 18

# Asymptotic Behaviour of Sums of Powers of Residuals in the Classic Linear Regression Model

Risto D.H. HEIJMANS

## 18.1 Introduction

In this paper, we study the asymptotic properties of sums of powers of residuals, a subject which, apart from some problems in *Econometric Theory,* has not received much attention in the econometric literature. We intend to show that some powerful results can be achieved. Furthermore, an inequality, not much used in econometrics, will be intensively applied—we refer to Markov's inequality. The Chebychev inequality, much more used, is a special form of the Markov inequality. The inequality simply states that, if

$$g(x) \geq 0 \quad \Rightarrow \quad P\left[g(X) \geq \theta\right] \leq \frac{E\left\{g(X)^{\alpha}\right\}}{\theta^{\alpha}}, \quad \alpha > 0,$$

where $E$ denotes expectation and $P$ denotes probability, see [2, p.74] . Using this inequality implies that certain consistency results can be proven much more easily as we are completely free in the choice of $\alpha$ as long as it is positive.

## 18.2 Set-up and Main Results

Let us consider the following model

$$y_i = x_{i.}\beta + \varepsilon_i, \quad i = 1, \ldots, n,$$

where $x_{i.}$ is a $(1, k)$ vector, $\varepsilon_i$ is a non-observable random variable with the following properties, the $\varepsilon_i$'s are i.i.d. with $E(\varepsilon_i^{2j+1}) = 0$, where $j \in \{0, \ldots, \}$, with $E(\varepsilon_i^2) = \sigma^2$ and with $E(\varepsilon_i^{2j}) = \mu_{2j}$. We further assume that the matrix $X = (x'_{1.}, \ldots, x'_{n.})'$ has full column rank $k$. Here the superscripted $'$ denotes transposition.

Another assumption usually made is that $\lim(n \to \infty)X'X/n = Q$ is a positive definite matrix.

The least-squares estimator of $\beta$, $\hat{\beta}$ say, of course obeys

$$\hat{\beta} = (X'X)^{-1}X'y = \beta + (X'X)^{-1}X'\varepsilon,$$

where $y = (y_1, \ldots, y_n)'$ and $\varepsilon = (\varepsilon_1, \ldots, \varepsilon_n)'$. Let us now define the vector of residuals by $e = y - X\hat{\beta}$ and let

$$S_{n,r} = \frac{1}{n}\sum_{i=1}^{n} e_i^r, \quad r \in \{1, 2, \ldots\}.$$

**Lemma 1.** Under the assumptions as mentioned above (except the last one), we have $S_{n,r} \xrightarrow{p} E(\varepsilon_1^r)$, where $\xrightarrow{p}$ denotes convergence in probability.

**Proof.**

$$\begin{aligned} S_{n,r} &= \frac{1}{n}\sum_{i=1}^{n}\left(y_i - x_{i.}\hat{\beta}\right)^r = \frac{1}{n}\sum_{i=1}^{n}\left\{y_i - x_{i.}\beta + x_{i.}\left(\beta - \hat{\beta}\right)\right\}^r \\ &= \frac{1}{n}\sum_{i=1}^{n}\left\{\varepsilon_i + x_{i.}\left(\beta - \hat{\beta}\right)\right\}^r \\ &= \frac{1}{n}\sum_{i=1}^{n}\varepsilon_i^r + \frac{1}{n}\sum_{i=1}^{n}\sum_{j=1}^{r}\binom{r}{j}\varepsilon_i^{r-j}\left\{x_{i.}\left(\beta - \hat{\beta}\right)\right\}^j. \end{aligned}$$

Now, as the $\varepsilon_i$'s are i.i.d., Khinchin's law of large numbers applies; and this implies that

$$\frac{1}{n}\sum_{i=1}^{n}\varepsilon_i^r \xrightarrow{p} E(\varepsilon_1^r) \quad \text{as} \quad n \to \infty.$$

Let us now turn to the second sum. Clearly, for Lemma 1 to obtain, we have to show that

$$\frac{1}{n}\sum_{i=1}^{n}\sum_{j=1}^{r}\binom{r}{j}\varepsilon_i^{r-j}\left\{x_{i.}\left(\beta - \hat{\beta}\right)\right\}^j \xrightarrow{p} 0 \quad \text{as} \quad n \to \infty.$$

To prove this, we proceed as follows:

$$\begin{aligned} &\frac{1}{n}\left|\sum_{i=1}^{n}\sum_{j=1}^{r}\binom{r}{j}\varepsilon_i^{r-j}\left\{x_{i.}\left(\beta - \hat{\beta}\right)\right\}^j\right| = \frac{1}{n}\left|\sum_{j=1}^{r}\binom{r}{j}\sum_{i=1}^{n}\varepsilon_i^{r-j}\left\{x_{i.}\left(\beta - \hat{\beta}\right)\right\}^j\right| \\ &\le \sum_{j=1}^{r}\binom{r}{j}\sqrt{\frac{1}{n^2}\sum_{i=1}^{n}\varepsilon_i^{2r-2j}\sum_{i=1}^{n}\left\{x_{i.}\left(\beta - \hat{\beta}\right)\left(\beta - \hat{\beta}\right)'x_{i.}'\right\}^j} \\ &\le \sum_{j=1}^{r}\binom{r}{j}\sqrt{\frac{1}{n}\sum_{i=1}^{n}\varepsilon_i^{2r-2j}\left\{n^{-1/j}\sum_{i=1}^{n}x_{i.}\left(\beta - \hat{\beta}\right)\left(\beta - \hat{\beta}\right)'x_{i.}'\right\}^j}, \end{aligned}$$

where we have applied the Cauchy–Schwarz inequality and made use of the fact that

$$\sum_{i=1}^{n} a_i^j \leq \left(\sum_{i=1}^{n} a_i\right)^j \quad \text{for } a_i \geq 0 \text{ and } j \in \{1, 2, \ldots\}.$$

Now

$$\begin{aligned}
&P\left[n^{-1/j}\sum_{i=1}^{n} x_{i.}\left(\beta-\hat{\beta}\right)\left(\beta-\hat{\beta}\right)' x_{i.}' > \delta\right] \\
&\quad = P\left[n^{-1/j}\sum_{i=1}^{n} x_{i.}(X'X)^{-1}X'\varepsilon\varepsilon' X(X'X)^{-1}x_{i.}' > \delta\right] \\
&\quad \leq E\left(n^{-1/j}\delta^{-1}\sum_{i=1}^{n} x_{i.}(X'X)^{-1}X'\varepsilon\varepsilon' X(X'X)^{-1}x_{i.}'\right) \\
&\quad = \left\{n^{-1/j}\delta^{-1}\sum_{i=1}^{n} x_{i.}(X'X)^{-1}X'\sigma^2 I_n X(X'X)^{-1}x_{i.}'\right\} \\
&\quad = \sigma^2 n^{-1/j}\delta^{-1}\text{trace}\left\{(X'X)^{-1}X'X(X'X)^{-1}\sum_{i=1}^{n} x_{i.}'x_{i.}\right\} \\
&\quad = k\sigma^2 n^{-1/j}\delta^{-1};
\end{aligned}$$

and, as $j$ remains finite and $n^{-1}\sum_{i=1}^{n}\varepsilon_i^{2r-2j} \xrightarrow{p} E\left(\varepsilon_1^{2r-2j}\right)$, we have just established that

$$\frac{1}{n}\sqrt{\sum_{i=1}^{n}\varepsilon_i^{2r-2j}\sum_{i=1}^{n}\left\{x_{i.}\left(\beta-\hat{\beta}\right)\left(\beta-\hat{\beta}\right)' x_{i.}'\right\}^j} \xrightarrow{p} 0 \quad \text{as} \quad n \to \infty.$$

We have used Markov's inequality and the fact that trace$AB$ = trace$BA$. The last convergence result we have derived implies, in its turn, that

$$\frac{1}{n}\sum_{i=1}^{n}\sum_{j=1}^{r}\binom{r}{j}\varepsilon_i^{r-j}\left\{x_{i.}\left(\beta-\hat{\beta}\right)\right\}^j \xrightarrow{p} 0 \quad \text{as} \quad n \to \infty,$$

which brings us to the end of the proof.

**Theorem 1.** If

$$d_{n,r}^2 = \mu_{2r} - \mu_r^2 + \frac{1}{n}s'X(X'X)^{-1}X's\left(r^2\mu_{r-1}^2\sigma^2 - 2r\mu_{r-1}\mu_{r+1}\right),$$

where $s$ is a vector consisting of ones, then we have

$$\sqrt{n}\left(\frac{S_{n,r}-\mu_r}{d_{n,r}}\right) \xrightarrow{d} N(0,1),$$

where $N(0,1)$ is the standard normal distribution and $\xrightarrow{d}$ denotes convergence in distribution.

**Proof.**

$$\begin{aligned}\sum_{i=1}^{n}\varepsilon_i^r &= \sum_{i=1}^{n}(y_i - x_{i.}\beta)^r = \sum_{i=1}^{n}\left\{y_i - x_{i.}\hat{\beta} + x_{i.}\left(\hat{\beta}-\beta\right)\right\}^r \\ &= \sum_{i=1}^{n}\left(y_i - x_{i.}\hat{\beta}\right)^r + r\sum_{i=1}^{n}\left(y_i - x_{i.}\hat{\beta}\right)^{r-1}x_{i.}\left(\hat{\beta}-\beta\right) \\ &\quad + \sum_{i=1}^{n}\sum_{j=2}^{r}\binom{r}{j}\left(y_i - x_{i.}\hat{\beta}\right)^{r-j}\left\{x_{i.}\left(\hat{\beta}-\beta\right)\right\}^j.\end{aligned}$$

Now

$$\begin{aligned}&r\sum_{i=1}^{n}\left(y_i - x_{i.}\hat{\beta}\right)^{r-1}x_{i.}\left(\hat{\beta}-\beta\right) \\ &\quad = r\sum_{i=1}^{n}\left\{\left(y_i - x_{i.}\hat{\beta}\right)^{r-1} - \mu_{r-1}\right\}x_{i.}\left(\hat{\beta}-\beta\right) + r\mu_{r-1}\sum_{i=1}^{n}x_{i.}\left(\hat{\beta}-\beta\right).\end{aligned}$$

Let us focus our attention on the first sum on the right-hand side of this expression:

$$\begin{aligned}&r\sum_{i=1}^{n}\left\{\left(y_i - x_{i.}\hat{\beta}\right)^{r-1} - \mu_{r-1}\right\}x_{i.}\left(\hat{\beta}-\beta\right) \\ &\quad = r\sum_{i=1}^{n}\left[\left\{\varepsilon_i + x_{i.}\left(\beta-\hat{\beta}\right)\right\}^{r-1} - \mu_{r-1}\right]x_{i.}\left(\hat{\beta}-\beta\right) \\ &\quad = r\sum_{i=1}^{n}\left(\varepsilon_i^{r-1} - \mu_{r-1}\right)x_{i.}\left(\hat{\beta}-\beta\right) \\ &\qquad + r\sum_{i=1}^{n}\sum_{l=1}^{r-1}\binom{r-1}{l}\varepsilon_i^{r-1-l}\left\{x_{i.}\left(\beta-\hat{\beta}\right)\right\}^l x_{i.}\left(\hat{\beta}-\beta\right).\end{aligned}$$

Since

$$\mathrm{var}\left\{n^{-1}\sum_{i=1}^{n}\left(\varepsilon_i^{r-1} - \mu_{r-1}\right)x_{i.}\right\} = n^{-2}\mathrm{var}\left(\varepsilon_i^{r-1}\right)\sum_{i=1}^{n}x_{i.}'x_{i.} \to 0 \quad \text{as} \quad n \to \infty,$$

and since

$$\sqrt{n}\left(\hat{\beta}-\beta\right) \xrightarrow{d} N(0, \sigma^2 Q^{-1}),$$

we have just established that

$$\frac{1}{\sqrt{n}}\sum_{i=1}^{n}\left(\varepsilon_i^{r-1} - \mu_{r-1}\right)x_{i.}\left(\hat{\beta}-\beta\right) \xrightarrow{p} 0 \quad \text{as} \quad n \to \infty.$$

Now

$$\frac{1}{\sqrt{n}}\left|\sum_{i=1}^{n}\varepsilon_i^{r-1-l}\left\{x_{i.}\left(\beta-\hat{\beta}\right)\right\}^l x_{i.}\left(\hat{\beta}-\beta\right)\right|$$

$$\leq \frac{1}{\sqrt{n}}\sum_{i=1}^{n}\left|\varepsilon_i^{r-1-l}\left\{x_{i.}\left(\beta-\hat{\beta}\right)\right\}^l x_{i.}\left(\hat{\beta}-\beta\right)\right|$$

$$\leq \sqrt{\frac{1}{n}\sum_{i=1}^{n}\varepsilon_i^{2r-2-2l}\sum_{i=1}^{n}\left\{x_{i.}\left(\beta-\hat{\beta}\right)\right\}^{2l}\left\{x_{i.}\left(\hat{\beta}-\beta\right)\right\}^2},$$

where we have applied the Cauchy–Schwarz inequality. Consider the second sum in the square root. There we have

$$\sum_{i=1}^{n}\left\{x_{i.}\left(\beta-\hat{\beta}\right)\right\}^{2l}\left\{x_{i.}\left(\beta-\hat{\beta}\right)\right\}^2$$

$$\leq \max_{1\leq i\leq n} x_{i.}\left(\beta-\hat{\beta}\right)\left(\beta-\hat{\beta}\right)' x_{i.}'\sum_{i=1}^{n}\left\{x_{i.}\left(\beta-\hat{\beta}\right)\left(\beta-\hat{\beta}\right)' x_{i.}'\right\}^l$$

$$\leq \max_{1\leq i\leq n} x_{i.}\left(\beta-\hat{\beta}\right)\left(\beta-\hat{\beta}\right)' x_{i.}'\left\{\sum_{i=1}^{n}x_{i.}\left(\beta-\hat{\beta}\right)\left(\beta-\hat{\beta}\right)' x_{i.}'\right\}^l.$$

But

$$\max_{1\leq i\leq n} x_{i.}\left(\hat{\beta}-\beta\right)\left(\hat{\beta}-\beta\right)' x_{i.}' \leq \max_{1\leq i\leq n} x_{i.}x_{i.}' \operatorname{trace}\left(\hat{\beta}-\beta\right)\left(\hat{\beta}-\beta\right)',$$

however, as $n \to \infty$, there is $n^{-1}\sum_{i=1}^{n} x_{i.}x_{i.}' \to \operatorname{trace}Q$ which implies $n^{-1}\max_{1\leq i\leq n} x_{i.}x_{i.}' \to 0$ as $n\to\infty$, see [1, Lemma 2.6.1]. Thus as the assumptions give

$$\sqrt{n}\left(\beta-\hat{\beta}\right)\xrightarrow{d} N(0,\sigma^2Q^{-1}) \quad \text{as} \quad n\to\infty.$$

We have shown that

$$\max_{1\leq i\leq n} x_{i.}\left(\beta-\hat{\beta}\right)\left(\beta-\hat{\beta}\right)' x_{i.}' \xrightarrow{p} 0.$$

Now, since

$$\sigma^{-4}\sum_{i=1}^{n}x_{i.}\left(\hat{\beta}-\beta\right)\left(\hat{\beta}-\beta\right)' x_{i.}' = \sigma^{-4}\operatorname{trace}\left(\hat{\beta}-\beta\right)\left(\hat{\beta}-\beta\right)' X'X,$$

and since this latter quantity converges in distribution to a $\chi_k^2$, we have established that

$$\frac{1}{\sqrt{n}}\sum_{i=1}^{n}\varepsilon_i^{r-1-l}\left\{x_{i.}\left(\beta-\hat{\beta}\right)\right\}^l x_{i.}\left(\hat{\beta}-\beta\right)\xrightarrow{p} 0$$

and thus also

$$\frac{1}{\sqrt{n}}\sum_{i=1}^{n}\sum_{l=1}^{r-1}\binom{r-1}{l}\varepsilon_i^{r-1-l}\left\{x_{i.}\left(\hat{\beta}-\beta\right)\right\}^l x_{i.}\left(\hat{\beta}-\beta\right)\xrightarrow{p}0.$$

This last result implies that the asymptotic behaviour of

$$\frac{1}{\sqrt{n}}r\sum_{i=1}^{n}\left(y_i-x_{i.}\hat{\beta}\right)^{r-1}x_{i.}\left(\hat{\beta}-\beta\right)$$

is the same as that of

$$\frac{1}{\sqrt{n}}r\mu_{r-1}\sum_{i=1}^{n}x_{i.}\left(\hat{\beta}-\beta\right).$$

Proceeding in the same manner as above, we can easily show that

$$\frac{1}{\sqrt{n}}\sum_{i=1}^{n}\sum_{j=2}^{r}\binom{r}{j}\left(y_i-x_{i.}\hat{\beta}\right)^{r-j}\left\{x_{i.}\left(\hat{\beta}-\beta\right)\right\}^j\xrightarrow{p}0.$$

This of course implies that

$$\frac{1}{\sqrt{n}}\sum_{i=1}^{n}\left\{\left(y_i-x_{i.}\hat{\beta}\right)^r-\mu_r\right\}$$

has the same asymptotic distribution as

$$\frac{1}{\sqrt{n}}\sum_{i=1}^{n}\left\{\varepsilon_i^r-\mu_r-r\mu_{r-1}x_{i.}\left(\hat{\beta}-\beta\right)\right\}.$$

Now, since

$$\begin{aligned}\sum_{i=1}^{n}x_{i.}\left(\hat{\beta}-\beta\right) &= \varepsilon'X(X'X)^{-1}\sum_{i=1}^{n}x'_{i.}\\ &= \sum_{j=1}^{n}\varepsilon_j x_{j.}(X'X)^{-1}\sum_{i=1}^{n}x'_{i.},\end{aligned}$$

we can focus on

$$\frac{1}{\sqrt{n}}\sum_{j=1}^{n}\left\{\varepsilon_j^r-\mu_r-r\mu_{r-1}\varepsilon_j x_{j.}(X'X)^{-1}\sum_{i=1}^{n}x'_{i.}\right\}=\frac{1}{\sqrt{n}}W_n.$$

$$\begin{aligned}\text{var}W_n &= n(\mu_{2r}-\mu_r^2)+r^2\mu_{r-1}^2\sigma^4\sum_{j=1}^{n}\left\{x_{j.}(X'X)^{-1}\sum_{i=1}^{n}x'_{i.}\right\}^2\\ &\quad -2r\mu_{r-1}\mu_{r+1}\sum_{j=1}^{n}x_{j.}(X'X)^{-1}\sum_{i=1}^{n}x'_{i.}.\end{aligned}$$

But

$$\sum_{j=1}^{n}\left\{x_{j.}(X'X)^{-1}\sum_{i=1}^{n}x'_{i.}\right\}^2 = \sum_{j=1}^{n}\left\{x_{j.}(X'X)^{-1}\sum_{i=1}^{n}x'_{i.}\sum_{i=1}^{n}x_{i.}(X'X)^{-1}x'_{j.}\right\}$$
$$= \text{trace}(X'X)^{-1}\sum_{i=1}^{n}x'_{i.}\sum_{i=1}^{n}x_{i.}(X'X)^{-1}(X'X)$$
$$= \sum_{i=1}^{n}x_{i.}(X'X)^{-1}\sum_{i=1}^{n}x'_{i.}.$$

So we finally arrive at

$$\text{var}W_n = n\left(\mu_{2r}-\mu_r^2\right)+s'X(X'X)^{-1}X's\left(r^2\mu_{r-1}^2\sigma^2-2r\mu_{r-1}\mu_{r+1}\right)=nd_{n,r}^2.$$

According to the Lindeberg–Feller theorem, see [3, p.31], this implies that

$$\sqrt{n}\left(\frac{S_{n,r}-\mu_r}{d_{n,r}}\right)\xrightarrow{d}N(0,1).$$

**Corollary 1.** Theorem 1 remains true if we replace $d_{n,r}$ with $a_{n,r}$, where

$$a_{n,r}^2=S_{n,2r}-S_{n,r}^2+\frac{1}{n}s'X(X'X)^{-1}X's\left(r^2S_{n,r-1}^2S_{n,2}^2-2rS_{n,r-1}S_{n,r+1}\right).$$

**Proof.** Lemma 1 shows that $a_{n,r}^2-d_{n,r}^2\xrightarrow{p}0$ as $n\to\infty$. Of course, readers should realise that $0\le n^{-1}s'X(X'X)^{-1}X's\le 1$, because $X(X'X)^{-1}X'$ is idempotent.

Usually, the probability law of the errors has one parameter only; so all moments are functions of that parameter. For example, in the case of the normal distribution, the one parameter is the variance and so one would have:

$$\mu_{2r}=\frac{\sigma^{2r}\,(2r)!}{r!2^r},\quad \mu_{2r+1}=0.$$

Invoking Slutsky gives

$$\frac{S_{n,2}^r\,(2r)!}{r!2^r}\xrightarrow{p}\mu_{2r},$$

and this property could be used to find another consistent estimator of $d_{n,r}$ in the case of normally distributed errors. Taking $r$ uneven could be of use in constructing a test on the normality of the errors. However, due to the pressure exerted by the editors, this author has not been able to pursue that particular avenue any further.

## References

[1] Anderson, T.W., (1971), *The Statistical Analysis of Time Series,* John Wiley and Sons, New York.

[2] Billingsley, P., (1986), *Probability and Measure, Second Edition,* John Wiley and Sons, Chichester.

[3] Serfling, R.J., (1980), *Approximation Theorems of Mathematical Statistics,* John Wiley and Sons, Chichester.

CHAPTER 19

# Matrix Methods for Solving Nonlinear Dynamic Optimisation Models

Hans M. AMMAN and David A. KENDRICK

## 19.1 Introduction

Nonlinear dynamic optimisation models are widely used in theoretical and empirical economic modelling, especially in the field of optimal growth and intertemporal macroeconomic modelling. In this chapter, we present a sequential quadratic programming algorithm for computing directly the steady state solution for a wide class of nonlinear dynamic optimisation problems in discrete time. The method is based on the method described in Amman and Neudecker [3] for solving the algebraic Riccati matrix equation.

Starting with the seminal work of Ramsey [18] and Hotelling [9], dynamic optimisation models have gained wide acceptance in both theoretical and empirical economics. The renewed interest in growth theory, Lucas [14], and the literature on overlapping generation models, Blanchard and Fisher [6], has attracted much attention to the computational issues connected to the solution of these models, Judd [10]. Especially the steady state properties of these optimisation models is an intriguing and interesting subject.

For finite horizon discrete time models, the standard mathematical programming methods can be applied to solve nonlinear models. In the infinite time horizon case, however, we have to apply dynamic programming techniques to obtain the steady state solution. In this chapter, we present such a method for computing the steady state solution, which can be applied to a broad class of optimisation models. The method is a combination of a sequential quadratic programming approach with the numerical solution of the algebraic Riccati equation as described in Amman and Neudecker [3]. In the next section, we will present our algorithm while, in the third section, we will apply our algorithm to an example.

The rational for this approach, in comparison to more direct method for solving the steady state, lies in the fact that in a dynamic programming framework the Riccati matrix equation comprises the first-order conditions of the optimisation problem. Having this Riccati equation available from the finite time solution, the steady state can easily be obtained from the algebraic matrix Riccati equations. Hence, the steady state solution comes with little additional cost.

## 19.2 A Nonlinear Optimisation Framework

In this section, we will present the algorithm for solving infinite horizon nonlinear dynamic optimisation problems. The algorithm works in two stages. In the first stage, a quadratic-linear approximation (QLA) is made of the optimisation problem at hand. In the second stage, this approximation is solved for the steady state solution. Once we have this steady state solution for the QLA, we go back to the first stage and iterate on the first stage and second stage until we have the solution of the nonlinear optimisation model.

Lets start out with the general form of our optimisation framework. The objective we want to minimise is

$$J = \sum_{t=0}^{\infty} \beta^t \Pi(x_t, u_t) \tag{1}$$

subject to the constraints

$$\begin{aligned} x_{t+1} &= f(x_t, u_t), \\ x_0 & \quad \text{given} \end{aligned} \tag{2}$$

with $x_t \in \Re^n$ as the state vector, and $u_t \in \Re^m$ as the control vector. $\Pi$ is twice continuous differentiable and $f$ is continuous and differentiable. This form of optimisation model is used in many branches of economics. See, for instance, Azariadis [4], Stockey and Lucas [20], Kendrick [12] and Sargent [19]. A simple way to solve the control model in equations (1)–(2) for a *finite time* horizon $T$, is to make a QLA of the form

$$\begin{aligned} J \approx & \sum_{t=0}^{T} \beta^t \Big\{ \Pi^j + \Pi_x^{'j}[x_t^{j+1} - x_t^j] \\ & + \Pi_u^{'j}[u_t^{j+1} - u_t^j] + \frac{1}{2}[x_t^{j+1} - x_t^j]' \Pi_{xx}^j [x_t^{j+1} - x_t^j] \\ & \times \frac{1}{2}[u_t^{j+1} - u_t^j]' \Pi_{uu}^j [u_t^{j+1} - u_t^j] + [x_t^{j+1} - x_t^j]' \Pi_{xu}^j [u_t^{j+1} - u_t^j] \Big\}, \end{aligned} \tag{3}$$

subject to the constraints

$$\begin{aligned} x_{t+1}^{j+1} &\approx x_{t+1}^j + f_x^j(x_t^{j+1} - x_t^j) + f_u^j(u_t^{j+1} - u_t^j). \\ x_0 & \quad \text{given} \end{aligned} \tag{4}$$

Examples of this approach can be found in Garbade [8] and Kendrick [12]. Starting with an initial guess for $x_t^0 \ \forall t \in [1,T]$ and $u_t^0 \ \forall t \in [0, T-1]$, the

QLA in equations (3)–(4) will surely convergence to an unique minimum if the functional in (1) is nice convex and the constraint (2) is concave, see Léonard and Van Long [13]. In this way, we obtain the optimal state vector $x_t^* \ \forall t \in [1,T]$ and $u_t^* \ \forall t \in [0, T-1]$. Quite often, it is explicitly assumed that the matrices $\Pi_{xx}$ and $\Pi_{uu}$ are positive definite, see Kendrick [12].[1]

The above QLA can be solved by backward substitution of the Riccati equation,[2] see Chow [7] and Kendrick [12],

$$\begin{aligned} X_t &= \Pi_{xx}^j + \beta f_x^{j'} X_{t+1} f_x^j \\ &\quad - (\beta f_x^{j'} X_{t+1} f_u^j + \Pi_{xu})(\Pi_{uu}^j + \beta f_u^{j'} X_{t+1} f_u^j)^{-1}(\beta f_u^{j'} X_{t+1} f_x^j + \Pi_{xu}^{j'}) \end{aligned} \tag{5}$$

and the tracking vector, which is recursive with respect to $X_{t+1}$,

$$\begin{aligned} p_t &= -[\beta f_x^{j'} X_{t+1} f_u^j + \Pi_{xu}][\beta f_u^{j'} X_{t+1} f_u^j + \Pi_{uu}^j]^{-1} \\ &\quad \times [\beta f_u^{j'} \{X_{t+1}(x_{t+1}^j - f_x^j x_t^j - f_u^j u_t^j) + p_{t+1}\} + \Pi_u^j - \Pi_{uu}^j u_t^j - \Pi_{xu}^{j'} x_t^j] \\ &\quad + \beta f_x^{'j} \{X_{t+1}(x_{t+1}^j - f_x^j x_t^j - f_u^j u_t^j) + p_{t+1}\} + \Pi_x^j - \Pi_{xx}^j x_t^j - \Pi_{xu}^j u_t^j, \end{aligned} \tag{6}$$

with the fixed end condition

$$X_T = \Pi_{xx}^j, \tag{7}$$

$$p_T = -\Pi_{xx}^j x_T^j. \tag{8}$$

Unfortunately, this form is not suitable for solving our infinite time horizon optimisation problem. Following Sargent [19], Amman and Neudecker [3] and Léonard and Van Long [13], the infinite time horizon steady state solution of the QLA, $T \to \infty$, leads to the algebraic Riccati matrix equation and a corresponding tracking vector. These two items can be derived from the value function for the infinite time horizon case. For the problem in equations (3) and (4), the value function has the form, see Léonard and Van Long [13],

$$\begin{aligned} V(x) &= \max_u \Big\{ \Pi^j + \Pi_x^{'j}[x_t^{j+1} - x_t^j] \\ &\quad + \Pi_u^{'j}[u_t^{j+1} - u_t^j] + \frac{1}{2}[x_t^{j+1} - x_t^j]' \Pi_{xx}^j [x_t^{j+1} - x_t^j] \\ &\quad \frac{1}{2}[u_t^{j+1} - u_t^j]' \Pi_{uu}^j [u_t^{j+1} - u_t^j] + [x_t^{j+1} - x_t^j]' \Pi_{xu}^j [u_t^{j+1} - u_t^j] \\ &\quad + \beta V(\tilde{x}) \Big\}. \end{aligned} \tag{9}$$

By assuming that the value function has a quadratic solution like

$$V = \frac{1}{2} x' X x + p' x + \gamma \tag{10}$$

[1]The stochastic version of the QLA may suffer from nonconvexities due to parameter uncertainty and learning. See Amman and Kendrick [2].

[2]The Riccati matrix in equation (5) is derived in Appendices A and B.

and that, in the infinite time horizon case, equation (4) has the form

$$x = f(x, u), \tag{11}$$

we can solve the QLA in equations (3)–(4) for the infinite time horizon case by determining $X$, $p$. The constant term $\gamma$ can be ignored because it does not play a role in the solution of the state and control vector. Dropping the iteration index $j$ for the derivative components, we get for $X$

$$\begin{aligned} X &= \Pi_{xx} + \beta f_x' X f_x \\ &\quad - (\beta f_x' X f_u + \Pi_{xu})(\Pi_{uu} + \beta f_u' X f_u)^{-1}(\beta f_u' X f_x + \Pi_{xu}'), \end{aligned} \tag{12}$$

and similarly for $p$

$$\begin{aligned} p &= -[\beta f_x' X f_u + \Pi_{xu}][\beta f_u' X f_u + \Pi_{uu}]^{-1} \\ &\quad \times [\beta f_u'\{X(x_\infty^j - f_x x_\infty^j - f_u u_\infty^j) + p\} + \Pi_u - \Pi_{uu} u_\infty^j - \Pi_{xu}' x_\infty^j] \\ &\quad + \beta f_x'\{X(x_\infty^j - f_x x_\infty^j - f_u u_\infty^j) + p\} + \Pi_x - \Pi_{xx} x_\infty^j - \Pi_{xu} u_\infty^j, \end{aligned} \tag{13}$$

which are almost identical to equation (5) and (6). The vectors $x_\infty^j$ and $u_\infty^j$ indicate the steady state values of the state vector and control vector from the previous iteration respectively. Now the next step is to solve the QLA for the steady state solution. This can be done by solving equation (12) for $X$ and equation (13) for $p$. Under quite weak conditions, these equations will have a solutions, see Bertsekas [5]. Let us concentrate on equation (12) and define

$$\begin{aligned} G(z) &= X - \Pi_{xx} - \beta f_x' X f_x \\ &\quad + (\beta f_x' X f_u + \Pi_{xu})(\Pi_{uu} + \beta f_u' X f_u)^{-1}(\beta f_u' X f_x + \Pi_{xu}') \end{aligned} \tag{14}$$

with $g = D^{+}\text{vec}\, G$ and $z = D^{+}\text{vec}\, X$. This gives

$$\begin{aligned} \frac{\partial g}{\partial z'} &= D^{+}\Big[I - \{(\beta f_x' X f_u + \Pi_{xu})(\Pi_{uu} + \beta f_u' X f_u)^{-1} f_u' - f_x'\} \\ &\quad \otimes \{(\beta f_x' X f_u + \Pi_{xu})(\Pi_{uu} + \beta f_u' X f_u)^{-1} f_u' - f_x'\}\Big] D, \end{aligned} \tag{15}$$

where the matrix $I$ is the identity matrix and $\otimes$ denotes the Kronecker product. The matrices $D$ and $D^{+}$ are the duplication matrix and inverse duplication matrix (c.f. Magnus and Neudecker [15]). These matrices "remove and restore" redundant upper triangular parts of the (symmetric) positive definite Riccati matrix $X$ and the function $G(z)$.

The problem here is to solve the multi-dimensional fixed-point problem $g(z) = 0$. There are several types of numerical methods to solve such a problem. With the help of equation (15), it is simple to construct an iterative scheme to solve our fixed-point problem, namely

$$z^{(k+1)} = z^{(k)} - \left(\frac{\partial g}{\partial z'}\right)^{-1} g, \tag{16}$$

with

$$z^{(0)} = D^{+}\text{vec}\,\Pi_{xx}$$

as an initial condition. This updating scheme is quite robust and will generally lead to a rapid solution up to some machine precision level.

In some particular cases, the updating scheme might not converge due to the fact that the initial $z^{(0)}$ is distant from a possible solution. In that case, we may use a more global method like the Levenberg–Marquardt scheme (c.f. Ralston and Rabinowitz [17]), which is of the form

$$z^{(k+1)} = z^{(k)} - \left[\left(\frac{\partial g}{\partial z'}\right)'\left(\frac{\partial g}{\partial z'}\right) + \lambda^{(k)}I\right]^{-1}\left(\frac{\partial g}{\partial z'}\right)' g, \qquad (17)$$

where $\lambda^{(k+1)} = \lambda^{(k)}/c^{k}$ with $c$ being an arbitrary constant $> 1$. If a solution exists for $g(z) = 0$ (c.g. Bertsekas [5], page 75), this method will converge for any initial $z^{(0)}$ when $\lambda^{(0)}$ is taken sufficiently large.[3]

The tracking equation (13) can be solved in the same manner. Once we have obtained the values of the Riccati matrix, $X^{*}$, and the tracking vector, $p^{*}$, we can compute the steady state solution of the state vector $x_{\infty}^{j+1}$ and control vector $u_{\infty}^{j+1}$ with the help of the equations,[4] see Amman [1] and Amman and Neudecker [3]

$$x_{\infty}^{j+1} = (I - f_x - f_u H)^{-1}(f_u h + x_{\infty}^{j} - f_x x_{\infty}^{j} - f_u u_{\infty}^{j}) \qquad (18)$$

$$u_{\infty}^{j+1} = H x_{\infty}^{j+1} + h, \qquad (19)$$

where

$$H = -(\beta f_u^{'} X^{*} f_u + \Pi_{uu})^{-1}(\beta f_u^{'} X^{*} f_x + \Pi_{xu}^{'}) \qquad (20)$$

$$h = -(\beta f_u^{'} X^{*} f_u + \Pi_{uu})^{-1}[\beta f_u^{'}\{X^{*}(x_{\infty}^{j} - f_x^{'} x_{\infty}^{j} - f_u^{'} u_{\infty}^{j}) + p^{*}\} + \Pi_u - \Pi_{uu} u_{\infty}^{j} - \Pi_{xu}^{'} x_{\infty}^{j}]. \qquad (21)$$

Now that we have an update of the steady state solution of the optimisation model, we can go back to the first stage of the algorithm and make a new QLA of the nonlinear model in equations (1)–(2) and repeat the above procedure. We can repeat these iterative steps until the following norm, $\Xi$, is met

$$\Xi = \frac{\|x_{\infty}^{j+1} - x_{\infty}^{j}\|}{\|x_{\infty}^{j}\|} < \epsilon_p^{1/2}, \qquad (22)$$

$\epsilon_p$ being the machine precision. The resulting $x_{\infty}^{*}$ and $u_{\infty}^{*}$ are the steady state values of the nonlinear optimisation model in equations (1)–(2).

[3] A GAUSS implementation of the algorithm can be obtained through email from the corresponding author.

[4] The derivations of equations (18)–(21) are described in Appendix C

## 19.3 An Example

Nonlinear dynamic optimisation schemes are frequently used in macroeconomics in the field of optimal growth. Lets take a simple growth model of the Ramsey [4] type[5]

$$\max \sum_{t=0}^{\infty} \left(\frac{1}{1+r}\right)^t \{\psi \ln(c_t) + (1-\psi)\ln(k_t)\}; \tag{23}$$

$$\begin{aligned} y_t^d &= c_t + i_t, \\ y_t^s &= \alpha\{\phi k^{-\rho} + (1-\phi)\}^{-1/\rho}, \\ k_{t+1} &= (1-\delta)k_t + i_t, \\ y_t^s &= y_t^d, \end{aligned} \tag{24}$$

where $y_t^d$ is total demand, $y_t^s$ total supply, $c_t$ consumption, $i_t$ investment, $k_t$ the capital stock per unit labour, $\delta$ the depreciation rate, $r$ the interest rate, $\rho$ the elasticity of substitution between capital and labour, $\psi$ the utility share and $\alpha$ and $\phi$ as technical constants. If we take the parameter values $r = 0.1$, $\psi = 0.5$, $\alpha = 1$, $\phi = 0.5$, $\rho = 2$, $\delta = 0.1$ and we take as initial values $x_\infty^0 = 1$ and $u_\infty^0 = 1$, we get as a steady state solution that $x_\infty^* = 4.805$, $u_\infty^* = 0.904$ and $V^* = 3.719$, which corresponds to the analytical solution.

Table 1 below gives the norm $\Xi$ at each iteration step.[6]

Table 1: Solution of the Optimal Growth model.

| Iteration $k$ | *Norm* $\Xi$ |
|---|---|
| 1 | 1.19e-00 |
| 2 | 8.09e-01 |
| 3 | 2.33e-01 |
| 4 | 1.73e-02 |
| 5 | 2.34e-03 |
| . | . |
| 19 | 3.70e-16 |

## 19.4 Summary

We have presented an algorithm for solving a broad class of nonlinear optimisation problems in discrete time. By making successive quadratic-linear approximations of the dynamic optimisation model, and computing the resulting

[5] We have used a very simple model with which it is possible to check the outcome of the algorithm with the analytical solution.

[6] Calculations were done in GAUSS on a 486 computer.

algebraic Riccati matrix iteratively, we have constructed a direct method for computing the steady state solution.

# Appendix A

## Derivation of the Riccati Equations in a Model with Discounting

The derivations of the Riccati equations for the model in the body of this chapter follow the procedures used in chapter 2 of Kendrick [12]; however, the model in the book is less general than the one used here. Both the model used in the book and the one used in this chapter are quadratic-linear optimal control models in tracking form. However, the model in the chapter includes discounting which is not included in the book. Therefore, this appendix generalises the derivation in the book to a model with discounting. In this, we follow the discussion in Sargent [19], pages 20–21. We begin with the model in equations (2-1)–(2-3) of the book which we write here as

$$\text{find} \quad (u_k)_{k=0}^{N-1}$$

to minimise the criterion

$$J = L_N(x_N) + \sum_{k=0}^{N-1} L_k(x_k, u_k), \tag{A-1}$$

where

$$L_N(x_N) = \frac{1}{2}x_N' W_N x_N + w_N' x_N + \tau_N, \tag{A-2}$$

$$L_k(x_k, u_k) = \frac{1}{2}x_k' W_k x_k + w_k' x_k + x_k' F_k u_k + \frac{1}{2}u_k' \Lambda_k u_k + \lambda_k' u_k + \tau_k, \tag{A-3}$$

subject to

$$x_{k+1} = f(x_k, u_k), \tag{A-4}$$

where

$$f(x_k, u_k) = A_k x_k + B_k u_k + c_k \tag{A-5}$$

with the initial conditions

$$x_0 \quad \text{given}, \tag{A-6}$$

where

$x_k$ = an $n$ element state vector for period $k$,
$u_k$ = an $m$ element control vector for period $k$.

The addition of discounting to the problem in equations (A-1)–(A-6) with $\rho$ as the discount rate and $\beta$ as the discount factor,

$$\beta = \frac{1}{(1+\rho)}, \tag{A-7}$$

results in the conversion of equation (A-1) to

$$J = \beta^N L_N(x_N) + \sum_{k=0}^{N-1} \beta^k L_k(x_k, u_k). \tag{A-8}$$

In the following, we will use the notation (DK x-y) to refer to equation (x-y) in Kendrick [12]. Using this notation and following equation (DK 2-22), we begin the derivation of the optimal cost-to-go with the terminal period cost-to-go, i.e.

$$J^*(x_N) = J_N^* = \frac{1}{2} x_N' \tilde{K}_N x_N + \tilde{p}_N' x_N + \tilde{\gamma}_N. \tag{A-9}$$

Then, from equation (A-8) and equation (A-2), the discounted cost which are incurred in the terminal period are

$$\frac{1}{2}\beta^N x_N' W_N x_N + \beta^N w_N' x_N + \beta^N \tau_N, \tag{A-10}$$

so that the comparison of equation (A-9) and equation (A-10) yields the terminal conditions for the Riccati equations, i.e.

$$\begin{aligned} \tilde{K}_N &= \beta^N W_N, \\ \tilde{p}_N &= \beta^N w_N. \end{aligned} \tag{A-11}$$

Sargent [19] uses the undiscounted or "current value" cost-to-go $V_N^*$ which is defined as

$$V_N^* = \beta^{-N} J_N^* \tag{A-12}$$

or, more intuitively, as

$$J_N^* = \beta^N V_N^*, \tag{A-13}$$

so that $J_N^*$ is the discounted cost-to-go. Analogous to equation (A-9), one can then write the current value cost-to-go as

$$V^*(x_N) = V_N^* = \frac{1}{2} x_N' K_N x_N + p_N' x_N + \gamma_N. \tag{A-14}$$

Substitution of equations (A-9) into equation (A-12) yields the current value cost-to-go for the terminal period, i.e.

$$V_N^* = \beta^{-N} J_N^* = \frac{1}{2}\beta^{-N} x_N' \tilde{K}_N x_N + \beta^{-N} \tilde{p}_N' x_N + \beta^{-N} \tilde{\gamma}_N. \tag{A-15}$$

Then the comparison of equations (A-14) and (A-15) yields

$$\begin{aligned} K_N &= \beta^{-N} \tilde{K}_N, \\ p_N &= \beta^{-N} \tilde{p}_N. \end{aligned} \tag{A-16}$$

Substitution of equation (A-11) into equation (A-16) then yields the terminal conditions for the current-value matrix and vector Riccati equations, namely

$$\begin{aligned} K_N &= W_N, \\ p_N &= w_N. \end{aligned} \tag{A-17}$$

Next proceed to period $N-1$ by using (DK 2-26), the optimal cost-to-go for period $N-1$. In the discounting case, this can be written as

$$J_{N-1}^{*} = \min_{u_{N-1}} \left\{ \beta^{N-1} L_{N-1}\left(x_{N-1}, u_{N-1}\right) + J_{N}^{*} \right\}. \tag{A-18}$$

Sargent [19], pages 18–20, has observed that the optimal cost-to-go in equation (A-18) can be replaced with an equivalent problem which is defined in terms of the undiscounted or "current value"elements of the cost which are discussed above. From equation (A-13), one can write the discounted optimal cost-to-go for period $N-1$ as

$$J_{N-1}^{*} = \beta^{N-1} V_{N-1}^{*}. \tag{A-19}$$

Then expressions (A-19) and (A-13) can then be substituted into equation (A-18) to obtain the optimal cost-to-go in terms of current values

$$\beta^{N-1} V_{N-1}^{*} = \min_{u_{N-1}} \left\{ \beta^{N-1} L_{N-1}\left(x_{N-1}, u_{N-1}\right) + \beta^{N} V_{N}^{*} \right\} \tag{A-20}$$

or as

$$V_{N-1}^{*} = \min_{u_{N-1}} \left\{ L_{N-1}\left(x_{N-1}, u_{N-1}\right) + \beta V_{N}^{*} \right\} \tag{A-21}$$

or, in general,

$$V_{N-j}^{*} = \min_{u_{N-j}} \left\{ L_{N-j}\left(x_{N-j}, u_{N-j}\right) + \beta V_{N-j+1}^{*} \right\}. \tag{A-22}$$

The optimal cost-to-go functions in equation (A-18) and equation (A-21) are the same except for the transformation of variables; therefore, one can solve the model using either equation (A-18) or equation (A-21). Here, we use equation (A-21) to solve the quadratic-linear model with discounting, i.e. equation (A-8) and equations (A-2) through (A-6). To do this, we use the terminal conditions in equation (A-17) and work backward in the cost-to-go to derive the Riccati equations. Begin with equation (A-21) and substitute equations (A-3) and (A-14) into it to obtain

$$\begin{aligned} V_{N-1}^{*} &= \min_{u_{N-1}} \left\{ \frac{1}{2} x_{N-1}^{'} W_{N-1} x_{N-1} + w_{N-1}^{'} x_{N-1} + x_{N-1}^{'} F_{N-1} u_{N-1} \right. \\ &\quad + \frac{1}{2} u_{N-1}^{'} \Lambda_{N-1} u_{N-1} + \lambda_{N-1}^{'} u_{N-1} + \tau_{N} + \frac{1}{2} \beta x_{N}^{'} K_{N} x_{N} \\ &\quad \left. + \beta p_{N}^{'} x_{N} + \beta \gamma_{N} \right\}. \end{aligned} \tag{A-23}$$

From equations (A-4) and (A-5), the system equation for the model for period $N$ is

$$x_{N} = A_{N-1} x_{N-1} + B_{N-1} u_{N-1} + c_{N-1}. \tag{A-24}$$

Substitution of equation (A-24) into equation (A-23) yields

$$\begin{aligned} V_{N-1}^{*} &= \min_{u_{N-1}} \left\{ \frac{1}{2} x_{N-1}^{'} \Phi_{N-1} x_{N-1} + \frac{1}{2} u_{N-1}^{'} \Theta_{N-1} u_{N-1} + x_{N-1}^{'} \Psi_{N-1} u_{N-1} \right. \\ &\quad \left. + \phi_{N-1}^{'} x_{N-1} + \theta_{N-1}^{'} u_{N-1} + \eta_{N} \right\}, \end{aligned} \tag{A-25}$$

where

$$\begin{aligned}
\Phi_{N-1} &= \beta A'_{N-1} K_N A_{N-1} + W_{N-1}, \\
\Theta_{N-1} &= \beta B'_{N-1} K_N B_{N-1} + \Lambda_{N-1}, \\
\Psi_{N-1} &= \beta A'_{N-1} K_N B_{N-1} + F_{N-1}, \\
\phi_{N-1} &= \beta A'_{N-1} \left( K'_N c_{N-1} + p_N \right) + w_{N-1}, \\
\theta_{N-1} &= \beta B'_{N-1} \left( K'_N c_{N-1} + p_N \right) + \lambda_{N-1}, \\
\eta_{N-1} &= \frac{1}{2} \beta c'_{N-1} K_N c_{N-1} + \beta p'_N c_{N-1} + \beta \gamma_N + \tau_{N-1}.
\end{aligned} \tag{A-26}$$

The application of the minimisation operator in equation (A-2) then yields

$$\Theta_{N-1} u_{N-1} + \Psi'_{N-1} x_{N-1} + \theta_{N-1} = 0. \tag{A-27}$$

Notice that it is assumed here that $\Theta$ is symmetric, which is the case in quadratic-linear tracking problems. Then solving equation (A-27) for $u_{N-1}$ yields

$$u_{N-1} = -\Theta_{N-1}^{-1} \Psi'_{N-1} x_{N-1} - \Theta_{N-1}^{-1} \theta_{N-1}, \tag{A-28}$$

which is the feedback rule

$$u_{N-1} = G_{N-1} x_{N-1} + g_{N-1}, \tag{A-29}$$

where

$$G_{N-1} = -\Theta_{N-1}^{-1} \Psi'_{N-1}, \tag{A-30}$$

$$g_{N-1} = -\Theta_{N-1}^{-1} \theta_{N-1}. \tag{A-31}$$

Substitution of the feedback rule equation (A-29) into the optimal cost-to-go of equation (A-25) yields

$$\begin{aligned}
V^*_{N-1} &= \frac{1}{2} x'_{N-1} \Phi_{N-1} x_{N-1} + \frac{1}{2} x'_{N-1} G'_{N-1} \Theta_{N-1} G_{N-1} x_{N-1} \\
&\quad + \frac{1}{2} g'_{N-1} \Theta_{N-1} g_{N-1} + g'_{N-1} \Theta_{N-1} G_{N-1} x_{N-1} \\
&\quad + x'_{N-1} \Psi_{N-1} G_{N-1} x_{N-1} + x'_{N-1} \Psi_{N-1} g_{N-1} + \phi'_{N-1} x_{N-1} \\
&\quad + \theta'_{N-1} G_{N-1} x_{N-1} + \theta'_{N-1} g_{N-1} + \eta_{N-1}
\end{aligned} \tag{A-32}$$

or

$$V^*_{N-1} = \frac{1}{2} x'_{N-1} K_{N-1} x_{N-1} + p'_{N-1} x_{N-1} + \gamma_{N-1}, \tag{A-33}$$

where

$$K_{N-1} = \Phi_{N-1} + G'_{N-1} \Theta_{N-1} G_{N-1} + 2 \Psi_{N-1} G_{N-1}, \tag{A-34}$$

$$p_{N-1} = \left( \Psi_{N-1} + G'_{N-1} \Theta_{N-1} \right) g_{N-1} + G'_{N-1} \theta_{N-1} + \phi_{N-1}, \tag{A-35}$$

$$\gamma_{N-1} = \frac{1}{2} g'_{N-1} \Theta_{N-1} g_{N-1} + \theta'_{N-1} g_{N-1} + \eta_{N-1}. \tag{A-36}$$

The matrix Riccati equations can then be obtained by substituting equation (A-30) into equation (A-34) to provide

$$K_{N-1} = \Phi_{N-1} + \Psi_{N-1}\Theta_{N-1}^{-1}\Psi_{N-1}^{'} - 2\Psi_{N-1}\Theta_{N-1}^{-1}\Psi_{N-1}^{'} \tag{A-37}$$

or

$$K_{N-1} = \Phi_{N-1} - \Psi_{N-1}\Theta_{N-1}^{-1}\Psi_{N-1}^{'}. \tag{A-38}$$

Substitution of equation (A-26) into equation (A-38) yields

$$\begin{aligned} K_{N-1} &= \beta A_{N-1}^{'} K_N A_{N-1} + W_N \\ &\quad - \left(\beta A_{N-1}^{'} K_N B_{N-1} + F_{N-1}\right)\left(\beta B_{N-1}^{'} K_N B_{N-1} + \Lambda_{N-1}\right)^{-1} \\ &\quad \times \left(F_{N-1}^{'} + \beta B_{N-1}^{'} K_N^{'} A_{N-1}\right). \end{aligned} \tag{A-39}$$

For the general time period, equation (A-39) becomes

$$\begin{aligned} K_k &= \beta A_k^{'} K_{k+1} A_k + W_k \\ &\quad - \left(\beta A_k^{'} K_{k+1} B_k + F_k\right)\left(\beta B_k^{'} K_{k+1} B_k + \Lambda_k\right)^{-1}\left(F_k^{'} + \beta B_k^{'} K_{k+1}^{'} A_k\right), \end{aligned} \tag{A-40}$$

with the terminal condition from equation (A-17)

$$K_N = W_N. \tag{A-41}$$

A comparison of equation (A-41) to the similar equation in Kendrick [12] which does not include discounting, i.e. equation (DK 2-53)

$$\begin{aligned} K_k &= A_k^{'} K_{k+1} A_k + W_k \\ &\quad - \left(A_k^{'} K_{k+1} B_k + F_k\right)\left(B_k^{'} K_{k+1} B_k + \Lambda_k\right)^{-1}\left(F_k^{'} + B_k^{'} K_{k+1}^{'} A_k\right) \end{aligned} \tag{A-42}$$

shows clearly the affect of the discounting. The difference between equation (A-41) and (A-43) is that each time the $K_{k+1}$ term appears in equation (A-41) it is multiplied by the discount factor $\beta$. The intuition for this is that the terms in the Riccati matrix are like Lagrangean multipliers in the sense that they carry information about the value of state variables. The discounting in equation (A-41) is therefore of the form

$$K_k = f\left(\beta K_{k+1}\right) = f\left(\frac{1}{1+\rho} K_{k+1}\right),$$

which shows that the value of state variables tend to be less in the next period than in the current period by the amount $1/(1+\rho)$.

Next consider the derivation of the Riccati vector, $p$. Begin by substituting equations (A-30) and (A-31) into equation (A-35) to obtain

$$\begin{aligned} p_{N-1} &= -\left(\Psi_{N-1} - \Psi_{N-1}\Theta_{N-1}^{-1}\Theta_{N-1}\right)\Theta_{N-1}^{-1}\theta_{N-1} \\ &\quad - \Psi_{N-1}\Theta_{N-1}^{-1}\theta_{N-1} + \phi_{N-1} \end{aligned} \tag{A-43}$$

or

$$p_{N-1} = -\Psi_{N-1}\Theta_{N-1}^{-1}\theta_{N-1} + \phi_{N-1}. \tag{A-44}$$

Substitution of equation (A-26) into equation (A-44) yields

$$\begin{aligned} p_{N-1} &= -\left(\beta A_{N-1}^{'} K_N B_{N-1} + F_{N-1}\right)\left(\beta B_{N-1}^{'} K_N B_{N-1} + \Lambda_{N-1}\right)^{-1} \\ &\quad \times \left(\beta B_{N-1}^{'}\left[K_N^{'} c_{N-1} + p_N\right] + \lambda_{N-1}\right) \\ &\quad + \beta A_{N-1}^{'}\left[K_N^{'} c_{N-1} + p_N\right] + w_{N-1} \end{aligned} \tag{A-45}$$

or, for the general time period,

$$\begin{aligned} p_k &= -\left(\beta A_k^{'} K_{k+1} B_k + F_k\right)\left(\beta B_k^{'} K_{k+1} B_k + \Lambda_k\right)^{-1} \\ &\quad \times \left(\beta B_k^{'}\left[K_{k+1}^{'} c_k + p_{k+1}\right] + \lambda_k\right) + \beta A_k^{'}\left[K_{k+1}^{'} c_k + p_{k+1}\right] + w_k. \end{aligned} \tag{A-46}$$

A comparison of equation (A-47) to equation (DK 2-54),[7] which is derived without discounting, reveals the difference in that both the $K_{k+1}$ and the $p_{k+1}$ terms on the right hand side are multiplied by $\beta$ in the discounting case but not in the case without discounting. Also, from equation (A-17), the terminal condition for equation (A-47) is

$$p_N = w_N. \tag{A-47}$$

In summary, the addition of discounting alters the Riccati equations in an intuitive fashion. The $K_{k+1}$ and $p_{k+1}$ terms on the right hand side of the Riccati matrix and vector equations are multiplied by the discount factor $\beta$.

# Appendix B

## Derivation of the Riccati Equations for an Extended Quadratic-Linear Tracking Model

As was discussed at the beginning of Appendix A, the derivations here follow those in Kendrick [12] chapter 2, but the model in the book is less general than the model in this chapter. In particular, the model in the chapter has discounting and that was considered in Appendix A where the derivations of the control model as a quadratic form was generalised to include discounting. However, there are a number of other differences between the model used in the book and the model used in this chapter. The additional differences are best seen by comparing the quadratic tracking model in the book to the quadratic tracking model in this chapter. In the book, the quadratic tracking model is converted to a quadratic form model before the Riccati equations are derived. In order to use that procedure here, we must first generalised the quadratic

[7]Not shown here.

tracking model of the book to include all the terms required by the model of this chapter. Then those results can be applied to obtain the Riccati equations for the model in the chapter.

We begin with the criterion function for the quadratic tracking model in chapter 2 of the book, i.e. equation DK(2-4)

$$\begin{aligned} J &= \frac{1}{2}\left[x_N - x_N^{\#}\right]' W_N^{\#}\left[x_N - x_N^{\#}\right] \\ &+ \frac{1}{2}\sum_{k=0}^{N-1}\left(\left[x_k - x_k^{\#}\right]' W_k^{\#}\left[x_k - x_k^{\#}\right] + \left[u_k - u_k^{\#}\right]' \Lambda_k^{\#}\left[u_k - u_k^{\#}\right]\right). \end{aligned} \tag{B-1}$$

To be useful in the derivations for this chapter the criterion function above needs to be extended to

$$\begin{aligned} J &= \lim_{N\to\infty}\sum_{k=0}^{N}\beta^k\Big(\frac{1}{2}\left[x_k - x_k^{\#}\right]' W_k^{\#}\left[x_k - x_k^{\#}\right] + \frac{1}{2}\left[u_k - u_k^{\#}\right]' \Lambda_k^{\#}\left[u_k - u_k^{\#}\right] \\ &+ \left[x_k - x_k^{\#}\right]' F_k^{\#}\left[u_k - u_k^{\#}\right] + w_k^{\#'}\left[x_k - x_k^{\#}\right] \\ &+ \lambda_k^{\#'}\left[u_k - u_k^{\#}\right] + \tau_k\Big). \end{aligned} \tag{B-2}$$

The modifications of equation (B-2) in equation (B-2) are

1. the time horizon is extended to infinity,
2. a discounting factor $\beta^k$ is introduced,
3. a cross term in $F_k$ is added,
4. linear terms in $w_k$ and $\lambda_k$ are added,
5. constant terms $\tau_k$ are added.

In order to make use of equation (B-2) in the derivation, it is useful to transform it from a tracking equation in terms like

$$\frac{1}{2}\left[x - x^{\#}\right]' W^{\#}\left[x - x^{\#}\right]$$

to a quadratic form in terms like

$$\frac{1}{2}x'W^{\#}x - x^{\#\prime}W^{\#}x + \frac{1}{2}x^{\#\prime}W^{\#}x^{\#}$$

by expanding it to

$$\begin{aligned} J &= \lim_{N\to\infty}\sum_{k=0}^{N}\beta^k\Big(\frac{1}{2}x_k'W_k^{\#}x_k + \frac{1}{2}u_k'\Lambda_k^{\#}u_k + x_k'F_k^{\#}u_k \\ &- x_k^{\#'}W_k^{\#}x_k - u_k^{\#'}\Lambda_k^{\#}u_k - x_k^{\#'}F_k^{\#}u_k - u_k^{\#'}F_k^{\#'}x_k \\ &+ \frac{1}{2}x_k^{\#\prime}W_k^{\#}x_k^{\#} + \frac{1}{2}u_k^{\#\prime}\Lambda_k^{\#}u_k^{\#} + x_k^{\#'}F_k^{\#}u_k^{\#} \\ &+ w_k^{\#'}x_k - w_k^{\#'}x_k^{\#} + \lambda_k^{\#'}u_k - \lambda_k^{\#'}u_k^{\#} + \tau_k\Big). \end{aligned} \tag{B-3}$$

Collection of terms in equation (B-3) yields

$$\begin{aligned} J &= \lim_{N\to\infty} \sum_{k=0}^{N} \beta^k \Big( \frac{1}{2} x_k' W_k^{\#} x_k - \left[ x_k^{\#'} W_k^{\#} + u_k^{\#'} F_k^{\#'} - w_k^{\#'} \right] x_k \\ &\quad + x_k' F_k^{\#} u_k + \frac{1}{2} u_k' \Lambda_k^{\#} u_k - \left[ u_k^{\#'} \Lambda_k^{\#} + x_k^{\#'} F_k^{\#} - \lambda_k^{\#'} \right] u_k \\ &\quad + \frac{1}{2} x_k^{\#'} W_k^{\#} x_k^{\#} - w_k^{\#'} x_k^{\#} + x_k^{\#'} F_k^{\#} u_k^{\#} + \frac{1}{2} u_k^{\#'} \Lambda_k^{\#} u_k^{\#} \\ &\quad - \lambda_k^{\#'} u_k^{\#} + \tau_k \Big). \end{aligned} \tag{B-4}$$

Then equation (B-4) can be written as a quadratic form

$$\begin{aligned} J &= \lim_{N\to\infty} \sum_{k=0}^{N} \beta^k \Big( \frac{1}{2} x_k' W_k x_k + w_k' x_k + x_k' F_k u_k \\ &\quad + \frac{1}{2} u_k' \Lambda_k u_k + \lambda_k' u_k + \eta_k \Big), \end{aligned} \tag{B-5}$$

where

$$\begin{aligned} W_k &= W_k^{\#}, \\ w_k &= -\left( W_k^{\#'} x_k^{\#} + F_k^{\#} u_k^{\#} - w_k^{\#} \right), \\ F_k &= F_k^{\#}, \\ \Lambda_k &= \Lambda_k^{\#}, \\ \lambda_k &= -\left( \Lambda_k^{\#'} u_k^{\#} + F_k^{\#'} x_k^{\#} - \lambda_k^{\#} \right), \\ \eta_k &= \frac{1}{2} x_k^{\#'} W_k^{\#} x_k^{\#} + x_k^{\#'} F_k^{\#} u_k^{\#} + \frac{1}{2} u_k^{\#'} \Lambda_k^{\#} u_k^{\#}, \\ &\quad - w_k^{\#'} x_k^{\#} - \lambda_k^{\#'} u_k^{\#} + \tau_k. \end{aligned} \tag{B-6}$$

The system equations for this model are

$$x_k = A_k x_k + B_k u_k + c_k \tag{B-7}$$

with initial conditions

$$x_0 \quad \text{given.}$$

The quadratic-linear model equations (B-5) through (B-7) is the same as the model (A-8) and (A-2) through (A-6) of Appendix A of this chapter except for the use of the infinite time horizon. The Riccati matrix equation for that model in equations (A-41) and (A-41) are

$$\begin{aligned} K_k &= \beta A_k' K_{k+1} A_k + W_k \\ &\quad - \left( \beta A_k' K_{k+1} B_k + F_k \right) \left( \beta B_k' K_{k+1} B_k + \Lambda_k \right)^{-1} \left( F_k' + \beta B_k' K_{k+1}' A_k \right) \end{aligned} \tag{B-8}$$

with the terminal conditions

$$K_N = W_N. \tag{B-9}$$

For the infinite horizon case, equations (B-8) and (B-9) remain the same.

The model in this chapter is in a tracking form, so it is necessary to substitute equation (B-7) into the Riccati matrix equation (B-8) to obtain

$$\begin{aligned} K_k &= \beta A_k^{'} K_{k+1} A_k + W_k^{\#} \\ &- \left(\beta A_k^{'} K_{k+1} B_k + F_k^{\#}\right)\left(\beta B_k^{'} K_{k+1} B_k + \Lambda_k^{\#}\right)^{-1} \\ &\times \left(F_k^{\#'} + \beta B_k^{'} K_{k+1}^{'} A_k\right). \end{aligned} \tag{B-10}$$

In order to apply this result to the model in the body of the chapter, consider the criterion function in equation (3)

$$\begin{aligned} J &\approx \sum_{t=0}^{\infty} \beta^t \Big\{ \Pi^j + \Pi_x^{j'} \left[x_t^{j+1} - x_t^j\right] + \Pi_u^{j'} \left[u_t^{j+1} - u_t^j\right] \\ &+ \frac{1}{2}\left[x_t^{j+1} - x_t^j\right]' \Pi_{xx}^j \left[x_t^{j+1} - x_t^j\right] + \frac{1}{2}\left[u_t^{j+1} - u_t^j\right]' \Pi_{uu}^j \left[u_t^{j+1} - u_t^j\right] \\ &+ \left[x_t^{j+1} - x_t^j\right]' \Pi_{xu}^j \left[u_t^{j+1} - u_t^j\right] \Big\}. \end{aligned} \tag{B-11}$$

A comparison of equation (B-11) with equation (B-2) yields

$$\begin{aligned} W_k^{\#} &= \Pi_{xx}^j, \\ F_k^{\#} &= \Pi_{xu}^j, \\ \Lambda_k^{\#} &= \Pi_{uu}^j, \\ w_k^{\#} &= \Pi_x^j, \\ \lambda_k^{\#} &= \Pi_u^j, \\ x_k &= x_t^{j+1}, \\ u_k &= u_t^{j+1}, \\ x_k^{\#} &= x_t^j, \\ u_k^{\#} &= u_t^j, \\ \tau_k &= \Pi^j. \end{aligned} \tag{B-12}$$

Also, in the body of the chapter, the symbol $X$ is used for the Riccati matrix which is labelled $K$ in Kendrick [12], so we have

$$K_k = X_t. \tag{B-13}$$

In a similar manner, a comparison of the system equations in the body of the chapter, i.e. equation (4)

$$x_{t+1}^{j+1} \approx x_t^j + f_x^j \left(x_t^{j+1} - x_t^j\right) + f_u^j \left(u_t^{j+1} - u_t^j\right) \tag{B-14}$$

with equation (B-7)

$$x_{k+1} = A_k x_k + B_k u_k + c_k \tag{B-15}$$

yields

$$\begin{aligned} A_k &= f_x^j, \\ B_k &= f_u^j, \\ c_k &= x_t^j - f_x^j x_t^j - f_u^j u_t^j, \\ x_{k+1} &= x_{t+1}^{j+1}, \\ x_k &= x_t^{j+1}, \\ u_k &= u_t^{j+1}. \end{aligned} \tag{B-16}$$

Substitution of equations (B-16), (B-18) and (B-21) into equation (B-12) yields

$$\begin{aligned} X_t &= \beta f_x^{j\prime} X_{t+1} f_x^j + \Pi_{xx}^j \\ &- \left(\beta f_x^{j\prime} X_{t+1} f_u^j + \Pi_{xu}^j\right) \left(\beta f_u^{j\prime} X_{t+1} f_u^j + \Pi_{uu}^j\right)^{-1} \left(\Pi_{xu}^{j\prime} + \beta f_u^{j\prime} X_{t+1} f_x^j\right) \end{aligned} \tag{B-17}$$

with the end condition from equation (B-9)

$$X_N = W_N. \tag{B-18}$$

This is the Riccati matrix expression in equation (5) in the body of the chapter with the terminal condition in equation (7). Next, derive the Riccati vector equation. Begin with the equation derived from a model with discounting in Appendix A, i.e. from equations (A-47) and (A-47)

$$\begin{aligned} p_k &= -\left(\beta A_k^{'} K_{k+1} B_k + F_k\right) \left(\beta B_k^{'} K_{k+1} B_k + \Lambda_k\right)^{-1} \\ &\times \left(\beta B_k^{'} \left[K_{k+1}^{'} c_k + p_{k+1}\right] + \lambda_k\right) + \beta A_k^{'} \left[K_{k+1}^{'} c_k + p_{k+1}\right] + w_k \end{aligned} \tag{B-19}$$

with the terminal condition

$$p_N = w_N. \tag{B-20}$$

The model in the body of this chapter is in a tracking form, while equation (A-47) above was derived from a quadratic form, so it is necessary to substitute equation (B-7) into equation (A-47). This yields

$$\begin{aligned} p_k &= -\left(\beta A_k^{'} K_{k+1} B_k + F_k^{\#}\right) \left(\beta B_k^{'} K_{k+1} B_k + \Lambda_k^{\#}\right)^{-1} \\ &\times \left(\beta B_k^{'} \left[K_{k+1}^{'} c_k + p_{k+1}\right] - \left[\Lambda_k^{\#\prime} u_k^{\#} + F_k^{\#\prime} x_k^{\#} - \lambda_k^{\#}\right]\right) \\ &+ \beta A_k^{'} \left[K_{k+1}^{'} c_k + p_{k+1}\right] - \left[W_k^{\#\prime} x_k^{\#} + F_k^{\#} u_k^{\#} - w_k^{\#}\right]. \end{aligned} \tag{B-21}$$

Next, convert equation (B-21) into the notation and form of the model in the body of this chapter by substituting equations (B-12), (B-13) and (B-16) into equation (B-21) to obtain

$$\begin{aligned} p_t &= -(\beta f_x^{j\prime} X_{t+1} f_u^j + \Pi_{xu}^j)(\beta f_u^{j\prime} X_{t+1} f_u^j + \Pi_{uu}^j)^{-1} \\ &\quad \times \Big(\beta f_u^{j\prime} \Big[X_{t+1}^{\prime} \Big\{x_{t+1}^j - f_x^j x_t^j - f_u^j u_t^j\Big\} + p_{t+1}\Big] \\ &\quad - \Big[\Pi_{uu}^{j\prime} u_t^j + \Pi_{xu}^{j\prime} x_t^j - \Pi_u^j\Big]\Big) \\ &\quad + \beta f_x^{j\prime} \Big[X_{t+1}^{\prime} \Big\{x_{t+1}^j - f_x^j x_t^j - f_u^j u_t^j\Big\} + p_{t+1}\Big] \\ &\quad - \Big[\Pi_{xx}^{j\prime} x_t^j + \Pi_{xu}^j u_t^j - \Pi_x^j\Big] \end{aligned} \tag{B-22}$$

with the end condition from equation (B-20)

$$p_N = w_N. \tag{B-23}$$

Equations (B-22) and (B-23) are the Riccati vector expression and terminal condition in equation (6) and equation (8) in the body of the chapter.

# Appendix C

### Derivation of the Steady State Equations for States and Controls

Here, we follow the pattern of the derivations in chapter 2 of Kendrick [12] as extended to models with discounting in Appendix A of this chapter. We begin from equations (A-29) through (A-31) for the feedback rule, i.e.

$$u_k = G_k x_k + g_k, \tag{C-1}$$

where

$$G_k = -\Theta_k^{-1} \Psi_k^{\prime}, \tag{C-2}$$

$$g_k = -\Theta_k^{-1} \theta_k. \tag{C-3}$$

Substitution of equation (A-26) for period k into equations (C-2) and (C-3) yields

$$G_k = -\left[\beta B_k^{\prime} K_{k+1} B_k + \Lambda_k\right]^{-1} \left[F_k^{\prime} + \beta B_k^{\prime} K_{k+1} A_k\right], \tag{C-4}$$

$$g_k = -\left[\beta B_k^{\prime} K_{k+1} B_k + \Lambda_k\right]^{-1} \left[\beta B_k^{\prime} \left(K_{k+1} c_k + p_{k+1}\right) + \lambda_k\right]. \tag{C-5}$$

Next, convert equations (C-4) and (C-5) to the equivalent equations for the more general tracking model as in Appendix B by substituting equation (B-7) into equations (C-4) and (C-5) to obtain

$$G_k = -\left[\beta B_k^{\prime} K_{k+1} B_k + \Lambda_k^{\#}\right]^{-1} \left[F_k^{\#\prime} + \beta B_k^{\prime} K_{k+1} A_k\right], \tag{C-6}$$

$$\begin{aligned} g_k &= -\left[\beta B_k^{\prime} K_{k+1} B_k + \Lambda_k^{\#}\right]^{-1} \\ &\quad \times \left[\beta B_k^{\prime} \left(K_{k+1} c_k + p_{k+1}\right) - \left(\Lambda_k^{\#} u_k^{\#} + F_k^{\#\prime} x_k^{\#} - \lambda k^{\#}\right)\right]. \end{aligned} \tag{C-7}$$

The next step is to convert equations (C-6) and (C-7) into the notation and form used in the model in the body of this chapter. We begin with equation (C-6) and with the notation

$$H_t = G_k \tag{C-8}$$

which is used in the body of the chapter. Then, substitution of equation (C-8) and equations (B-16) through (B-21) into equation (C-6) yields

$$H_t = -\left[\beta f_u^{j\prime} X_{t+1} f_u^j + \Pi_{uu}^j\right]^{-1}\left[\Pi_{xu}^{j\prime} + \beta f_u^{j\prime} X_{t+1} f_x^j\right]. \tag{C-9}$$

The infinite horizon version of equation (C-9) is equation (20) in the body of the chapter.

Next, consider the feedback gain vector $g$. Convert equation (C-7) into the notation and form used in the body of the chapter by using the notation

$$h_t = g_k \tag{C-10}$$

and by substituting equations (B-12) through (B-16) into equation (C-7) to obtain

$$\begin{aligned} h_t \quad = \quad & -\left[\beta f_u^{j\prime} X_{t+1} f_u^j + \Pi_{uu}^j\right]^{-1} \\ & \times \left[\beta f_u^{j\prime}\left(X_{t+1}\left\{x_{t+1}^j - f_x^j x_t^j - f_u^j u_t^j\right\} + p_{t+1}\right)\right. \\ & - \left.\left(\Pi_{uu}^j u_t^j + \Pi_{xu}^j x_t^j - \Pi_u^j\right)\right]. \end{aligned} \tag{C-11}$$

The infinite horizon version of equation (C-11) is equation (21) in the body of the chapter.

We turn now to the derivation of the steady state system equations and feedback rule. Begin with the system equations and feedback rule in chapter 2 of Kendrick [12], i.e. equations (DK 2-2) and (DK 2-5):

$$x_{k+1} = A_k x_k + B_k u_k + c_k, \tag{C-12}$$

$$u_k = G_k x_k + g_k. \tag{C-13}$$

Substitution of equation (C-13) into equation (C-12) yields

$$x_{k+1} = A_k x_k + B_k G_k x_k + B_k g_k + c_k. \tag{C-14}$$

In the steady state, equation (C-14) becomes

$$x = (A + BG)\,x + Bg + c \quad \text{or} \tag{C-15}$$

$$x = (I - A - BG)^{-1}(Bg + c). \tag{C-16}$$

Convert the steady state system equation above to the notation and form used in the model in the body of the chapter by substituting equations (B-12), (B-16) and (C-8) into equation (C-16) to obtain

$$x^{j+1} = \left(I - f_x^j - f_u^j H\right)^{-1}\left(f_u^j h + x^j - f_x^j x^j - f_u^j u^j\right). \tag{C-17}$$

The infinite horizon version of this equation is equation (19) in the body of the chapter. Substitution of equations (C-8) and (C-10) into equation (C-13) above provides the steady state feedback rule

$$u^{j+1} = Hx^{j+1} + h, \qquad \text{(C-18)}$$

and the infinite horizon version of this equation is equation (20) in the chapter.

## References


[1] Amman, H.M., (1995), Numerical Methods for Linear-Quadratic Models, pp. 579–618 in H.M. Amman, D.A. Kendrick and J. Rust (eds.), *Handbook of Computational Economics,* North-Holland, Amsterdam.

[2] Amman, H.M., and D.A. Kendrick, (1995), Nonconvexities in Stochastic Control Models, *International Economic Review,* 36, 455–475.

[3] Amman, H.M., and H. Neudecker, (1997), Solving the Algebraic Riccati Matrix Equation, *Journal of Economic Dynamic and Control,* 21, 363–370.

[4] Azariadis, C., (1993), *Intertemporal Macroeconomics,* Blackwell, Oxford.

[5] Bertsekas, D.P., (1976), *Dynamic Programming and Stochastic Control,* Academic Press, New York.

[6] Blanchard, O.J., and S. Fisher, (1989), *Lectures on Macroeconomics,* MIT Press, Cambridge Massachusetts.

[7] Chow, G.C., (1975), *Analysis and Control of Dynamic Economic Systems,* John Wiley, New York.

[8] Garbade, K.D., (1975), On the Existence and Uniqueness of Solutions of Multiperiod Linear Quadratic Control Problems, *International Economic Review,* 17, 719–732.

[9] Hotelling, H., (1931), The Economics of Exhaustible Resources, *Journal of Political Economy,* 39, 137–175.

[10] Judd, K., (1998), *Numerical Methods in Economics,* MIT Press.

[11] Kendrick, D.A., and L. Taylor, (1981), Numerical Solutions of Nonlinear Planning Models, *Econometrica,* 38, 453–467.

[12] Kendrick, D.A., (1981), *Stochastic Control for Economic Models,* McGraw-Hill, New York.

[13] Léonard, D., and N. Van Long, (1992), *Optimal Control Theory and Static Optimisation in Economics,* Cambridge University Press, Cambridge UK.

[14] Lucas, R.E., (1988), On the Mechanics of Economic Development, *Journal of Monetary Economics,* 22, 3–42.

[15] Magnus, J.R., and H. Neudecker, (1994), *Matrix Differential Calculus with Applications in Statistics and Econometrics,* John Wiley, New York.

[16] McGratten, E.R., (1994), A Note on Computing Competitive Equilibria in Linear Models, *Journal of Economic Dynamics and Control,* 18, 149–160.

[17] Ralston, A., and P. Rabinowitz, (1986), *A First Course in Numerical Analysis,* McGraw-Hill, Singapore.

[18] Ramsey, F.P., (1928), A Mathematical Theory of Saving, *Economic Journal,* 38, 543–559.

[19] Sargent, T.J., (1987), *Dynamic Macroeconomic Theory,* Harvard University Press, Cambridge.

[20] Stockey, N.L., and R.E. Lucas, (1989), *Recursive Methods in Economic Dynamics,* Harvard University Press, Cambridge.

CHAPTER 20

# Computers, Multilinear Algebra and Statistics

D.S.G. POLLOCK

## 20.1 Introduction

We are sometimes surprised to discover that the same intellectual puzzle is confronting us in two quite separate endeavours which we happen to be pursuing at the same time. Our sense of bemusement vanishes when we grasp the solution; and it may occur to us that, in reality, we have been pursuing different aspects of the same problem. Then we are liable to feel that it was our own obtuseness which was preventing us, at first, from recognising the connection.

This paper concerns some puzzles which arose in the context of computer graphics and on the interface between multilinear algebra and multivariate statistics. The problems were amenable to a common solution. But this was not obvious at first.

## 20.2 Problems with the Computer Screen

The difficulties which accompanied the first generation of personal computers, which had so many limitations and inadequacies, are already fading from our memories. It is hard to believe that, as recently as 1984, the best-selling brand of home computer was endowed with only 64 kilobytes of memory which was managed by an 8-bit microprocessor. (See, for example, Jarrett [5]).

A large part of this memory was preempted by what the manufacturers were proud to describe as a high-resolution graphical display. This consisted of a grid of 1000 cells arranged in 40 columns and 25 rows. In each of the cells, one could place a character or a graphical symbol, or one could leave the cell blank. A character cell was itself a small grid of eight rows and eight columns; and the pixels (i.e. picture elements) within this grid corresponded to individual bits (i.e. binary digits) within the computer's memory.

The grid of character cells constituted the low-resolution graphics. The much finer grid of pixels constituted the so-called high-resolution graphics. There is a hint of a salesman's deception in this distinction.

An example of how a character was composed of bits is provided by the matrix $A$ below. The elements of the matrix, which is described as a bitmap, provide a crude rendition of the letter A, with enough space on one side of it and above to prevent its clashing with the adjacent symbols:

$$A = \begin{bmatrix} 0 & 0 & 0 & 0 & 0 & 0 & 0 & 0 \\ 0 & 0 & 0 & 0 & 1 & 0 & 0 & 0 \\ 0 & 0 & 0 & 0 & 1 & 0 & 0 & 0 \\ 0 & 0 & 0 & 1 & 0 & 1 & 0 & 0 \\ 0 & 0 & 0 & 1 & 0 & 1 & 0 & 0 \\ 0 & 0 & 1 & 1 & 1 & 1 & 1 & 0 \\ 0 & 0 & 1 & 0 & 0 & 0 & 1 & 0 \\ 0 & 1 & 1 & 1 & 0 & 1 & 1 & 1 \end{bmatrix}. \tag{1}$$

The computer screen was used primarily to display text in uppercase characters whose bitmaps were stored permanently in a read-only memory bloc; and the cells of the low-resolution display were numbered accordingly, following the usual way of reading from left to right and from one line of the text to the next.

Statisticians are able to describe the operation of numbering the cells in terms of a concise notation of their own. Let the character cells of the computer screen correspond to the elements of a matrix $X$. Then the operation of joining successive lines of a textual display in a long string corresponds to the operation of forming the row vector $\mathrm{rvec}(X)$. Numbering the cells of the display is a matter of indexing the elements of $\mathrm{rvec}(X)$ in the usual way.

The bits within the character cells were numbered according to a scheme which is analogous to the scheme for numbering the character cells. Thus, if the elements of the matrix $A$ above correspond to the pixels of a character, then the numbering of the pixels corresponds to the indexing of $\mathrm{rvec}(A)$.

Let $i_m$ and $i_n$ be column vectors containing $m = 25$ units and $n = 40$ units respectively. Imagine that nothing but the letter A is printed on the computer screen and that it fills every available cell. Then the result corresponds the matrix

$$X = i_m i_n' \otimes A, \tag{2}$$

where $\otimes$ is a tensor product which statisticians commonly call the Kronecker product and which some would prefer to call the Zeyfuss product.

From the point of view of the computer's memory, the graphical bits were numbered in a single sequence beginning with the first bit in the first character cell in the top left corner of the screen. The sequence passed from the final bit in the first character cell (number 64 in the lower right corner) to the first bit of the second character of the line. Thereafter, it passed through successive characters of the line. When the end of the line was reached, it passed to the line below.

In the notation of equation (2), the arrangement of the graphics memory can be represented by an rvec within an rvec:

$$\mathrm{rvec}(i_h i_w') \otimes \mathrm{rvec}(A). \tag{3}$$

The ambition of some owners was to use their home computer for plotting scientific graphs. Then it was necessary to gain access to the individual pixels to turn them on or off (i.e. to set their values to 1 or to 0). The problem was to find the precise locations of the corresponding binary digits within the computers's graphics memory of 64,000 bits.

In tackling this problem, it is hardly helpful to contemplate equation (2). The reason is that the symbols of that equation are designed to conceal the very details with which the problem is concerned. It is easier to tackle the problem in the terms of a programming language.

Let us adopt the following system of indexing and numeration:

$$\begin{aligned} i &= 1,\ldots,m; && \text{character lines,} \\ j &= 1,\ldots,n; && \text{character columns,} \\ k &= 1,\ldots,h; && \text{pixel rows,} \\ l &= 1,\ldots,w; && \text{pixel columns.} \end{aligned} \tag{4}$$

The following procedure, which is coded in the Pascal language, runs through the entire graphics memory bit by bit, pausing only to mark the end of the pixel rows and character lines and the passage from one character cell to another:

(5)

```
procedure MemoryScan;

var
  i, j, k, l : integer;
  m, n, h, w : integer;
  p : longint;

begin
m := 25;
n := 40;
h := 8;
w := 8;

Writeln('Begin a new screen');
for i := 1 to m do
  begin {i}
    Writeln('Begin a new line');
    for j := 1 to n do
      begin {j}
        Writeln('Begin a new character');
        for k := 1 to h do
          begin {k}
            Writeln('Begin a new row of pixels');
            for l := 1 to w do
              begin {l}
```

```
                p := l + n * ((k - 1) + m * ((j - 1) + w * (i - 1)));
                Writeln(p : 6);
              end; {l}
            end; {k}
          end; {j}
        end; {i}

      end; {MemoryScan}
```

This procedure depends upon the fact that the generic pixel has a location in the graphics memory which is indexed by

$$\begin{aligned} p(i,j,k,l) &= l + (k-1)w + (j-1)wh + (i-1)whn \\ &= l + w[(k-1) + h\{(j-1) + n(i-1)\}]. \end{aligned} \tag{6}$$

The first expression on the LHS is a readily intelligible rendition of the rule for counting the pixels. The second expression represents a computationally efficient formula which is incorporated in the following computer routine:

(7)

```
function P(i, j, k, l, m, n, h, w : integer) : longint;
begin
  P := n * (i - 1);
  P := h * (P + j - 1);
  P := w * (P + k - 1);
  P := P + l;
end;
```

The problem which we face is how to find, within the graphics memory, a pixel which has the screen coordinates $(x, y)$, where $x \in [1, 320]$ is a number of pixels representing the horizontal location, measured from the left edge of the screen, and $y \in [1, 200]$ is the vertical location in terms of a pixel count starting from the top of the screen. This is a matter of finding the indices $i, j, k$ and $l$ which correspond to the coordinates $x$ and $y$. Once the indices have been found, the function $p(i, j, k, l)$ can be used to find the memory location.

The two row indices $i$ and $k$ are obtained from the vertical coordinate $y$ using the identity

$$\begin{aligned} y &= h \times (y \text{ div } h) + (y \text{ mod } h) \\ &= h \times (i-1) + k. \end{aligned} \tag{8}$$

The column indices $j$ and $l$ are obtained from the horizontal coordinate $x$ using the identity

$$\begin{aligned} x &= w \times (x \text{ div } w) + (x \text{ mod } w) \\ &= w \times (j-1) + l. \end{aligned} \tag{9}$$

The function which finds the memory location for the pixel with the coordinates $(x, y)$ is as follows:

(10)
```
function Pxy(x, y, m, n, h, w : integer) : longint;
  var
      i, j, k, l : integer;
begin
  i := (y div h) - 1;
  k := y mod h;
  j := (x div w) - 1;
  l := x mod w;
  Pxy := P(i, j, k, l, m, n, h, w);
end;
```

## 20.3 An Index Notation for the Computer Screen

A third-generation computer programming language, such as Pascal, is not the best means of expressing the simple relationships which lie at the heart of the problem of the computer screen. The computer language, which conveys all of the operational details, is too explicit for the purpose. On the other hand, there is too much concealment in the matrix notation of equation (2) for this to be of much use. Something midway between the two is required.

In pursuit of an appropriate notation, we might think of representing equation (2) in the form of

$$X = i_m \otimes i_n' \otimes \left\{ \sum_k \sum_l a_{kl}(e_k \otimes e_l') \right\}, \tag{11}$$

where the values $a_{kl}$ are the elements of the matrix $A$ of (1). Here the $e_k$ is a basis vector of order $h$ with a unit in the $k$th position and with zeros elsewhere, whereas $e_l$ is a basis vector of order $w$.

It will be recalled that equation (2) represents nothing but a computer screen full of the repeated letter A. To create a more interesting effect, we require every pixel of the screen to be determined independently. The general expression for a screen of arbitrary pixel values would be

$$X = \sum_i \sum_j \sum_k \sum_l x_{ijkl} \left( e_i \otimes e_j' \otimes e_k \otimes e_l' \right). \tag{12}$$

This notation can benefit from some compression. Let us write $e_i^j = e_i \otimes e_j'$ for the matrix of order $m \times n$ with a unit in the $i$th row and the $j$th column and with zeros elsewhere. Let $e_k^l = e_k \otimes e_l'$ be a similar matrix of order $h \times w$. A natural extension of the notation gives us

$$e_{ik}^{jl} = e_i^j \otimes e_k^l = e_i \otimes e_j' \otimes e_k \otimes e_l'. \tag{13}$$

This can be used to advantage in equation (12), but there remains the problem of the summation signs. These are redundant; and we can write the equation simply as

$$X = (x_{ijkl} e_{ik}^{jl}). \tag{14}$$

Here it is the parentheses which imply the fourfold summation. Stripping them away gives the element $x_{ijkl}e_{ik}^{jl}$, which is a matrix of order $mh \times nw$ with an arbitrary scalar in the $ik$th row and the $jl$th column and with zeros elsewhere.

In so far as they relate to the rows and columns of the high-resolution (i.e. pixel resolution) screen display, the composite indices or words $ik$ and $jl$ obey a lexicographic scheme. As in a dictionary, the leading letters of the words vary slowly and the following letters vary more rapidly. In this case, a lexicographic interpretation of the sequence $ijkl$, which indexes the scalar $x_{ijkl}$, also leads to the correct memory location for the pixel which corresponds to the arbitrary element of the matrix $x_{ijkl}e_{ik}^{jl}$.

However, such a result is not to be relied upon in general. It is the indexing of the basis vectors which determines the ordering of the scalar elements; and it is a fortuitous circumstance when the scalar elements display their indices in the order which is taken by the basis vectors.

To understand the role of the indices better, it is appropriate to consider the matrix

$$X = (x_{ij}e_i^j), \tag{15}$$

of which the elements correspond to the characters of a low-resolution textual display on the computer screen. The row vector rvec($X$) can be represented by

$$X^r = (x_{ij}e_i^j)^r = (x_{ij}e^{ij}). \tag{16}$$

The corresponding column vector, which is the more familiar one in statistics and which is commonly denoted by vec($X$), is represented by

$$X^c = (x_{ij}e_i^j)^c = (x_{ij}e_{ji}). \tag{17}$$

Notice that, in this case, the elements $x_{ij}$ acquire an ordering within the vector $X^c$ which is the reverse of a lexicographic ordering. A further operation which naturally accompanies the two vectorisation operations is that of transposition:

$$X' = (x_{ij}e_i^j)' = (x_{ij}e_j^i). \tag{18}$$

It follows immediately from (16), (17) and (18) that

$$X^r = X'^{c\prime} \qquad \text{and} \qquad X^c = X'^{r\prime}. \tag{19}$$

The operations under (16) to (19) may be described as tensor commutations. (See Pollock [11] and Magnus and Neudecker [7] for example).

The operation which maps the matrix of the pixels of the computer console into the bits of the graphics memory is more complicated than any of the foregoing commutations, and there is no concise symbolism to denote it. It can be represented as follows:

$$(x_{ijkl}e_{ik}^{jl}) \longrightarrow (x_{ijkl}e_{ijkl}). \tag{20}$$

This mapping is characterised by the way in which the superscript column indices $jl$ have been interspersed with the subscript row indices $ik$ to create a specific order for the basis elements of the set of long row vectors which comprise $mnhw$ elements.

Based on the four indices, there are 24 similar schemes which are available for ordering the basis vectors; and the one which is reflected in equation (20) is the natural one within the context of the computer screen.

## 20.4 The Index Notation Applied to Matrix Differential Calculus

The theory of matrix differential calculus is an important adjunct of multivariate statistical analysis. This body of knowledge has been used with increasing effectiveness in finding the values which optimise statistical likelihood functions of ever-increasing complexity. The field was surveyed by Magnus and Neudecker [8] who have also been responsible for some major contributions to the subject, beginning with a seminal paper by Neudecker [10].

For a long time, matrix differential calculus was bedevilled by a lack of consensus over the appropriate definition of one of the essential objects of the study, which was the derivative of a matrix-valued function $Y = Y(X)$ with respect to its matrix argument $X$.

There were three alternative definitions; and now it sees to be generally agreed that only one of them is truly viable (See, for example, Magnus and Neudecker [8, p. 173] and Pollock [11], [12]). The other definitions, which have been widely used in multivariate statistical analysis, are not consistent with the classical representation of linear algebra via matrix theory, and they lead to serious practical difficulties which do not arise when the appropriate definition is adopted.

It is worthwhile to use the index notation of the previous section to examine the alternative definitions of the derivative and to establish relationships amongst them. The task of determining which of the definitions was to be favoured began to demand attention insistently at the time when the home computer which has been described in the previous sections first appeared in the market.

According to the first definition, the derivative of $Y$ with respect to $X$ is a partitioned matrix $[\partial Y/\partial x_{ij}]$ whose $ij$th partition is derived from the matrix $Y$ by replacing each element $y_{kl}$ by its derivative $\partial y_{kl}/\partial x_{ij}$. Thus the elements of the matrix derivative have the same disposition as the elements $x_{ij}y_{kl}$ of the product $X \otimes Y = (x_{ij}e_i^j) \otimes (y_{kl}e_k^l)$, which we can also write as

$$X \otimes Y = (\{x_{ij}y_{kl}\}e_{ik}^{jl}). \tag{21}$$

This definition has been studied by Rogers [13] amongst others; and he has ascribed to it the notation $\varepsilon Y/\varepsilon X$. Graham [3] has also adopted it as the basic definition in his textbook.

According to the second definition, the derivative of $Y$ with respect to $X$ is a partitioned matrix $[\partial y_{kl}/\partial X]$ whose $kl$th partition is derived from the matrix

$X$ by replacing the elements $x_{ij}$ by the derivatives $\partial y_{kl}/\partial x_{ij}$. The elements of the matrix derivative therefore have the same disposition as the elements of

$$Y \otimes X = (\{x_{ij}y_{kl}\}e_{ki}^{lj}). \tag{22}$$

This is probably the most commonly used definition. It has been employed extensively by both Rogers [13] and Balestra [1] in their treatises of matrix differential calculus; and it was adopted by MacRae [6] in a seminal article to which many authors have referred.

From another point of view, the two definitions already considered follow directly from the definitions of Dwyer and MacPhail [2] who considered the forms $\partial Y/\partial x_{ij}$ and $\partial y_{kl}/\partial X$ without arranging them into partitioned matrices.

According to the third definition, the derivative of $Y$ with respect to $X$ is a matrix $\partial Y^c/\partial X^c$ whose elements $\partial y_{kl}/\partial x_{ij}$ have the same arrangements as the elements $y_{kl}x_{ij}$ in the product

$$Y^c \otimes (X^c)' = (\{x_{ij}y_{kl}\}e_{lk}^{ji}). \tag{23}$$

This final definition, which we may call the vectorial definition, has been used by Neudecker [8] and by the present author—Pollock [9]—amongst many others. Nel [9] has ascribed to this derivative the notation $\partial \text{vec}Y/\partial \text{vec}'X$.

The relationships between the three definitions are revealed by juxtaposing their expressions:

$$\begin{aligned}
\left[\frac{\partial Y}{\partial x_{ij}}\right] &= \left(\frac{\partial y_{kl}}{\partial x_{ij}}\, e_{ik}^{jl}\right), \qquad (24)\\
\left[\frac{\partial y_{kl}}{\partial X}\right] &= \left(\frac{\partial y_{kl}}{\partial x_{ij}}\, e_{ki}^{lj}\right),\\
\left[\frac{\partial Y^c}{\partial X^c}\right] &= \left(\frac{\partial y_{kl}}{\partial x_{ij}}\, e_{lk}^{ji}\right).
\end{aligned}$$

The first two derivatives are seen to differ from each other only in respect of the orders within the pair of column indices $j$, $l$ and within the pair of row indices $k$, $i$. The third derivative $\partial Y^c/\partial X^c$ differs from the other two in a more complicated way which requires the conversion of the basis vectors $e_i$ and $e^l$ into $e^i$ and $e_l$ respectively.

## 20.5 Chain Rules

To illustrate the contention that the vectorial definition is the appropriate one, we shall consider the problem of defining a chain rule for matrix derivatives.

Let $X = X(Z)$ and $Y = Y(X)$ be two matrix transformations whose composition is $Y = Y(Z)$. Then the vectorial definition gives rise to a rule in the form of

$$\frac{\partial Y^c}{\partial Z^c} = \frac{\partial Y^c}{\partial X^c}\frac{\partial X^c}{\partial Z^c}, \tag{25}$$

which entails nothing more than the multiplication of the forms $\partial Y^c/\partial X^c$ and $\partial X^c/\partial Z^c$ according to the ordinary rules of matrix algebra. Chain rules obeying the normal algebra of matrix compositions are not available for the other definitions which we have considered.

To provide the simplest example of the chain rule under (25), let us consider the following matrix equations:

$$\begin{aligned} Y &= AXB' \quad \text{or} \quad (y_{kl}e^l_k) = (a_{ki}e^i_k)(x_{ij}e^j_i)(b_{lj}e^l_j), \\ X &= CZD' \quad \text{or} \quad (x_{ij}e^j_i) = (c_{if}e^f_i)(z_{fu}e^u_f)(d_{ju}e^j_u). \end{aligned} \tag{26}$$

In forming such matrix products, we must invoke some further rules of the index notation. In the first place, we notice that the act of matrix multiplication is accompanied by the cancellation of a superscript index within the leading factor with a subscript index in the following factor. Thus, in forming the product $(a_{ki}e^i_k)(x_{ij}e^j_i) = (\{a_{ki}x_{ij}\}e^j_k)$, we are cancelling the index $i$. This leaves the index $i$, which is to be found on the RHS only within the scalar product $\{a_{ki}x_{ij}\}$, without an accompanying basis vector.

The second rule deals with such unassociated indices. Thus, if adjacent indices of a product within braces contain a common index, as in the product $\{a_{ki}x_{ij}\}$, then we must take a sum in respect of the index. Thus $\{a_{ki}x_{ij}\} = \sum_i a_{ki}x_{ij}$.

The composition of the two mappings of (26) gives

$$Y = (AC)Z(BD)' \quad \text{or} \quad (y_{kl}e^l_k) = (\{a_{ki}c_{if}\}e^f_k)(z_{fu}e^u_f)(\{b_{lj}d_{ju}\}e^l_u). \tag{27}$$

The vectorised versions of the two equations under (26) are given by

$$\begin{aligned} Y^c &= (B\otimes A)X^c \quad \text{or} \quad (y_{kl}e_{lk}) = (\{b_{lj}a_{ki}\}e^{ji}_{lk})(x_{ij}e_{ji}), \\ X^c &= (D\otimes C)Z^c \quad \text{or} \quad (x_{ij}e_{ji}) = (\{d_{ju}c_{if}\}e^{uf}_{ji})(z_{fu}e_{uf}), \end{aligned} \tag{28}$$

and that of their composition under (27) is given by

$$Y^c = (BD\otimes AC)Z^c \quad \text{or} \quad (y_{kl}e_{lk}) = (\{b_{lj}d_{ju}\}\{a_{ki}c_{if}\}e^{uf}_{lk})(z_{fu}e_{uf}). \tag{29}$$

Next, by referring to the definition under (24), we find that $\partial Y^c/\partial X^c = (\{\partial y_{kl}/\partial x_{ij}\}e^{ji}_{lk}) = (\{b_{lj}a_{ki}\}e^{ji}_{lk})$. In this manner, we can easily confirm that

$$\begin{aligned} \frac{\partial Y^c}{\partial X^c} &= B\otimes A, \\ \frac{\partial X^c}{\partial Z^c} &= D\otimes C, \\ \frac{\partial Y^c}{\partial Z^c} &= BD\otimes AC. \end{aligned} \tag{30}$$

Finally, by confirming that

$$\begin{aligned} (B\otimes A)(D\otimes C) &= (\{b_{lj}a_{ki}\}e^{ji}_{lk})(\{d_{ju}c_{if}\}e^{uf}_{ji}) \\ &= (\{b_{lj}d_{ju}\}\{a_{ki}c_{if}\}e^{uf}_{lk}) \\ &= (\{b_{lj}d_{ju}\}e^u_l)\otimes(\{a_{ki}c_{if}\}e^f_k) = BD\otimes AC, \end{aligned} \tag{31}$$

we verify the chain rule in question.

To obtain chain rules for the alternative definitions, we are obliged to invent special operations of composition which do not accord with the usual matrix algebra. Consider the following derivatives:

$$\begin{aligned}\left[\frac{\partial y_{kl}}{\partial X}\right] &= (\{b_{lj}a_{ki}\}e_{ki}^{lj}) = (A')^c \otimes B^r, \\ \left[\frac{\partial x_{ij}}{\partial Z}\right] &= (\{d_{ju}c_{if}\}e_{if}^{ju}) = (C')^c \otimes D^r, \\ \left[\frac{\partial y_{kl}}{\partial Z}\right] &= (\{a_{ki}c_{if}\}\{b_{lj}d_{ju}\}e_{kf}^{lu}) = ([AC]')^c \otimes (BD)^r.\end{aligned} \tag{32}$$

A chain rule is obtained by defining a star product of the derivatives such that

$$\left[\frac{\partial y_{kl}}{\partial Z}\right] = \left[\frac{\partial y_{kl}}{\partial X}\right] * \left[\frac{\partial x_{ij}}{\partial Z}\right]. \tag{33}$$

In terms of our example, this becomes

$$\begin{gathered}([A']^c \otimes B^r) * ([C']^c \otimes D^r) = ([AC]')^c \otimes (BD)^r \quad \text{or} \\ (\{b_{lj}a_{ki}\}e_{ki}^{lj}) * (\{d_{ju}c_{if}\}e_{if}^{ju}) = (\{a_{ki}c_{if}\}\{b_{lj}d_{ju}\}e_{kf}^{lu}).\end{gathered} \tag{34}$$

In place of the ordinary convention of matrix multiplication, which implies that

$$(e_{lk}^{ji})(e_{ji}^{uf}) = (e_{lk}^{uf}), \tag{35}$$

we have a new convention of star products to the effect that

$$(e_{ki}^{lj}) * (e_{if}^{ju}) = (e_{kf}^{lu}). \tag{36}$$

This star product is a generalisation of a product which MacRae [6] used in defining a rule for the composition of a derivative $\partial y/\partial X$ of a scalar function $y = y(X)$ with the derivative $[\partial x_{ij}/\partial Z]$ of a matrix function $X = X(Z)$.

The generalised star product has the manifest disadvantage that it cannot be extended in any simple way to accommodate the composition of multiple tensor products. By contrast, the usual rules for matrix manipulation extend easily to such cases. For example, for triple products or matrices, we have the simple rule that

$$(A \otimes B \otimes C)(D \otimes E \otimes F) = AD \otimes BE \otimes CF. \tag{37}$$

## References

[1] Balestra, P., (1976), *La Dérivation Matricielle,* Sirey, Paris.

[2] Dwyer, P.S., and M.S. MacPhail, (1948), Symbolic Matrix Derivatives, *Annals of Mathematical Statistics,* 19, 517–534.

[3] Graham, A., (1981), *Kronecker Products and Matrix Calculus with Applications,* Ellis Horwood, Chichester.

[4] Henderson, H.V., and S.R. Searle, (1981), The vec-Permutation Matrix, the vec Operator and Kronecker Products: A Review, *Linear and Multilinear Algebra,* 9, 271–288.

[5] Jarrett, D., (1984), The Complete Commodore 64, Hutchinson and Co., London.

[6] MacRae, E.C., (1974), Matrix Derivatives with an Application to an Adaptive Linear Decision Problem, *Annals of Statistics,* 2, 337–346.

[7] Magnus, J.R., and H. Neudecker, (1979), The Commutation Matrix: Some Properties and Applications, *Annals of Statistics,* 7, 381–394.

[8] Magnus, J.R., and H. Neudecker, (1988), *Matrix Differential Calculus,* John Wiley and Sons, Chichester.

[9] Nel, D.G., (1980), On Matrix Differentiation in Statistics, *South African Statistical Journal,* 14, 137–193.

[10] Neudecker, H., (1969), Some Theorems on Differentiation with Special Reference to Kronecker Matrix Products. *Journal of the American Statistical Association,* 64, 953–963.

[11] Pollock, D.S.G., (1979), *The Algebra of Econometrics,* John Wiley and Sons, Chichester.

[12] Pollock, D.S.G., (1985), Tensor Products and Matrix Differential Calculus, *Linear Algebra and its Applications,* 67, 169–193.

[13] Rogers, G.S., (1980), *Matrix Derivatives,* Marcel Dekker, New York.

# Author Index

# Subject Index

## Advanced Studies in Theoretical and Applied Econometrics

1. J.H.P. Paelinck (ed.): Qualitative and Quantitative Mathematical Economics. 1982 ISBN 90-247-2623-9
2. J.P. Ancot (ed.): Analysing the Structure of Econometric Models. 1984 ISBN 90-247-2894-0
3. A.J. Hughes-Hallett (ed.): Applied Decision Analysis and Economic Behaviour. 1984 ISBN 90-247-2968-8
4. J.K. Sengupta: Information and Efficiency in Economic Decision. 1985 ISBN 90-247-3072-4
5. P. Artus and O. Guvenen (eds.), in collaboration with F. Gagey: International Macroeconomic Modelling for Policy Decisions. 1986 ISBN 90-247-3201-8
6. M.J. Vilares: Structural Change in Macroeconomic Models Theory and Estimation. 1986 ISBN 90-247-3277-8
7. C. Carraro and D. Sartore (eds.): Developments of Control Theory for Economic Analysis. 1987 ISBN 90-247-3345-6
8. D.P. Broer: Neoclassical Theory and Empirical Models of Aggregate Firm Behaviour. 1986 ISBN 90-247-3412-6
9. A. Italianer: Theory and Practice of International Trade Linkage Models. 1986 ISBN 90-247-3407-X
10. D.A. Kendrick: Feedback. A New Framework for Macroeconomic Policy. 1987 ISBN 90-247-3593-9; Pb 90-247-3650-1
11. J.K. Sengupta, and G.K. Kadekodi (eds.): Econometrics of Planning and Efficiency. 1988 ISBN 90-247-3602-1
12. D.A. Griffith: Advanced Spatial Statistics Special Topics in the Exploration of Quantitative Spatial Data Series. 1988 ISBN 90-247-3627-7
13. O. Guvenen (ed.): International Commodity Market Models and Policy Analysis. 1988 ISBN 90-247-3768-0
14. G. Arbia: Spatial Data Configuration in Statistical Analysis of Regional Economic and Related Problems. 1989 ISBN 0-7923-0284-2
15. B. Raj (ed.): Advances in Econometrics and Modelling. 1989 ISBN 0-7923-0299-0
16 A. Aznar Grasa: Econometric Model Selection. A New Approach. 1989 ISBN 0-7923-0321-0
17. L,R. Klein and J. Marquez (eds.): Economics in Theory and Practice. An Eclectic Approach. Essays in Honor of F.G. Adams. 1989 ISBN 0-7923-0410-1
18. D.A. Kendrick: Models for Analyzing Comparative Advantage. 1989 ISBN 0-7923-0528-0
19. P. Artus and Y Barroux (eds.): Monetary Policy. A Theoretical and Econometric Approach. 1990 ISBN 0-7923-0626-0
20. G. Duru and J.H.P. Paelinck: Econometrics of Health Care. 1990 ISBN 0-7923-0766-6

**Kluwer Academic Publishers—Dordrecht / Boston / London**

## Advanced Studies in Theoretical and Applied Econometrics

21. L. Phlips (ed.): Commodity, Futures and Financial Markets. 1990 ISBN 0-7923-1043-8
22. H.M. Amman, D.A. Belsley and L.F. Pau (eds.): Computational Economics and Econometrics. 1991 ISBN 0-7923-1287-2
23. B. Raj and J. Koerts (eds.): Henri Theil's Contributions to Economics and Econometrics. Volume 1: Econometric Theory and Methodology. 1992 ISBN 0-7923-1548-0
24. B. Raj and J. Koerts (eds.): Henri Theil's Contributions to Economics and Econometrics. Volume 2: Consumer Demand Analysis and Information Theory. 1992 ISBN 0-7923-1664-9
25. B. Raj and J. Koerts (eds.): Henri Theil's Contributions to Economics and Econometrics. Volume 3: Economic Policy and Forecasts, and Management Science. 1992 ISBN 0-7923-1665-7

    Set (23-25), ISBN 0-7923-1666-5
26. P. Fisher: Rational Expectations in Macroeconomic Models. 1992 ISBN 0-7923-1903-6
27. L. Phlips and L.D. Taylor (eds.): Aggregation, Consumption and Trade. Essays in Honor of H.S. Houthakker. 1992 ISBN 0-7923-2001-8
28. L. Mátyás and P. Sevestre (eds.): The Econometrics of Panel Data. Handbook of Theory and Applications. 1992 ISBN 0-7923-2043-3
29. S. Selvanathan: A System-Wide Analysis of International Consumption Patterns. 1993 ISBN 0-7923-2344-0
30. H. Theil, D. Chen, K.W. Clements and C. Moss: Studies in Global Econometrics. 1996 ISBN 0-7923-3660-7
31. P.J. Kehoe and T. J. Kehoe (eds.): Modeling North American Economic Integration. 1995 ISBN 0-7923-3751-4
32. C. Wells: The Kalman Filter in Finance. 1995 ISBN 0-7923-3771-9
33. L. Mátyás and P. Sevestre (eds.): The Econometrics of Panel Data. A Handbook of the Theory with Applications. Second Revised Edition. 1996 ISBN 0-7923-3787-5
34. B. Assarsson, D.L. Edgerton, A. Hummelmose, I. Laurila, K. Rickertsen and P.H. Vale: The Econometrics of Demand Systems with Applications to Food Demand in the Nordic Countries. 1996 ISBN 0-7923-4106-6
35. D.A. Griffith, C. Amrhein and J-M. Huriót (eds.): Econometric Advances in Spatial Modelling and Methodology Essays in Honour of Jean Paelinck. 1998 ISBN 0-7923-4915-6
36. R.D.H. Heijmans, D.S.G. Pollock and A. Satorra (eds.): Innovations in Multivariate Statistical Analysis. 2000 ISBN 0-7923-8636-1

**Kluwer Academic Publishers—Dordrecht / Boston / London**

MIX
Papier aus verantwortungsvollen Quellen
Paper from responsible sources
FSC® C105338

If you have any concerns about our products,
you can contact us on
**ProductSafety@springernature.com**

In case Publisher is established outside the EU,
the EU authorized representative is:
**Springer Nature Customer Service Center GmbH**
**Europaplatz 3, 69115 Heidelberg, Germany**

Printed by Libri Plureos GmbH
in Hamburg, Germany